EXPERIMENTAL HYDRODYNAMICS OF FAST-FLOATING AQUATIC ANIMALS

EXPERIMENTAL HYDRODYNAMICS OF FAST-FLOATING AQUATIC ANIMALS

Viktor V. Babenko

ELSEVIER

ACADEMIC PRESS
An imprint of Elsevier

Academic Press is an imprint of Elsevier
125 London Wall, London EC2Y 5AS, United Kingdom
525 B Street, Suite 1650, San Diego, CA 92101, United States
50 Hampshire Street, 5th Floor, Cambridge, MA 02139, United States
The Boulevard, Langford Lane, Kidlington, Oxford OX5 1GB, United Kingdom

British Library Cataloguing-in-Publication Data
A catalogue record for this book is available from the British Library

Library of Congress Cataloging-in-Publication Data
A catalog record for this book is available from the Library of Congress

ISBN: 978-0-12-821025-3

For Information on all Academic Press publications
visit our website at https://www.elsevier.com/books-and-journals

Publisher: Mica Haley
Acquisitions Editor: Anna Valutkevich
Editorial Project Manager: Liz Heijkoop
Production Project Manager: Omer Mukthar
Cover Designer: Victoria Pearson

Typeset by MPS Limited, Chennai, India

Working together
to grow libraries in
developing countries

www.elsevier.com • www.bookaid.org

Contents

Annotation

The basic principle of the optimization of living organisms is the development, as a result of the centuries-old evolution, of the minimum expenditure of energy to support the process of vital activity. In this regard, in **Part I** of this monograph, the structural features of the systems of the organism of aquatic animals and their interaction in the process of movement, aimed at reducing energy consumption, are investigated. On the example of dolphins and other high-speed aquatic animals, the features of the morphology and physiology of the skeleton, muscles, skin covers, blood and lymphatic systems, and innervations are considered. Since the movement is carried out in the aquatic environment, when considering these systems, the power impact of the environment on the body is taken into account. The influence on the body's systems of swimming speed, nonstationary movement, an unconventional way to create thrust, and a specific body structure are considered. Some features of the hydrodynamic effects on the body during movement in the aquatic environment are considered.

In accordance with these features, the specific structures of these systems of the body of aquatic animals are analyzed. A detailed description of the structure of the skeleton and the location of the nodes of the nervous system are given, as well as layer-by-layer arrangement along the body thickness of dolphins and along the body length of the motor muscles. The structure of the skin is particularly detailed. Features of the structure of the circulatory system are also considered. With the help of the developed devices, temperature measurements were made on the surface of the skin of the body, the distribution of elasticity, and the damping properties of dolphin skin. The results of direct and indirect measurements of other mechanical parameters of the skin, in particular, the phase velocity of the development of disturbances on the skin surface are given. The work of interacting systems of the body and the mechanisms regulating the mechanical properties of the skin are considered, showing ways to reduce body resistance.

In **Part II**, modeling of individual systems and the detected features of the morphological structure of fast-swimming aquatic organisms is performed. Special devices, experimental installations, and experimental techniques on analogs of aquatic animals have been developed.

Numerous physical and integral measurements have been carried out, and new theoretical generalizations also have been developed.

All this allowed us to discover new phenomena and patterns that resulted in the form of discoveries and inventions in **Part III** of this monograph. Hydrodynamic inventions on analogs of fast-swimming aquatic animals allowed the development of new technical solutions and technologies. Numerous examples of the implementation of the results of experimental hydrodynamic studies are given.

Preface

Democritus (460–370 BCE) and Aristotle (384–322 BCE) pointed to the imitation of nature by man. At the beginning of the 15th century, J. Battista Danti designed and manufactured wings, with the help of which flew over Lake Perudjy, in Italy.

However, Leonardo da Vinci can be considered the founder of the new scientific direction of bionics. He developed a bionic research method: "If you want to talk about such things, you should in the first part determine the nature of air resistance; in the second part—the structure of the bird and its feathers; in the third part—the action of this plumage during various movements; in the fourth part—the role of the wings and tail." In 1490, he designed and built, perhaps, the first model of an aircraft. This model had wings similar to those of a bat. Using the muscles of the arms and legs, the person had to fly. He established the advantage of planning (soaring) flight. He developed a plan for the book, *Treatise on Birds*, consists of four books, of which the first would be about their flight with the help of wings flapping, the second about flying without flapping with the help of the wind, the third about flight in general, that is, birds, bats, fish, animals, insects, and the latter about flying with the help of mechanisms. He developed a helicopter project in which the Archimedes spiral was "a screw machine, which, if rotated at high speed, is screwed into the air and rises." He conducted tests in this direction from 1490 to 1519. Finally, he formulated the definition of bionics: "A bird is an instrument acting in accordance with mathematical law, which can be made in human power with all its movements. The instrument built by man lacks the soul of the bird, which in a particular case must be replaced by the soul of man."

Giovanni Alfonso Borelli (1608–1679), a pupil of Galileo, wrote a book, *On the Movement of Animals*, which was published in 1680. This book contains sections on the structure, shape, and action of the muscles of humans and some animals, the contraction of process muscles, heart, blood circulation, and digestion. There is a chapter on the flight of birds, which was published many times independently.

Lomonosov M., Galvani L., Hermann von Helmholtz, and other scientists tried to learn not only the laws of nature, but also to use their knowledge in practice. J. Cayley (1773–1857) was engaged in the study of the flight of birds and the construction of their models. He believed that in his time, first of all, it was necessary to conduct experiments and first copy nature. In 1809, he formulated a very important postulate: the shape of the stern of the body of rotation is more important than the bow in the formation of aerodynamic drag. He developed a drawing based on the cross-sections of the trout body. The profile shape of the trout body corresponds exactly to the modern laminarized profiles of the type of NASA, N.E. Zhukovsky, and Chaplygins profiles.

The famous Russian physicist N.A. Umov (1846–1915) carried out a systematic analysis of the general laws of living organisms and technical devices. Much earlier, N. Wiener formulated an important bionic postulate: "The action of a living material system can be replaced by the action of some inanimate material system."

The general principles of the functioning of living organisms and machines were investigated by N.E. Zhukovsky (1847–1921). He developed mathematical analogies between two phenomena—between the actions of living organisms and similar technical devices. He performed an exhaustive analysis of the flight of birds. First, he performed an analysis of the processes observed in nature, for example, during the flight of birds. He then introduced theoretical concepts explaining the action of animals and used in the design of mechanical devices, in particular, aircraft and propellers. He argued that with an external difference in the work of living organisms and mechanical devices there is a complete internal analogy.

Professor of the University of Liverpool, Hele-Shaw (1854–1951), as early as 1897–1898 studied the flow of a viscous fluid near a streamlined body. He developed a hydrodynamic apparatus and studied the effect of the mucous membrane of fish on reducing resistance.

In 1962, J. Sickmann published an article on the movement of swimming animals [273], and in 1975 a book about swimming and flying in nature was published [289].

Fundamental research was performed by M.J. Lighthill, who in 1969 published the articles *Hydrodynamics of aquatic animal propulsion* [188], in 1975 *Aerodynamic Aspects of the Flight of Animals* [189], and in 1983 the monograph *Mathematical Biofluid Dynamics* [190].

Professor of the Mining Institute of Dnipro, Ukraine, Ya.I. Grdina (1871–1931) made a great contribution to the development of biomechanics. From 1910 to 1916, several of his monographs were published, in particular, *The Dynamics of Living Organisms*. He established a classification of relations, dynamic equations of motion and methods of their integration, variational principles, and basic theorems of the dynamics of living organisms.

Academician V.V. Shuleikin (1895–1979) published a monograph *Physics of the Sea*, in the section on *Biological Physics of the Sea* systematized the results of many years of research into the hydrodynamics of swimming in fish and dolphins [271].

Academician M.A. Lavrentiev (1900–1980) developed a theoretical model for swimming of fish and other aquatic animals. He first applied the mathematical apparatus: the method of flat cross sections.

Academician G.V. Logvinovich [192–194] developed the method of M.A. Lavrent'ev.

Later, the American researcher T. Wu [300] applied the general principles of the theory of the thin body and the theorem on the conservation of momentum and energy, and developed the theory of swimming fish. The theory of G.V. Logvinovich modified those of L.F. Kozlov [171] and E.V. Romanenko [251–253] for aquatic animals, in particular for swimming dolphins.

In 1932–1936, British physiologist James Gray completed a series of studies on the bioenergetics of dolphin swimming [117]. If we assume that the dolphin's musculature develops a specific power the same as that of a well-trained athlete, then when swimming at a speed of 10 m/s for dolphins, the power of the body does not allow such swimming speeds to be developed. According to his calculations, at this speed, the dolphin must develop power, which should be seven times greater than that which its muscles can develop. In 1949, this discrepancy was called the "Gray paradox." For a long time in various countries and in different directions, a search was made for facts that could explain this paradox. A number of scientists called this paradox erroneous, but at the same time they did not present any arguments against it.

In 1938, M.O. Kramer received a patent, which simulates the structure of the hair of aquatic animals and the cover of birds. After the Second World War, he moved from Germany to the United States. On the way across the ocean, he watched swimming dolphins. In 1957, he performed a series of experiments on elastic coatings, the structure of which imitated the structure of the upper layer of dolphin skin. When towing on a boat, a model in the form of a body of revolution, in which the cylindrical part was covered with its elastic materials, a decrease in resistance of 57% was obtained [177–180]. He received three US patents for his coating designs. This result partially explained the "Gray paradox" and aroused considerable worldwide interest in bionic research and in a new method of reducing resistance. At the same time, Thomps's reliable results were obtained: when a small amount of water-soluble polymer powder WSR-301 was added to the water, the hydraulic resistance in pipes was reduced by 80%!

In 1964, the book *The Man and the Dolphin* [191] by J. Lily was published, and attracted more attention to the study of various aspects of the biology and hydrodynamics of dolphins.

Since 1941, E. Holst published the results of studies that simulated the structure and aerodynamics of the wings of birds, including prehistoric birds, as well as insects, such as dragonflies. Based on a study of the flight of dragonflies, models of helicopters with a coaxial three-blade propeller were developed.

In 1948, G.A. Woodcock studied dolphin swimming.

In 1955, M.M. Sleptsov published a monograph *Cetaceans of the Far Eastern Seas* [274].

In 1963, Hertel H. published a monograph in which he examined various parts of bionics: structural bionics, principles for constructing light structures in nature and technology, cybernetic bionics, and others [126,127]. Hertel H.'s book is a classic book on bionics, especially when analyzing movement in air and water. Many sections of bionics discussed in this book continue to be studied at present.

Professor L.F. Kozlov, the head of the department "Boundary Layer Control and Hydrobionics," who constantly developed numerous scientific and practical problems for many years, was the founder of the new research direction "Hydrobionics" and the flow of elastic surfaces in Ukraine. Since 1965, systematic studies of bionics began in Ukraine and the Russian Federation in various organizations, including various institutes of the Academy of Sciences of Ukraine. Professor L.F. Kozlov (1927–1987) organized hydrobionic research in Ukraine at the Institute of Hydromechanics of the Academy of Sciences of Ukraine. From 1967 to 1998, the journal *Bionika* was published, the editor-in-chief of which was Kozlov. For several years, Kozlov also coordinated hydrobionic studies in Ukraine, published numerous articles on various issues of bionics and a number of monographs. In particular, in 1983 the monograph *Theoretical Bio-hydrodynamics* was published [175]. Kozlov organized a modern experimental basis for carrying out new scientific research. Under his leadership, a large number of experimental facilities and equipment were designed and manufactured. In the Institute of Hydromechanics of the Academy of Sciences of Ukraine, Kiev, the DISA Elektronic thermo-anemometric equipment and the Brul & Cer multichannel tape recorder were purchased. He instructed V.V. Babenko to master a new method of measuring the characteristics of the boundary layer based on the tellurium method of W.X. Wortman, and I.P. Ivanov to develop a laser anemometer, with the help of which numerous experimental studies of the characteristics of the boundary layer were carried out.

L.F. Kozlov, V.M. Shakalo [165,166,168,265,266] and Ye.V. Romanenko [250–253], when performing independent experimental studies on dolphins moving in the channel, found a corresponding decrease in the pulsation characteristics of speed and pressure in the boundary layer on the body of dolphins on various phases of nonstationary movement.

V.A. Rodionov discovered the specific structure of the musculature. An increased amount of myoglobin was found in the muscles of the cetaceans, allowing accumulation of oxygen in the body during deep diving.

S.V. Pershin, A.G. Tomilin, and A.S. Sokolov, in 1971, received a diploma for discovery No. 95 on the specific structure of the cetacean caudal fins: "The phenomenon of self-regulation of the hydroelasticity of cetacean fins" [226].

V.V. Babenko discovered that the whole body of a dolphin is an adaptive system that has automatic mechanisms for changing the shape of the body, its geometric dimensions, the mechanical characteristics of the external covers during diving, and depending on the speed of movement. In 1969, he discovered that the skin of dolphins is an active, self-adjusting system depending on driving patterns. In 1983, V.V. Babenko, V.E. Sokolov, L.F. Kozlov, S.V. Pershin, A.G. Tomilin, and O.B. Chernyshev received a diploma for opening No. 265 "The property of cetacean skin is to actively regulate hydrodynamic resistance to swimming by controlling the local interaction of the skin with the flowing stream" [37].

In 1979, S.V. Pershin published a monograph that deals with the bionic questions of navigation and flight in nature. In 1988 he published a book *Basics of Hydrobionics* [231], in which the main results are published by him and various authors, mainly in the journal *Bionics*. In 1988, Pershin formulated the concept of "hydrobionics" [231]: "Hydrobionics is a new scientific direction, one of the independent search directions in bionics, which is distinguished by its studied living objects and habitat, by its goals, objectives and research methods. Hydrobionics is a new integrated scientific direction for studying the principles of structure and functions of the organs of locomotion of aquatic animals (hydrobionts) with the aim of improving the propulsion and technical maneuverability means of movement in the aquatic environment, as well as to create fundamentally new systems."

In 1968, G.K. Herzog published a monograph in which the anatomy of birds was studied and an aerodynamic analysis of the structure of the wings and tail presented. Also presented were the results of a bird flight simulation performed by E. Holst.

In 1970, the fundamental monograph by V.V. Ovchinnikov [215] on physiology and some issues in bionics of swordfish and sailfish was published.

In 1970, the fundamental monograph by G.B. Agarkov and others was published, *Dolphin Morphology* [4], on the morphology and innervation of the skin of dolphins.

In 1972, the fundamental monograph by A.V. Yablokov and others, *Whales and Dolphins* [311], was published.

In 1973, Academician V.E. Sokolov published the monograph *Skin of Mammals*, and in 1987, *Morphology and Ecology of Marine Mammals* [281,282].

In 1973, the monograph of V.A. Matyukhin on bioenergy and the physiology of swimming fish were published [199].

In 1975, the book *Swimming and Flying in Nature* [289] was published.

In 1976, a monograph by Yu.G. Aleev was published, which deals with the hydrodynamic aspects of swimming fish.

In 1997 published a book by Angelo Mojetta et al. in which numerous photographs are given and some morphological features of various species of sharks are explained [122].

In 1978 a monograph by V.R. Protasov and A.G. Staroselskaya was published, in which drawings of various fish are presented.

The word "bionics"—a symbiosis of the concepts "biology" and "technique"—was suggested.

United States Air Force Major J.E. Steele introduced this concept in 1960 at a congress in Dayton (Ohio, United States). By this he illustrated that this was a science that explores systems that "copy natural systems with the same qualities, or are their analogues."

In Germany, in 1993, a commission of experts initiated by the technological center of the Union of German Engineers of Düsseldorf proposed the following definition: "Bionics as a scientific discipline that systematically deals with the technical transformation and application of structures, methods and evolutionary principles for the development of biological systems. The tasks of bionics are the assessment of various, deeply analyzed phenomena of nature for possible technical implementation. The range of applications extends from the production of materials to issues of processes, for example, systems of main traffic arteries." A prerequisite for the development of bionics is biological basic research with appropriate methods of analysis used in physics and chemistry. At the same time, nature offers an unlimited reservoir for structures and methods that, in the process of evolution, led to an optimal state. The technical interpretation of biological systems is designated as "technical biology" and is a section of classical biology. "Technical biology" is, for example, the study of the structure of the skin of fast-swimming sharks, which have thin long grooves on their skin that can reduce flow resistance; bionics applies here only if the transformation of this knowledge leads, for example, to the development of a decrease in the resistance of artificial structures of the grooves.

W. Nachtigal also investigated the classification of the structure of bionics. He singled out 12 groups: (1) historical, (2) bionics structures, (3) construction bionics, (4) climate bionics, (5) structures bionics, (6) motion bionics, (7) device bionics, (8) anthropobionics, (9) sensors bionics, (10) bionics of nerves, (11) bionics of processes, and (12) bionics of evolution [208].

L.F. Kozlov also investigated chemical and pharmaceutical bionics.

R. Isenmann investigated the connection between economics and ecology with bionics. He developed eight basic environmental principles modeled on a picture of nature.

R. Reiner also investigated the connection and analogy between biological systems and sociological systems—this is sociological bionics [248]. He reviewed three analogies that were investigated earlier in the book by E.V. Zerbst [306]. Direct analogy: this consists of the development of a completely isomorphic model. For example, the biomechanics of the human hand is completely transferred to the mechanical arm of the robot. Functional analogy: this consists of developing the isomorphism of the functions of the original and its model. An example would be the wind concentrator of I. Rechenberg and R. Bannasch. Structural analogy: this is the agreement between the original and the model, in which the structure and properties are rapidly studied using computer calculations. An example is the evolutionary strategy of I. Rechenberg.

E.V. Zerbst developed in the monograph *Bionics, Biological Functional Principles and their Technical Application* methodology, functions and tasks of bionics and its various directions [306]. He considers three components of bionics: general, systematic, and applied bionics. Systematic bionics consists of four parts: structural, energy, functional, and organizational. The book developed the scientific foundations of bionics. Considered one of the most important parts of science, these are the principles of modeling and methods of similarity. The scientific foundations of the theory of dimensionality in relation to bionics and methods of analogies, in particular, electromechanical analogies, have been developed. An organism is considered as an open thermodynamic system.

In 2007 a collection of works edited by David Wagg and others was published, in which the results of studies of engineering applications of adaptive structures found in various examples of nature are presented. In particular, there is a chapter on bionic adaptive structures.

In 2007, the book *Wrapping Phenomenon in Nature* was published in two volumes in the form of a collection of works edited by R.J. Liebe, in which the results of studies on the flow of birds and insects in the air (1st volume) and the flow of various fish and sea animals (2nd volume) were published.

In 2010, a collection of papers edited by M.S.De.F. Guilak and R.K. Mofrad, which presented the results of computer modeling studies in biomechanics, a collection of works edited by N. Kato and S. Kamimura, which presents the results of studies on the biomechanics of navigation and flight, and a collection of works edited by G. Taylor, M.S. Triantafillou, and C. Tropea, which presents the results of studies on the movement of aquatic animals, were published.

In the monographs L.F. Kozlov [175] and S.V. Pershin [231], as well as in their articles published even earlier, the features of the shape and structure of the hydrobiont fins and the specific features of the creation of thrust are considered. Similar issues are discussed in the article by M.S. Triantafillou, G.S. Triantafillou, and R. Gopalakrishnan (1991) and at a conference (see Proc. Conf. Paper. Nov. 2015).

In 1967, at the Scientific and Technical Conference VVMIOLU, Leningrad, Associate Professor S. Krolenko demonstrated for the first time a model of a self-propelled dolphin. In fact, the results of these studies were the first in a new scientific direction for the development of biorobots: hydrobionts. A direction was created on the study of biobots/hydrobionts and the design of underwater gliders. Examples include the works of M.J. Wolfgang et al. [299], P.R. Bandyopadhyay [74–76], Kumph John Muir, M.S. Triantafillou [182], V.V. Babenko [237,238], and others.

Much attention has been paid in China to the study of robots, for example, in 2005, an article by Yu Jing-hu and others was published on the management of the tail section of robot-fish, and in 2008 an article by Chao Zhou and others was published on designs of robot-fish. Most articles in this direction in China are published in Chinese in various journals.

In 2010, a collection of papers edited by D. Floreano, J. Zufferey, M.V. Srinivasan, and C. Ellington, *Flight of Insects and Robots*, was published, in 2014, along with a doctoral dissertation: Feitian Zhang *Modeling, Construction and Control of a Planning Robot Fish*, and in 2015, a book by George V. Lauder and James L. Tangorra *The Locomotion of Fish: Biology and Robots of the Body and Movement of the Tail*.

The first part of this monograph presents the results of experimental studies on hydrobionics, obtained by the author and other researchers who conducted their research in various institutes of the National Academy of Sciences of Ukraine (NASU) and other organizations within Ukraine and the Russian Federation. The author expresses his deep gratitude for their creative and dedicated work.

The author is deeply grateful to Academician V.T. Grinchenko, Director of the Institute of Hydromechanics of the National Academy of Sciences of Ukraine, Kiev, for supporting and discussing the materials in this book.

In 1993—1994 the investigations of Professor V.V. Babenko were sponsored by the J. Soros International Science Foundation (the number of the head of research is UAW000 and UAW200).

In 1995—1997, Professor V.V. Babenko conducted research under a contract between the Institute of Hydromechanics of the National Academy of Sciences of Ukraine, Kiev, Ukraine, and Cortana Corporation, President K.J. Moore, United States (Task 11 and 12, DARPA Delivery Order 0011 of Contract MDA972-92-D-0011). The results of this collaboration were not only scientific reports. Professor Babenko received 11 US patents. In natural conditions on a high-speed vessel, a study was conducted of the effectiveness of some patents on combined methods of drag reduction. In 1995, Cortana Corporation provided financial support for participation in a conference, which was held in Johns Hopkins University, Baltimore, United States, and the president of Cortana Corporation, K.J. Moore, provided great support in discussing and preparing three reports at this conference. These three reports systematized the main scientific results of experimental investigations, which formed the basis for writing the first part of this monograph.

In 1998 and 1999, Babenko conducted hydrobionic investigations and their systematization at Wissenschaftskollegs zu Berlin (Institute for Applied Research, Berlin).

In 1999, Babenko V. V. received financial support for participation in a conference in

Newport (grant N00014-98-1-4040, Office of the Marine International Research Unit—European special programs, Assistant Code 240 223, Old Maryellend Road, support program visits to the Naval Undersea Center) [51—54].

In 2005—2006, Professor Babenko worked as a visiting professor at the Engineering Center for Applied Research of the Ship (ASERC, Busan, Korea) at the National University of Pusan (Director Professor Ho Hwan Chun). The results of this work were published in a book in English in 2012 [70].

All this made it possible to continue research on the problems of hydrobionics in different directions. As a result, the main research directions were formulated and the results obtained in this book were systematized.

We offer sincere thanks to the individuals and organizations that provided financial support for this research. We also thank K.J. Moore, President of Cortana Corporation, for very important creative collaboration and creative discussions.

Further reading

Paul W. Webb, Hydrodynamics and Energetics of Fish Propulsion. Department of the Environment Fisheries and Marine Service, (1975) p. 158.

A.H. Woodcock studied dolphin swimming (The swimming of dolphins. 1948 Nature, 161, pp.602).

List of symbols

x, y, z	longitudinal, normal and transverse axis of a Cartesian coordinate system
x_i	measurement location coordinate
$\bar{y} = y/\delta$	dimensionless vertical coordinate
b	width
l, L	length
d, D	diameter
$\lambda = l^2/S = l/b$	wing extension
l	wingspan
S	wing area
b	wing chord
R	radius; resistance force
δ	boundary layer thickness
δ^*	theoretical value of displacement thickness
δ^{**}, δ_2	impulse loss thickness
$k = 2\pi f/u = 2\pi/\lambda$	wave number
$\Delta y, \Delta z$	transverse and transversal amplitudes of disturbing oscillations
$\varepsilon = \dfrac{\sqrt{\frac{1}{3}\left(\bar{u}'^2 + \bar{v}'^2 + \bar{w}'^2\right)}}{U_\infty}; T_u$	degree of turbulence
$E = \sqrt{u'^2}/U_\infty$	maximum value of speed pulsation
t	time
U, V и W, u, v и w	longitudinal, normal and transversal components of avelocity of the basic flow
u', v', w'	longitudinal, normal and transversal components of the disturbing motion
$\sqrt{\bar{u}'^2}, \sqrt{\bar{v}'^2}, \sqrt{\bar{w}'^2}, \overline{u'v'}$	time-averaged components of velocity pulsations
$\dfrac{2\overline{u'v'}}{U_\infty^2}$ —	Reynolds stresses
τ_0	shear stress on the wall
$q = \rho V^2/2$	dynamic pressure
p	basic flow pressure, $\sqrt{\overline{p'^2}} = 0.5\rho U_\infty^2\, Re^{-0.3}$
P	resulting pressure
$\bar{p} = \dfrac{2(p - p_o)}{\rho U^2}$	dimensionless pressure
$\dfrac{2\sqrt{\left(p'\right)^2}}{\rho U_\infty^2} = \alpha_\tau \lambda$	pulse pressure energy p'
$C_p, \ \overline{P}_i = \dfrac{P - P_\infty}{\frac{1}{2}\rho U_\infty^2}, \ \bar{p} = \dfrac{p}{\rho u_H^2}$	dimensionless pressure; surface pressure distribution coefficients
$\alpha = 2\pi/\lambda$	wave number of disturbing oscillations
$k = 2\pi f/u = 2\pi/\lambda$	wave number
f и λ	frequency and length of the wave harmonics
u	average velocity at the measuring point
λ	oscillation wavelength
λ_x	disturbing wavelength of longitudinal vortices in the direction of the x axis
$\tilde{\lambda}_x = \lambda_x/\delta^*$	dimensionless wavelength oscillation

λ_z — three-dimensional disturbing wavelength in the direction of the axis z

$\beta = \beta_r + i\beta_i$ — complex frequency of disturbing oscillations

$\beta_r = 2\pi n$ — circular frequency

$n = 0.159 \frac{\beta_r \nu}{U_\infty^2} \frac{U^2}{\nu}$ — disturbing oscillation frequency

Δn — frequency range of disturbing oscillations

$T = 1/n$ — vibration cycle; period

β_i — amplification coefficient

$c = \beta / \alpha = c_r + i c_i$ — complex velocity of disturbing movement

c_r — velocity of propagation of a wave of disturbing movement

c_i — amplification coefficient

$\alpha\delta$ — dimensionless wave number of disturbing oscillation

$\beta_r \nu / U_\infty^2 ; \omega_r$ — dimensionless frequency of disturbing oscillation

c_r / U_∞ — dimensionless velocity of distribution of disturbing movement dynamic viscosity coefficient

$\nu = \mu/\rho$ — kinematic viscosity coefficient

$\rho = \gamma/g$ — fluid density (mass per unit volume)

$\mathbf{Re}_1 = U_\infty / \nu$ — single Reynolds number

$\overline{\tau} = \frac{\tau}{\rho u_H^2}$ — dimensionless shear stress

τ — shear stress (force per unit area)

τ_w — shear stress on the wall, $\tau_w = \mu \left(\frac{\partial u}{\partial y}\right)_0$

$\tau_w / \rho U^2 \sim \sqrt{\frac{\mu}{\rho U_\infty l}} = \frac{1}{\sqrt{\mathrm{Re}_l}}$ — dimensionless shear stress on the wall for Blasius profile

σ — stress of a viscoelastic material

$E = \sigma / \varepsilon$ — elastic modulus of elastic material

ε — elastomer elongation; relative lateral deformation of polymeric materials

t_i, h — flexible surface thickness

$C_m = \sqrt{(T/M)}$ — phase velocity of oscillation propagation, the velocity of forced oscillations on the dolphin's skin surface

T_F — tension of the elastic material element

M, m — oscillating mass of the element of elastic material

$S, E' = E / t_i$ — stiffness of elastic material

$\mathbf{Re} = UL/\nu$ — Reynolds number

$Fr = U / \sqrt{gL}$ — Froude number

$Sh = UT / L$ — Strouhal number

$Eu = 2p / \rho U^2$ — Euler number

c_a, C_L — lift coefficient

$C_F = 2P/S\rho U^2, C_F = \frac{1}{x}\int_0^x C_f dx' = 2\frac{\Theta}{x}$ — full coefficient of friction or average coefficient of friction resistance

$C_{fr} c_{fr} \, c_{w}, C_{xi}$ — friction coefficient

$\lambda, \, C_f = 2\tau_w / \rho U^2_\infty$ — local coefficient of friction on the wall

$c_\tau = 2\tau_w / \rho U^2_\infty$ — local coefficient of friction

$c_w = \frac{1.328}{\sqrt{\mathrm{Re}_l}}$ — local coefficient of friction for profile

$\xi = 2R / \rho U^2 S$ — drag coefficient

$\delta \sim \sqrt{\frac{\mu l}{\rho U}} = \sqrt{\frac{\nu l}{U}}$ — boundary layer thickness

$\delta_{\text{лам.}} = 5\sqrt{\frac{\nu l}{U}}$ — thickness of the laminar boundary layer

$k = Y/Q$ — aerodynamic quality (the ratio of lift to frontal drag)

ρ — mass density of medium

S — characteristic body area

$Q = c_x \rho U^2 S/2$ — body drag

$Y = c_y \rho U^2 S/2$ — lifting force

$Z = c_z \rho U^2 S/2$ — lateral force

$M_x = m_x \rho U^2 SL/2$ moment relative to longitudinal coordinates x
$M_y = m_y \rho U^2 SL/2$ moment relative to the vertical coordinate y
$M_z = m_z \rho U^2 SL/2$ moment relative to transversal coordinate z

Abbreviations

BL	boundary layer
TBL	turbulent boundary layer
T-S;	Tollmin - Schlichting
CVS;	coherent vortices structures
lateral	side
ventral	abdominal
caudal	tail

Method of investigation into the problems of hydromechanics on the basis of the laws of swimming aquatic animals

1.1 Introduction

Experimental hydrobionics is a new area of research, which studies the dynamic interactions of aquatic animals and their analogues with the environment in order to simulate the discovered new phenomena and patterns that can then be applied in engineering. Experimental hydrobionics investigates the hydrodynamics of the movement of various hydrobionts, which significantly differ in size and shape of the body, speed, degree and type of adaptation to the environment, etc., as well as hydrodynamic features of the flow around their analogues. The development of experimental hydrobionics has been due to the connection with such sciences as theoretical and experimental hydrodynamics, biology, physiology, electronics, physics, and chemistry. Below are described some results of the developments of techniques for experimental investigation into hydrobionics from studying high-speed aquatic animals, which will be used as generalizations in the future for hydrobionts. In 1995, at a conference in Johns Hopkins University (Baltimore, United States) we made three presentations, in particular, we proposed a hydrobionic study of the problems of hydromechanics (Table 1.1) and the structure of experimental hydrobionics, which consists of biological and technical problems (Tables 1.2 and 1.3).

The hydrobionic research method (Table 1.1) can be explained by the example of solving a specific technical problem—the development of an effective method for drag reduction of vehicles moving under water. The first stage of the proposed method is to search for the corresponding hydrobionic objects. Depending on the required range of Reynolds numbers for the analysis of hydrobionts, either high-speed dolphin species, sharks, or swordfish, etc., can be chosen. The next step is to determine the hydrodynamic parameters characterizing the laws of drag reduction. After assessing the significance of the determining parameters, a plan and procedure for conducting hydrobionic experimental

TABLE 1.1 Method of hydrobionic research of hydromechanics problems.

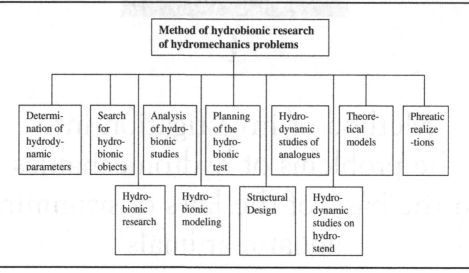

investigations is developed. Based on the analysis of the obtained results, the stage of hydrobionic modeling is developed. After that, an appropriate model experiment is planned. The basic parameters for the development of appropriate models for investigation in hydrodynamic laboratories are then determined. At the same time, constructive developments are carried out to create models of the phenomenon under study. The next stage is to conduct experimental studies on models—hydrobionic analogues of the outer covers of hydrobionts.

Hydrodynamic studies can be of two types: the study of integral characteristics and the study of the physical mechanisms of the phenomena under consideration. In this example, the first type of research was carried out in a high-speed towing tank [58,63,86], and the second type on a low- turbulence hydrodynamic test bench specially designed and manufactured by the author [170].

On the basis of the obtained results of a model experiment on hydrobionic analogues, new theoretical models of the hydrodynamic problems under consideration are being modernized or developed. This allows us to develop new technologies and technical solutions. All stages are carried out taking into account the detected specific patterns of interaction of living organisms with the environment and the principles of hydrobionts. The indicated method includes biological and technical problems (Tables 1.2 and1.3).

Biological problems consist of three main groups. The first group is the analysis and selection of the object of study, which, in turn, includes two characteristics. The first is classification by hydrobionic characteristics, which includes three characteristics. The first of these is the habitat. For example, there are only deep-water hydrobionts and hydrobionts living in the aquatic environment when moving at different depths, hydrobionts that periodically move in air or, on the contrary, periodically move from air to water. The second characteristic of hydrobionts differs in characteristic Reynolds numbers. When analyzing a

TABLE 1.2 Biological problems.

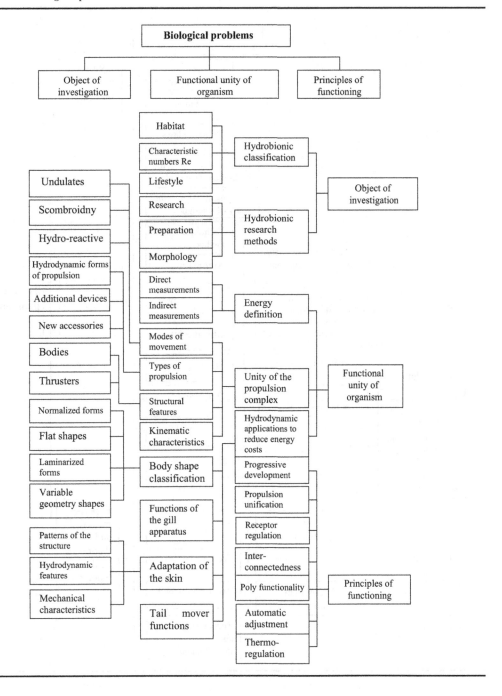

TABLE 1.3   Technical problems.

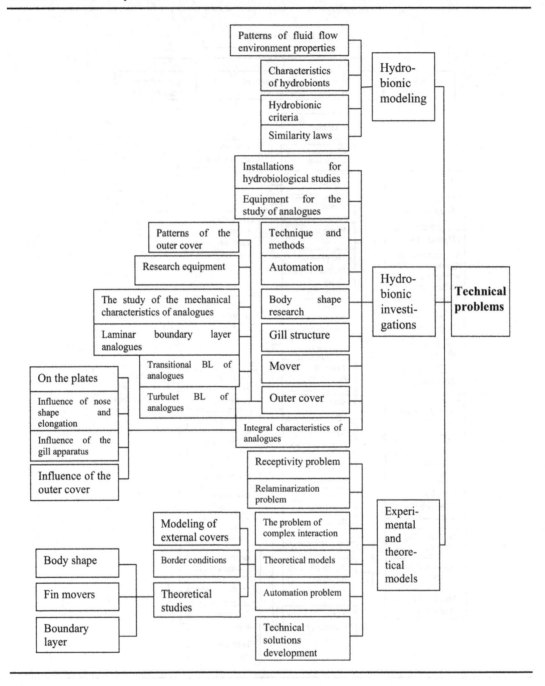

hydrobiont, it is necessary to take into account its speed and size. From the value of the Reynolds numbers, it is possible to predict the flow regime in the boundary layer. The third characteristic is the way of life, which makes it possible to understand the features of some hydrodynamic properties of the body.

Hydrobionic research methods are also divided into three areas. Methods for the study of living hydrobionts should be based on the gradual habituation of limited space and contact with humans. The research process begins with the capture. Here, and later, we will talk about dolphins as an example to explain the table. Catching is done using nets, and it is very important not to injure the animal when removing it from the net. Transportation is made in special sealed boxes, which are filled with water and covered on the inside with soft material. During transport, the protruding part of the body is constantly covered with water so as not to cause a skin burn due to exposure to the air. During the transportation of dolphins, we measured the elasticity and temperature of the skin along the length of the body. After being caught, the animals are excited and the data obtained correspond to the maximum swimming speeds. Caught animals were first placed in an open-air cage, restricted by nets, so that they got used to their new living conditions and trainer. Then the dolphins were moved to an aquarium, where they were trained and prepared for experimental research. Parts of these experiments were carried out with the help of an apparatus placed on the body in front of the vertical fin. Some experiments were carried out with dolphins placed on cradles in special boxes.

Methods of preparation and analysis performed for morphological studies should take into account the principle of interconnectedness of the body systems. Methods of morphological analysis are performed in the nodes of the grid proposed by V.V. Babenko, which takes into account the peculiarity of the body shape and the fins located on the body, taking into account the modern view of the formation of the boundary layer on the body of revolution and the features of coherent vortex structures arising from the movement of bodies [59,60,70].

Further, the features of the organism are investigated—this is the identification of the functional unity of the organism and its systems. The hydrobiont's energy is determined by direct or indirect measurements. Direct measurements are carried out by the mass of the motor muscles or by the composition of exhaled air, and indirect measurements are, for example, the characteristics of jumps in the air. However, it is necessary to make amendments taking into account the knowledge of the unity and interconnectedness of the work of various body systems.

The unity of the motor-propulsion complex consists of the adaptation of the organism to the simultaneous operation of the mover during the creation of an emphasis and its influence on drag reduction. Unity means that the mover works in accordance with the oscillations of the whole body, and that the detected musculature is associated with the skin muscle [287].

There are a number of main ways of movement: undulating; scombroidny, when only the tail part of the body works; bodywork when the fins located in various places of the body, work; flapping, when a pair of fore (penguins) or hind (seals) fins are used; and use of a hydrojet (squid). The types of movers are characterized by the hydrodynamic form and devices on the body, increasing the efficiency of the main mover. The morphological features of the structure of the body are associated primarily with the habitat, lifestyle, and characteristic Reynolds numbers when swimming. Kinematic characteristics allow us to estimate the energy of swimming, to calculate the hydrodynamic characteristics of the body when moving in water, and how these correspond to the way of life for the animal.

In the area of hydrodynamic adaptations to reduce energy consumption, the role of body shape, which can be normal, flat, laminarized, and with variable geometry, for example, in the squid, is distinguished. An important role in the adaptive functions of the organism is played by the shape of the nasal part of the body and the gill apparatus, especially in fast-swimming aquatic organisms. Adaptations in the skin are aimed at reducing drag friction and are determined by the hydrodynamic regularities of the morphological structure, the hydrodynamic features of swimming, and the mechanical characteristics of the skin.

The role of the mover is not only to create support, but also to reduce drag. In particular, this occurs due to the occurrence of suction force in the feed and the organization of the propulsion wave along the body, which changes the frequency characteristics of the boundary layer.

The most important feature for understanding the structure of hydrobionts and their systems, as well as for modeling these features in technology, is the study and systematization of the principles of the functioning of hydrobionts, which will be discussed below.

Technical problems consist of tasks of hydrobionic modeling, hydrobionic investigations, the development of experimental facilities and devices, and the development of theoretical models. Hydrobionic modeling begins with an analysis of the basic laws of habitat fluid flow and their properties. In the first approximation, the same concepts are used as in the case of studying the motion of a rigid body. However, the simulation should take into account that the laws of flow around a living deformable body during unsteady flow may make some corrections. Based on the modeling of the movement of hydrobionts, they have been characterized by the same dimensional values and dimensionless parameters as solids, for example, speed, linear dimensions, Reynolds, Strouhal, Froude, etc. However, the obtained quantitative values for hydrobionts can differ significantly from those for solids.

When studying the properties of the skin, a model equation of the motion of the cover element is compiled and dimensionless π-parameters are determined in the usual way. The laws of similarity can be refined in the process of conducting hydrobionic experimental studies on hydrobionts and their analogues.

For hydrobionic studies, special hydrobionic installations and devices have been developed, the designs of which takes into account the principles of hydrobionts. An important section of hydrobionic research is the study of the features of kinematics and power characteristics of fin thrusters using electrical and mechanical methods.

From the point of view of energy saving, it is essential to study the role of the outer covers of hydrobionts and their analogues for drag reduction. For proper modeling of the outer covers, it is first necessary to develop mechanical models of the skin and equipment, and methods for measuring their mechanical properties. When modeling and developing analogues, one should take into account the properties of the covers, which include the fact that the total tension of the skin is equal to the total frictional resistance force of a body at a given swimming speed. Experiments and calculations have shown that the degree of skin tension varies along the body. When traveling at different speeds, the level of skin tension changes. In this regard, when conducting experiments on analogues, it is necessary to consider which section along the body length the tension of the external surface of the analogues corresponds to.

The study of the mechanical characteristics of analogues should be carried out using standard instruments, and specially designed instruments for carrying out measurements on hydrobionts.

To clarify the mechanisms of interaction of the skin of hydrobionts with streamlined flow, skin analogues were developed, on which the boundary layer was studied in various hydro-aerodynamic installations under various flow conditions [44,70,111,170]. In the case that the analogue of the skin had a greater number of signs similar to the skin of hydrobiont, the characteristics of the boundary layers under different flow conditions differed significantly from the characteristics of the boundary layer measured on a rigid plate.

The results of physical investigations were tested on the developed models for the study of integral characteristics. Studies were carried out on plates and longitudinally streamlined cylinders. The role of the nasal contours and various methods of control for the coherent vortex structures of the boundary layer were also investigated.

Based on the study of biological and related technical problems, new tasks in fluid mechanics were developed and a number of promising new problems were formulated for solving by experimental and theoretical methods.

1.2 Principles of hydrobionics

The above features of the interactions of hydrobionts with the habitat are due to specific principles of the structure of the organisms, which are necessary to understand the characteristics of the body structure of hydrobionts and their systems, as well as to simulate these features in technology—the study and systematization of the principles of hydrobionts. Below the basic functional principles of hydrobionts are listed.

Ekkerhard W. Zerbst, 1987 [306]

1. The blood vessel system is a bionic model for optimal shaping of branching pipe systems.
2. Bionic aspects of the principle of counterflow of the circulatory system (A.G. Tomilin discovered and formulated this principle of counterflow in 1947).
3. Bionic aspects of biological gas exchange and transport.

Rolf Reiner, 1992 [248]

1. Accident (mutations).
2. Selection.
3. The principle of striving for a collective lifestyle.
4. The principle of modulation.
5. The principle of hierarchical interactions.
6. The principle of self-organization:
    a. Open systems (exchange of energy, matter, information with the environment).
    b. Is found far enough from the equilibrium (heat).

   c. Operational closure is detected (a prerequisite for maintaining feeding—the exchange between macrostructures and microrelations).

Werner Nachtigal, 1998 [209]

1. Integration instead of an additional construction (first formulated by V.V. Babenko in 1975).
2. Optimization of the whole instead of maximizing the individual element.
3. Polyfunctionality instead of monofunctionality (first formulated by V.V. Babenko in 1978).
4. Fine tuning in relation to the environment.
5. Saving energy instead of wasting it.
6. Direct and indirect use of solar energy.
7. Time limit instead of excessive quality durability.
8. Total recycling instead of reducing accumulation.
9. "Weave" net connections instead of linearization.
10. Development of the sequence "experience—error—process."

S.V. Pershin (1967, 1988) [223,231]

1. *The principle of balance strength buoyancy*
   The average density of the body of hydrobionts is close to the density of water, so they are balanced by the static Archimedean buoyancy force. Positive or negative buoyancy is neutralized when moving by hydrodynamic resultant forces of the hull and fins.
2. *Biological principle of convergence (similarity of features)*
   Animals of various species under long-term and consistently uniform living conditions, under the influence of natural selection, approach each other according to certain characteristics directly related to environmental exposure.
3. *The biological principle of divergence (divergence of signs)*
   Within one species there are no exactly identical organisms. Initially, their homogeneous groups in the process of natural selection in somewhat different environmental conditions specialized in different directions. As a result, species and new species appeared with more pronounced differences in structure, function, and lifestyle. For example, the elasticity of the skin of dolphins swimming at different speeds is significantly different.
4. *The biological principle of embryogenesis*
   In the process of embryonic development, the general characteristics of a large group of animals are detected earlier than special ones. Therefore, in the embryo, the sequence of the appearance of various traits basically corresponds to the historical sequence of the appearance of these traits in the ancestors of this animal, and they can be traced in embryogenesis.
5. *The principle of progressive swimming*
   With an increase in the size of hydrobionts and an increase in their level of organization, the maximum swimming speed increases substantially. This is due to the fact that the effective power of hydrobionts is proportional to the cube of linear

dimensions, and the hydrodynamic resistance to movement is proportional to the area of the wetted surface to the square of the linear dimensions of the body.

6. *The principle of cyclic movement*

    Swimming of aquatic animals is always unsteady, usually periodic, and close to harmonic. Active swimming is always alternated with inertia movement. This principle is due to the vital biological properties of the regulation of parameters of living tissues, due to the cycles of energy exchange, respiration, blood circulation, and the mechanical efficiency of unsteady swimming.

7. *The principle of unification of the mover*

    Among hydrobionts, the most common are wavy movers with elastic flexural-oscillatory complexes of various structures, which, when increasing the size and speed of swimming, are localized in the tail part of the body.

8. *The principle of relative hydrodynamic compliance*

With an increase in the Reynolds number, the areas of hydrobionts with regularly changing characteristics of propulsion and navigation methods are consistently localized in accordance with the hydrodynamic flow regimes in the boundary layers and vortex wakes. However, the characteristic Reynolds numbers are significantly different from those for solids due to the specific flexible skin and nonstationarity of flow.

V.V. Babenko (1969, 1981) [14,36,47−50,53,56,62,67].

1. *The basic principle of living organisms*

    In the process of evolution, all living organisms have developed the property of minimal energy expenditure to insure the process of life activity.

2. *The principle of receptor regulation*

    During the motion of a solid body, various hydrodynamic forces act on its surface. The solid body is not sensitive to these forces, because the designers preset such a safety margin so that the body does not deform. In all aquatic organisms and, in particular, in cetaceans, the nerve endings are located in the skin very close to the streamlined surface—at a distance of up to several tens of microns from the surface of the body. Calculations showed that the pressure and velocity of the boundary layer are felt very well by such receptors. Since the surface of hydrobionts is innervated, the force effect of the hydrodynamic field of the external environment, and especially the *gradients* of hydrodynamic loads, is sensitively perceived by a living organism, along any vortex disturbance the animal feels. This principle means that as a result of centuries-long evolution, the organism of a hydrobiont has developed such adaptations in order to eliminate the painful sensations arising under the influence of hydrodynamic and hydrostatic forces arising in the environment.

3. *The principle of interconnectedness*

    All systems in a living organism act only interconnectedly. For example, there is a functional connection between the skeleton of a dolphin and its moving ribs with the motor muscles, skin muscle, and all skin, including the circulatory and lymphatic systems, and innervation. During the oscillatory work of the propulsion unit, the mobility of the body is provided by a specific skeleton structure. At the same time, the skin muscle and membrane strands in the skin are activated so that a propulsive wave runs over the body surface. The outer structure of the skin is such that it generates coherent vortex structures of various shapes into the boundary layer. Interaction of the skin-generated vortex structures with the vortex structures of the boundary layer occurs.

4. *The principle of polyfunctionality*

Most body systems have more than one function. For example, when the driving mode changes, the oscillation frequency of the tail propulsion device changes. As shown above, this changes the mechanical characteristics of the skin and the diameter of the whole body, which affects the resistance of the whole body. Thus, in addition to the direct functions of creating a thrust, the motive muscles affect the drag reduction and maneuverability. Another example of polyfunctionality is the work of the gill apparatus of hydrobionts, in which the biopolymer can dissolve in water and form specified coherent vortex structures in the boundary layer, which affects the characteristics of the boundary layer and body resistance, and during maneuvering flow behind the gills is prevented.

5. *The principle of combined adaptive systems*

To achieve the greatest efficiency, not only one body system works, but several. For example, when flowing around tuna, a drag reduction is achieved due to the laminarized body shape, the stabilizing properties of the skin, and the presence of mucous cover on the body parts.

6. *The principle of automatic control*

All devices in the body act automatically. For example, in cetaceans, the sectional structure of the circulatory and lymphatic systems and innervation is found. In addition, motor plaques were found in each section—nerve formations that serve as sensors for static and dynamic pressure. Thus, in each section there are systems necessary for automatic regulation with a feedback system.

7. *The principle of thermoregulation*

Hydrobionts have mechanisms for regulating heat flows and, moreover, body heating during movement is directed not at the organization of the thermal boundary layer, but at regulating the mechanical characteristics of the skin, aimed at controlling the coherent vortex structures of the boundary layer and the dynamics of the body oscillations.

8. *The principle of interaction with the physical fields of the habitat*

As a result of evolution in the body, adaptations and systems have been developed that interact with the physical fields of the habitat: thermodynamic, hydrostatic and hydrodynamic, electrical, geomagnetic, etc.

9. *The principle of unity of the engine—mover complex*

The engine and the mover are interconnected as one in the body function and contribute to the automatic regulation and adjustment of ways to reduce drag.

10. *The principle of the unity of simultaneous functioning of all principles*

The higher the organization of the animal, the more principles implemented in its body. The organism functions optimally with the simultaneous functioning of all the principles of vital activity it uses.

1.3 Methods of experimental investigations and criteria for dynamic similarity

It has become obvious that, in the study of hydrobionts, the basic laws of classical hydromechanics with fluid flowing around bodies are applicable only in the first approximation. For example, experimental studies on hydrobionts have confirmed that when solving the equations of the boundary layer, it is necessary to determine new boundary conditions. To determine the true picture of fluid flow in the boundary layer of hydrobionts, it is necessary to carry out physical studies of the boundary layer on hydrobionts and their analogues. Measurements of the velocity and pressure pulsations on the body surface of a living hydrobiont performed in Refs. [165,251] showed a significant difference in the pulsation characteristics from similar parameters on a rigid body of the same length and shape. Similar measurements on analogues of the outer covers of hydrobionts also revealed differences in the physical flow pattern as compared with the hard surface in the laminar [170], transition [111], and turbulent [44,70] boundary layers.

For a closer approximation of the analogue to the outer covers of hydrobionts, blood circulation modeling was performed and the boundary layer was studied on such analogues. The mechanical characteristics of the analogue changed upon heating, and therefore, its stabilizing properties changed as compared with in the absence of a thermal boundary layer. The results of measurements of the kinematic characteristics of the boundary layer [44,70,111,170] indicate that the characteristics of the boundary layer of the analogue of the outer cover of a hydrobiont differ from those of the boundary layer in the flow around a rigid surface.

Based on this and the results of investigations conducted in Refs. [3,14,19,26,29, 30,132,146,164,165,169−173,194,228,246,247,282,291−293,309−311, etc.], it becomes obvious that the pattern of fluid flow in the boundary layer of hydrobionts differs from that for a rigid body due to the nonstationarity of movement, the operation of the propulsion complex, creating an additional suction force in the tail part of the body, the specific structure of the skin, various adaptive functions of aquatic organisms, and hydrodynamic features as noted above.

In experimental hydrobionics, there is a method for studying living hydrobionts, a method for morphological studies of the body, and a method for studying analogues of hydrobionts. The first, in turn, consists of a method investigation in natural conditions [165,223,251] and a research method in hydrodynamic installations [139,175,264]. The study of living hydrobionts should, first, be carried out with the maximum approximation to their natural life conditions. Second, it is important to determine whether a hydrobiont is trained. Research on a hydrobiont unprepared for experience in conditions unfamiliar to it causes a reaction that distorts the actual characteristics. At the same time, experiments on a trained hydrobiont allowed him to "tolerate" the sensors and made it possible to obtain reliable research results. Third, it is necessary, if possible, to obtain a "pure" measured value. Since a hydrobiont is a single mechanism in which the energetics of swimming, the propulsion complex, and the work of various systems of the body are interconnected and multifunctional, under ideal experimental conditions, it is very difficult to isolate any useful signal in its pure form.

All this leads to the need to develop additional equipment and devices in order to obtain reliable measurement results. These include various types of film boxes for underwater observations [143], the use of stereophotography, telemetry measurements, and the placement of various sensors on the body of hydrobionts [167,251]. It also provides for the measurement of the hydrodynamic characteristics of a hydrobiont as it moves by inertia.

The method of morphological studies of hydrobionts includes determining the standard geometric mesh applied to the body to determine the points of morphological analysis [14,251], the hydrodynamic analysis of the structure of the outer covers of the body, as well as its individual elements and systems.

The method of research on analogues of hydrobionts is that it is almost impossible to fulfill the laws of complete similarity for living objects. Therefore individual features of hydrobionts are distinguished and analogues are developed for studying individual functions of hydrobionts, if possible taking into account the functioning of this element in conjunction with other systems of the body of the hydrobiont.

Initially, in the study of hydrobionts, it was natural to determine the hydrodynamic characteristics typical for technical objects—the coefficients of resistance, lift, and moments. However, given the above, even these characteristics of living objects are quite difficult to determine in their pure form, as performed with standard hydrodynamic studies.

Measurements of hydrodynamic characteristics are carried out on living hydrobionts under seminatural conditions or in hydrobionic installations [15,16,18,144,146,166,167], on dead aquatic organisms, and also on their models tested in hydrodynamic installations. The following hydrodynamic characteristics are of cognitive and practical interest: coefficients of forces and moments, patterns of pressure distribution along the body during different phases of oscillatory motion, determination of hydrodynamic characteristics of the propulsion unit, as well as the body of the hydrobiont, its body elements and their interaction, measurement of the velocity of movement and characteristics of the boundary layer determination of the effectiveness of the hydrodynamic effect of the skin, and specific features of the body's hydrostatics during deep-water diving. In addition to the external hydrodynamic parameters of a hydrobiont, its internal hydrobionics is also of great interest, for example, the study of the specific features of the structure of the circulatory system and the patterns of blood flow in the vessels.

Despite these difficulties, it was possible to obtain some hydrodynamic characteristics of hydrobionts. In particular, new phenomena unknown in the technique were discovered. These include specific properties of fin propulsion, hydrobiont skin, and some body structure features, for example, concerning the hydrodynamics of the swordfish rostrum. It is believed that the rostrum allows redistribution of tangential stresses on the body to reduce hydrodynamic resistance and a painful effect on the body of hydrobionts [215]. Based on the principles of interconnectivity and the combined effect of hydrobiont devices on the flow, it was suggested that the hydrobiont rostrum is a polyfunctional organ. Along with the already-mentioned properties, the rostrum, in combination with the specific configuration of the head part of the body, forms a stream for its optimal interaction with the mover at high swimming speeds.

In the study of hydrodynamic forces, moments, and structure of flows arising in the process of movement of bodies, it is necessary to know the basic parameters that allow comparing the hydrodynamic characteristics of various bodies, regardless of their size and speeds, and the physical properties of the environment. It is especially important to know the general parameters characterizing the fluid flow when modeling the flow around bodies. The main similarity criteria and the results of experimental studies of some similarity criteria required when conducting hydrobionic studies are given in Refs. [16−19,34, etc.]. Some of the important similarity criteria are listed below [261].

The Reynolds number characterizes the ratio of inertial and viscous forces in a fluid flow:

$$Re = \frac{UL}{\nu} \tag{1.1}$$

The Froude number characterizes the ratio of inertial forces and gravity forces in fluid flows:

$$Fr = \frac{U}{\sqrt{gL}} \tag{1.2}$$

The Froude number characterizes the process of wave formation and the force of the wave nature. Sometimes, to calculate the Froude number, the expression is taken as a linear quantity.

$\sqrt[3]{V}$, where $V = D/\gamma$ is the volumetric displacement of the body at rest. Then

$$Fr_V = \frac{U}{\sqrt{g\sqrt[3]{V}}} \tag{1.3}$$

Both expressions for the Froude number are related as follows:

$$Fr_V = Fr\sqrt{\frac{L}{\sqrt[3]{V}}} \tag{1.4}$$

The Strouhal number characterizes periodic motion in a fluid:

$$Sh = \frac{UT}{L} \tag{1.5}$$

This number allows you to compare the forces and processes of flow around moving bodies in a fluid with acceleration.

The Euler number characterizes the absolute hydrodynamic pressure p at a given point of fluid flow:

$$Eu = \frac{p}{\frac{\rho U^2}{2}} \tag{1.6}$$

Often, the similar ratio expressed below is used in terms of excess pressure, $p - p_o$:

$$\bar{p} = \frac{p - p_o}{\frac{\rho U^2}{2}}, \tag{1.7}$$

where p_o is the static pressure in the fluid.

An important parameter is also the local coefficient of friction resistance, which is also expressed through dynamic head:

$$c_T = \frac{\tau_0}{\frac{\rho U^2}{2}} \tag{1.8}$$

The coefficient of hydrodynamic resistance is:

$$\xi = \frac{R}{\frac{\rho}{2} U^2 S}, \tag{1.9}$$

where R is the water resistance to the movement of the hydrobiont, U is the swimming speed of the hydrobiont, S is the wetted surface of the hydrobiont, and ρ is the density of water.

The thickness of the boundary layer has the form:

$$\delta \sim \sqrt{\frac{\mu l}{\rho U}} = \sqrt{\frac{\nu l}{U}} \tag{1.10}$$

The thickness of the laminar boundary layer is calculated by the formula:

$$\delta_{\text{лам.}} = 5\sqrt{\frac{\nu l}{U}} \tag{1.11}$$

If this expression is divided by the plate length, then we obtain the dimensionless thickness of the boundary layer:

$$\frac{\delta}{l} = \frac{5}{\sqrt{Re_l}}, \tag{1.12}$$

where $Re_l = \frac{Ul}{\nu}$.

Based on Newton's law, the shear stress on the wall is:

$$\tau_w(x) = \mu \left(\frac{\partial u}{\partial y}\right)_w, \tag{1.13}$$

where the index w means that the value on the wall is taken, on which $y = 0$. Since

$$\left(\frac{\partial u}{\partial y}\right)_w \sim U_\infty / \delta, \text{ то } \tau_w \sim \frac{\mu U}{\delta}.$$

Substituting the expression for the thickness of the boundary layer, we get:

$$\tau_w(x) \sim \mu U_\infty \sqrt{\frac{\rho U_\infty}{\mu x}} = \sqrt{\frac{\mu \rho U_\infty^3}{x}} \tag{1.14}$$

Consequently, the shear stress on the wall arising due to friction is proportional to the free-stream velocity U_∞ to a power of 2/3. Dividing the stress $\tau_w(x)$ by $\rho U_\infty{}^2$, we obtain the dimensionless shear stress on the wall:

$$\frac{\tau_w}{\rho U^2} \sim \sqrt{\frac{\mu}{\rho U_\infty l}} = \frac{1}{\sqrt{Re_l}}. \tag{1.15}$$

For the laminar boundary layer, we obtain the coefficient of tangential resistance:

$$c_f \frac{\tau_w(x)}{\frac{\rho}{2} U_\infty^2} = \frac{0.664}{\sqrt{Re_x}} \sqrt{\frac{l}{x}} \tag{1.16}$$

The total friction resistance can be obtained from the distribution along the plate of the local tangential resistance $\tau_w(x)$ by integrating the local averaged values of the friction resistance. For one side of a flat plate of width b and length l, we have frictional resistance:

$$W = b \int_0^l \tau_w(x) dx \tag{1.17}$$

For the profile of Blasius, we have [44]:

$$c_w = \frac{1328}{\sqrt{Re_l}} \tag{1.18}$$

1.4 Modeling in experimental investigations of swimming aquatic animals

The definition of similarity criteria in experimental hydrobionics is complicated by the fact that it is necessary to first determine the laws of similarity between different species of hydrobionts and other living organisms, then determine whether there are unambiguous patterns for hydrobionts in vivo and in vitro, and also define the criteria for similarity between hydrobionts and their analogues in whole or separate systems. The principles of similarity and modeling in the study of hydrobionts are underutilized. Studies conducted in Refs. [17,29,143,224,227,228,251, etc.] confirm that the basic similarity criteria used in traditional hydrodynamics can be used as a first approximation in experimental studies of hydrobionts.

In determining the similarity criteria between a hydrobiont and its analogue, the laws of complete, partial, and conditional similarity can be observed [261]. Full similarity testing of the hydrobiont and its analogue can be carried out with identical geometrical, kinematic, and dynamic conditions of their functioning. If not all physical quantities are similar, then partial similarity is realized. If it is impossible to carry out even a partial similarity, then the analogy of a hydrobiont with its model is judged on the basis of the general laws of change of the studied parameters depending on the selected independent variables. At present, it is almost impossible to realize complete similarity in experimental hydrobionics.

There are three known methods for determining the similarity criteria: analysis of the relation of forces, dimensions, and the governing equations of the system [261,283]. As an example, we consider the flow around the outer covers of hydrobionts using the theory of

TABLE 1.4 Independent variables.

Variable parameter	Designation	Dimension	Parameter characteristic
Instantaneous value average speed	$\bar{u}$	LT^{-1}	Dependent variable
Characteristic length	L	L	Geometrical parameters of the body and flow
Diameter of body cross section	D	L	
The thickness of the skin	H	L	
Longitudinal radius of body curvature	R	L	
The mutual position of the protruding parts of the body	l_i	L	
Boundary layer thickness	δ	L	
Oscillation wavelength in the boundary layer	λ	L	
The density of the skin	ρ_M	FT^2L^{-4}	Kinematic parameters of the skin and flow
Oscillating mass of skin	M	FT^2L^{-2}	
Skin viscosity (damping)	η	FTL^{-2}	
Flow viscosity	μ	FTL^{-2}	
Fluid density	ρ	FT^2L^{-4}	
Flexural rigidity	$G_{\text{н}}$	FT^{-2}	
The modulus of skin elasticity	E	FL^{-2}	Dynamic skin and flow parameters
Skin shear modulus	G	FL^{-2}	
Poisson's ratio	σ	$F^\circ L^\circ T^\circ$	
Frequency group of skin oscillations	ω	T^{-1}	
Group phase velocity of skin oscillations	c_M	LT^{-1}	
Skin tension	T	FL^{-2}	
Pressure distribution along the body, pressure module	P	FL^{-3}	
Phase angle of pressure pulsations	$\lvert p \rvert$	FL^{-3}	
Main frequency spectrum of pressure pulsations	φ_n	$F^\circ L^\circ T^\circ$	
	β	T^{-1}	
Body frequency	n or f	T^{-1}	
Time	T	T	
Skin temperature	t^o_M	Θ	Thermodynamic parameters of the skin and flow
Flow temperature	t^o	Θ	
Coefficient of thermal conductivity	k	$LM^{-1}T^{-3}\theta^{-3}$	

TABLE 1.5 List of π-parameters.

π-parameters	Characteristic π-parameters
$\pi_1 = \bar{u}/U_\infty \quad \pi_2 = l/D \quad \pi_3 = h/\delta \quad \pi'_3 = h/D$	Geometrical parameters of the body and flow
$\pi_4 = l_i/l$	Reynolds number
$\pi_5 = U_\infty l/\nu \quad \pi'_5 = U_\infty/\nu \quad \pi''_5 = U_\infty \delta/\nu$	Wave number
$\pi_6 = 2\pi\delta/\lambda \quad \pi'_6 = \frac{U_\infty R}{\nu}(\lambda/R)^{3/2}$	Kinematic parameters of the body and flow
$\pi_7 = \rho_M/\rho \quad \pi'_7 = \rho_M h/\rho\delta$	
$\pi_8 = \eta/\mu \quad \pi'_8 = \eta/\rho U_\infty$	
$\pi_9 = \pi'_5 M/\rho \quad \pi'_9 = M/\rho l^3$ $\pi_{10} = |p|h/G_u$	
$\pi_{11} = \rho U_\infty^2/E \quad \pi_{12} = E/h$	Cauchy number
$\pi''_{12} = E/h\mu U_\infty \pi'_5 2 = \nu/hU_\infty \pi_{11}$ $\pi_{13} = G/E$	
$\pi_{14} = (\pi_{12}/M)^{1/2}\delta/U_\infty$	Dynamic parameters of skin and flow
$\pi'_{14} = (\pi_{12}/M)^{1/2}U_\infty^{-1}\pi'_5 = \pi_{14}/\pi''_5 = (\pi''_{12}/\pi_9)^{1/2}$	
$\pi_{15} = c_M/U_\infty = (T/M)^{1/2}U_\infty^{-1}$	
$\pi_{16} = (\pi_{15}^2 + \omega^2/\pi_6^2)^{1/2} = (\pi_{15}^2 + \pi'_{14}\pi'_5 2/\pi_6^2)^{1/2}$	
$\pi_{17} = \sigma \quad \pi'_{17} = \Delta/\pi \quad \pi''_{17} = \pi''_5\pi'_8/\pi_6\pi_9\pi_{16}$	
$\pi_{18} = T/\mu U_\infty = \pi_9\pi_{15}^2 \quad \pi_{19} = D\sqrt{\omega\rho/\mu}$	
$\pi_{20} = \varphi_n$	
$\pi_{21} = nl/U_\infty$	Strouhal number
$\pi_{22} = U_\infty/\sqrt{gl} \quad \pi'_{22} = U_\infty/\sqrt{g^3\sqrt{D}}$	Froude number
$\pi_{23} = \delta/U_\infty \quad \pi'_{23} = U_\infty t/l$	
$\pi_{24} = t_M^o/t^o \quad \pi_{25} = (t^o - t_M^o)k/h$	Thermodynamic parameters

dimensions [261,283]. To determine the independent variables and π-parameters, we use the skin element scheme given in Refs. [14,17]. A list of independent variables is given in Table 1.4, and of the parameters in Table 1.5. The instantaneous value of the average velocity $\bar{u}$ is taken as the dependent variable. Some of the similarity criteria were described earlier [16–18,29,44].

In connection with the above features of hydrobionts, let us explain the influence of some parameters on the value $\bar{u}$. Parameter D is introduced due to the fact that the mobility of the skeleton of hydrobionts allows them to change the diameter of the body during swimming, and the parameter h due to the fact that the thickness of the skin along the body varies. When swimming, the body bends in the longitudinal direction with a variable curvature along it. Calculations and experiments on analogues show that the formation of longitudinal vortices in the tail part of the body significantly increases the stability of the profile of the transition boundary layer. In this regard, the important parameter is the value of R. In hydrobionts, the relative position of the fins and the body varies considerably, the lateral fins are very mobile, and in some hydrobionts they can be folded into special indentations on the body [215], therefore the parameter l_i is introduced.

The kinematic characteristics of the boundary layer along the skin analogue differ from the same characteristics when flowing around a hard surface, which indicates a change in the ratios of the characteristic thicknesses of the boundary layer. To account for these features, the parameter δ is introduced. Alternatively, the displacement thickness δ^* or impulse loss thickness δ^{**} can be applied. Justification of kinematic skin and flow parameters is given in Refs. [16,44,70].

Due to the fact that in high-speed hydrobionts the traction is mainly realized by the caudal fin, the body of the hydrobiont can be represented as a multilayer oscillating beam of variable cross section, clamped at one end. The parameter G_i allows you to take into account the effect of the amount $\bar{u}$ of bending stiffness of the body or skin. Additional parameters can include the moment of inertia, the moment of resistance, and the flexibility of such a beam.

The choice of parameters E, G, T, ω, and c_M is justified in Refs. [19,29,44,70]. On the parameter ω, we note the following. In Ref. [33], a photograph of transverse microfolds on the skin of a living dolphin is shown, which was previously shown on a section of dolphin skin in vitro [14]. During swimming, when a body bends along a hydrobiont, a wave of variable amplitude advances [264]. On the concave side, in this case, a strip of microfolds with wavelength λ is formed. The frequency caused by these microfolds of longitudinal nearwall pulsations in the boundary layer can be determined by the formula $f = c/\lambda$, where c is the traveling wave velocity [143]. The frequency f also depends on the curvature of the body, determined by the speed of swimming.

The values of p, $|p|$, φ_n, and β characterize the field of pressure pulsations formed in the boundary layer with unstated oscillations of the skin and body of a hydrobiont. Measurements of pressure pulsations on the surface of hydrobiont showed [146] that, with slow swimming, the frequency of pressure pulsations of the skin was 130–140 Hz, and during hops it was 115–230 Hz. When a single pulse is applied, vibrations with a frequency of 600 Hz propagate on the surface of the plexiglass, 130 Hz on the surface of the skin analogue, and 157 Hz when it is temperature controlled. By changing the mechanical parameters of the skin, hydrobionts are able to influence the field of pressure pulsations in the boundary layer, which must be taken into account in modeling.

As shown in Ref. [15], there is no thermal boundary layer on the surface of the body of hydrobionts. However, with the help of blood circulation it is possible to change the mechanical parameters of the skin, and, consequently, all parameters that depend on temperature. To account for these factors, the parameters t_M^o, t^o, and k, are entered.

The characteristics of a series of π-parameters (Table 1.5) are evident from the above and Table 1.4. The parameter π_6 in the interaction of the skin with the laminar boundary layer characterizes the wave number of unstable oscillations in the boundary layer [44,69], and in other flow regimes—the main energy-carrying vortices. In addition, π_6 can be determined by the wavelength of microfolds and longitudinal irregularities in the skin of a hydrobiont. The parameters $\pi_7-\pi_9$ are explained in Refs. [16,44]. The parameter π_{10} characterizes the distribution of the pressure field on the flexible body and depends on the controlled characteristics of the body [19,227] and environmental parameters. Parameters $\pi_{11}-\pi_{14}$ characterize the effect of skin elasticity on the boundary layer [44], π_{15} and π_{16} the propagation velocity of disturbances in the skin of hydrobionts, π_{17} the damping properties of the skin, π_{18} the effect of tension in the skin on the boundary layer, and π_{19} the ratio of inertial properties of the oscillating pressure field to viscous flow properties. In connection with the oscillatory motion of the body, parameters π_{21} and π_{22} are introduced.

Simulation in experimental hydrobionics is as follows. Initially, the values and types of π-parameters of hydrobionts in ontogenesis and phylogenesis are determined. Then, on the basis of the developed hydrodynamic and morphological methods, the peculiarities of hydrobionts are investigated and the corresponding analogues are developed. Finally, modeling is performed between the hydrobiont and its analogue, taking into account complete, partial, and conditional similarities.

To perform a complete similarity between a hydrobiont and its analogue, it is necessary that the equality $\pi_{1h} = \pi_{1a}$ be fulfilled, if the relations

$$\pi_{2h} = \pi_{2a}, \pi_{3h} = \pi_{3a}, \ldots, \pi_{25h} = \pi_{25a}.$$

With partial similarity, it is necessary that only part of the relations be fulfilled, and with conditional similarity, the π-parameters can be approximately equal for a hydrobiont and its analogue. As can be seen from Table 1.5, π-parameters characterize the geometry, properties, and forces. Measurements on in vivo hydrobionts of some characteristics showed [15,16,29,146] that in ontogenesis and phylogenesis these characteristics may differ significantly, and their changes along the body or depending on length and body weight are determined by power dependencies [29,171]. Considering the above, it is clear that in order to perform correct modeling, it is necessary to investigate the π-parameters of adult, healthy aquatic organisms at characteristic swimming speeds, in which the corresponding systems of the body are worked out and tuned to the process of evolution.

With this simulation, it turned out that π-parameters become constant and are within the following limits: $\pi_3 \approx 1 \div 2$; $\pi'_6 \approx 16$; $\pi_7 = 1$; $\pi'_7 = 1 \div 2$; $\pi_9 = (1 \div 20) \cdot 10^4$; $\pi''_{12} = (0.2 \div 6.5) \cdot 10^{-5}$; $\pi'_{14} = (0.2 \div 5) \cdot 10^{-5}$. Some other π-parameters measured also turned out to have constant values. To determine the remaining π-parameters, a series of complex and thorough researches on hydrobionts needs to be performed.

However, on the basis of the measurements already made, it can be expected that the hydrobionts and parameter π_1 will be constant. This gives grounds to perform the correct modeling of a hydrobiont. For example, the basis of geometric and kinematic π-parameters [14,17,18] were first developed in simplified [29], and then more complex [22,23,27,35,57] analogues of the skin of hydrobionts. As can be seen from Refs. [44,136,143,165,170], compliance with some dynamic π-parameters allowed us to obtain new results.

In experimental hydrobionics, it is more economical to perform simulations when performing partial or conditional similarity. The development of a technical solution, with the manufacture and study of an analogue of the skin of a hydrobiont with the help of specially developed installations and devices [44,70,170] significantly save time and allowed real devices to be obtained that gave a positive effect. In addition, the positive results of experiments on analogues [44,70,103,170] designed in accordance with the laws of partial similarity allow us to state that the morphofunctional patterns discovered do exist in nature.

Some aspects of the hydrodynamics of fast-swimming aquatic organisms are considered in Refs. [61,95,97].

When modeling in experimental hydrobionics, it is necessary to know the basic positions of the boundary layer [91,95,128,148,152,153,256,262,285, etc.].

Kinematic and dynamic parameters of swimming aquatic animals

2.1 Basic hydrodynamic forces and moments

When analyzing the main forces and moments arising on the body of floating hydrobionts, we will use the accumulated knowledge in aero-hydrodynamics [96,97,106,181,198,240,298]. In the general case, the aero-hydrodynamic force R and the aero-hydrodynamic moment M act on the moving body, which are usually decomposed into components along the axes of the velocity in a relative coordinate system. The force R is applied at the center of gravity (CG) of the moving body, and the moment M rotates the body relative to the axes passing through the CG. A rectangular coordinate system is considered. In the velocity coordinate system, the projections of the force R and moment M are denoted by X, Y, Z, M_x, M_y, and M_z. At the same time, for a positive value (by sign) of a moment about an axis, such a moment is considered which tends to turn the body around the axis counter-clockwise. Since the x axis is directed toward the movement of the body, the projection of the force X will have negative values. Introduce the symbol of the force $Q = -X$.

The projections of the aero-hydrodynamic force on the axis of the velocity coordinate system have the following names: the projection on the opposite direction of the x axis (Q) is called the drag force; the projection of force R on the direction perpendicular to the speed of movement and directed upwards along the axis of coordinate y is called the lifting force Y. The third projection of force R is directed in the transverse direction along the axis of coordinate z and is called the lateral force (Z). In the English literature, frontal drag is denoted by D (drag), and the lifting force is denoted by L (lift), and in German, respectively, by W (Widerstand) and A (Auftrieb). The ratio of lift to drag is called aerodynamic quality:

$$k = \frac{Y}{Q} \tag{2.1}$$

Quality is the tangent of the angle of inclination of the force R to the direction of the incident flow. Under conditions of normal horizontal movement, one can consider the drag force approximately equal to the thrust force T, and the lifting force Y is balanced by the weight of the moving body G. The aerodynamic quality is equal to the ratio G/T. For the best modern forms of bearing surfaces at low speeds, the aerodynamic quality is 35–40, for airplanes it is 10–20, and for easy-flowing forms of rotation bodies it is 3–5.

The formula for body drag is:

$$Q = -\frac{\rho_\infty V_\infty^2}{2} S \int \left[\bar{p} \cos(p,x) + \bar{\tau} \cos(\tau,x) \right] \frac{d\Sigma}{S}, \tag{2.2}$$

where ρ is the mass density of the medium, V_∞ is the velocity of the undisturbed flow, S is the characteristic area of the body, for example, the area of the maximum cross section, the area of the wetted surface and others, $\bar{p}$ and $\bar{\tau}$ are nonmeasured normal and tangential stresses acting on the elementary surface area of the body area $d\Sigma$ corresponding to the normal body surface area.

If we denote the integral in the above formula in terms of the corresponding coefficient, for example, in c_x, then the considered components of the forces will be written in the following form:

$$Q = c_x \frac{\rho_\infty V_\infty^2}{2} S, \tag{2.3}$$

$$Y = c_y \frac{\rho_\infty V_\infty^2}{2} S, \tag{2.4}$$

$$Z = c_z \frac{\rho_\infty V_\infty^2}{2} S. \tag{2.5}$$

The dimensionless coefficients c_x, c_y, and c_z are called the coefficient of drag, lift and lateral force, respectively.

Based on the obtained expressions, it can be stated that the aero-hydrodynamic force R when the body moves in the medium is proportional to the dynamic pressure $\frac{\rho_\infty V_\infty^2}{2}$, squared linear dimensions of the body S and depends on the dimensionless coefficient c, corresponding to the shape of the body and the flow conditions:

$$R = c \frac{\rho_\infty V_\infty^2}{2} S, \tag{2.6}$$

where $c = \sqrt{c_x^2 + c_y^2 + c_z^2}$.

Expressions for body moments are written in the same way:

$$M_x = m_x \frac{\rho_\infty V_\infty^2}{2} SL, \tag{2.7}$$

$$M_y = m_y \frac{\rho_\infty V_\infty^2}{2} SL, \tag{2.8}$$

$$M_z = m_z \frac{\rho_\infty V_\infty^2}{2} SL. \tag{2.9}$$

The dimensionless coefficients m_x, m_y, and m_z are the coefficient of moment of heel, moment of yaw, and moment of pitch, respectively. The expression for the total moment will be:

$$M = m \frac{\rho_\infty V_\infty^2}{2} SL, \qquad (2.10)$$

where $m = \sqrt{m_x^2 + m_y^2 + m_z^2}$.

The aero-hydrodynamic moment is proportional to the dynamic pressure, the cube of linear dimensions, and is characteristic for the shape of a particular body and the conditions of its flow around a certain dimensionless moment coefficient. Thus, to determine the aero-hydrodynamic characteristics of each body, it is necessary to determine the above three aero-hydrodynamic force coefficients and three aero-hydrodynamic moment coefficients.

Determining the indicated drag coefficients and moments for a given body is a difficult task. In practice, these coefficients of individual elements of the body and then in the assembly of a specific body are determined on models in modern aero-hydrodynamic installations. The results obtained are verified in real-life experiments. The accumulated statistics are used in the preliminary design. As science progresses, the data obtained are refined in the light of new modern knowledge.

2.2 Components of hydrodynamic resistance of bodies

The main components of the resistance of moving hydrodynamic bodies are expressed in general form by the equation:

$$R = R_{inductive} + R_{form} + R_{viscous} + R_{ledges} + R_{interference} + R_{spray} + R_{aerodynamic} + R_{vortex} + R_{non\text{-}stationary}$$

$$(2.11)$$

The physical substantiation of some components of resistance is shown in Fig. 2.1 [297], which shows how when moving near or on the free surface of a liquid, the body creates a system of waves in the liquid depending on the speed of movement, the shape of the body and the distance from the surface of the liquid (Fig. 2.1A). The process of formation of waves by the body leads to a change in the velocity field and pressure, including along its surface, to varying degrees when the body moves under a solid boundary (Fig. 2.1A, on the left) or in an unlimited fluid (Fig. 2.1A, on the right).

As a result, a pressure force arises, the projection of which on the direction of motion is called the wave resistance R_{wave}. The work created by the body to overcome this force is spent on creating the energy for formation of waves. Wave resistance is a characteristic part of the resistance of bodies moving near or on the free surface. In addition, the wave resistance occurs in the case of the existence of wind waves on the surface of the liquid.

If a moving body could deform in a similar way, as well as the induced wave motion on the water surface, then depending on the phase shift of the induced wave and the wavelike motion of the body, it would be possible for the body to accumulate some of the energy of the induced wave motion of the water. In other words, in this way, the wave resistance of the body can be significantly reduced. If the length of the body is

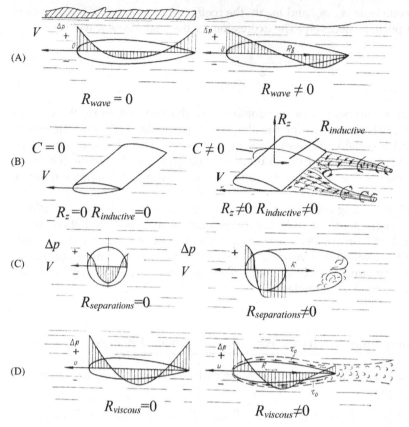

FIGURE 2.1 The scheme of the main components of the resistance: (A) wave; (B) inductive; (C) separations; (D) viscous resistance [297].

small compared with the length of the induced or existing wave motion, then the wave resistance can also be reduced. Dolphins, for example, have learned to use the energy of the induced wave motion: they either glide waves on the surface along the slope of the wave, or when moving near the water surface they use the wave energy by setting the tail and side fins at the appropriate angle of attack. The problems of wave energy utilization are discussed in detail in the papers [174,232]. Wave resistance significantly depends on the speed of movement and the shape of the body contours. Wave resistance tends to zero for small or large Froude numbers. Therefore the transition to movement in these areas may lead to its reduction or complete elimination. With an increase in the depth of motion below the surface of the water, the wave resistance decreases significantly.

Some vehicles have wings of low elongation for various purposes, including steering wheels. It is known that, according to the theorem of N.E. Zhukovsky, the lifting force rises when a profile flows around as a result of circulation C of velocity around a profile (circulating a vector field along a given closed loop G). Fig. 2.1B, on the left, shows the axisymmetric profile of the wing at a zero angle of attack. When flow is around it, there is no circulation of velocity around the profile, so there is also no magnitude of lift and inductive resistance. The presence of circulation C means that the wing can be replaced by a system of vortices, which continue behind the wing in the form of chords, called free vortices (Fig. 2.1B, to the right). The system of vortex chords is pulled together in a pair of longitudinal large vortices located in the direction of the edge of the wing. Work on overcoming the force of inductive resistance is spent on creating the energy of these vortices. Inductive resistance is inversely proportional to the wingspan and depends on the shape of the wing in the plan. The structure of the fins in aquatic animals makes it possible to have a variable and controlled distribution of the angles of attack along the wingspan, depending on the kinematics of its oscillation. On certain parts of the wing's trajectory, the angles of attack are zero.

The elongation of the wing of the small elongation is expressed by the formula:

$$\lambda = \frac{l^2}{S} = \frac{l}{b}, \tag{2.12}$$

where λ is the elongation, l is the wingspan, S is the wing area, and b is the wing chord. According to the theory of large wing elongation

$$c_{pi} = (\pi\lambda)^{-1} c_L^2 (1 + \delta); \tag{2.13}$$

$$c_L = (\pi\lambda)^{-1} \frac{2\pi\lambda\alpha}{2 + \sqrt{\lambda^2 + 4}}, \tag{2.14}$$

where c_{pi} is the coefficient of inductive resistance, c_L is the coefficient of lift, α is the angle of attack of the fins, and $(1 + \delta)$ is a factor that takes into account the deviation of the geometric shape of the wing in terms of the optimal shape in the sense of the smallest inductive losses. For rectangular wings, δ usually does not exceed 0.15. The minimum value of δ for wings with an elliptical circulation distribution (δ) is zero. Therefore it is rational to design the wing by changing the angle of the chords (twist) and the shape of the profile along the span to obtain the required elliptic law [198].

In Fig. 2.1C shows a diagram of the formation of components of resistance, depending on the form of the body of the R_{form} using the example of a flow around a circular transverse cylinder. The left of the figure shows the pressure distribution along the cylinder at a low flow velocity at which there is no flow separation. With an increase in the flow velocity, a separation occurs behind the cylinder, depending on the flow velocity. Form resistance is determined by the flow in the wake behind the wing, the emerging nature of the flow in the boundary layer. For an oscillating wing, the separation behind the wing is alternately discharged into the stream. In a moving apparatus, the form resistance is mainly determined by the shape of the apparatus and is determined by the components given in Section 2.1.

Fig. 2.1D provides a diagram of the formation of the viscous component of the resistance of a moving body. The viscosity of the fluid leads to the formation of a boundary layer along the surface of the moving body, as well as the body trace behind a streamlined body. As a result, the structure of the velocity field in front of the body and behind it in a viscous fluid is different, and the structure of the jet in a trace consists of either a turbulent flow or vortex depending on the shape of the body and the speed of movement. For example, at low speeds, eddies characteristic of a laminar or transition boundary layer are formed on a streamlined body. If the shape of the stern of the body has a shape, behind which a separation is formed, then large vortices develop in the track. In all cases, this leads to energy costs and a decrease in pressure in the stern of the body. The part of the resistance that occurs due to tangential stresses on the surface of the body is called friction resistance. The part of the resistance that characterizes the poorly flowing rear part of the body, leading to the formation of separation vortex, is called form resistance. Work related to overcoming the components of friction resistance and form resistance is expended on the creation of a boundary layer and a jet in the wake of the body. Ultimately, the development of vortex structures in a wake causes dissipation of the vortex energy. All the energy formed in the wake turns into heat. Since the nature of the components of friction resistance and form is in the presence of viscosity in the flow, these two components are usually combined into one and are called viscosity resistance:

$$R_{viscous} = R_{friction} + R_{form} \qquad (2.15)$$

In the viscosity component, the main contribution is friction resistance. The coefficient of longitudinal fullness φ characterizes the distribution of the underwater volume along the length of the body. The smaller the φ, the more volume is concentrated in the middle part of the body, and the pointedness of the extremities of the body increases. This is advantageous when the Froude numbers do not exceed 0.3–0.35. In Fig. 2.2 it is shown that, depending on the speed of movement, the structure of the boundary layer may be different. At low speeds of movement along the body, only a laminar boundary layer is formed. With increasing speed, a transition boundary layer is also formed. At high speeds of movement on a greater extent of the body there is a turbulent boundary layer (TBL). V. V. Babenko performed numerous experimental studies of the vortex structure of the transition and TBL. In Refs. [44,70], the corresponding models of the shape of the disturbing motion are given for different flow regimes in the boundary layer. The methods of controlling coherent vortex structures (CVS) under various flow regimes in the boundary layer have been experimentally investigated.

Resistance R_{ledges} is determined by all protruding parts, streamlined air, and water flows. The interference component of the resistance $R_{interference}$ arises due to the interaction of individual parts of the body, for example, the hydrofoil and the struts on which the wing is fixed. This component is determined by the vortices that occur in the corners of the articulation.

Spray resistance component R_{spray} occurs in vehicles moving on a water surface when splashes from the body or protruding parts downstream hit the body of the device. This component also comes from splashes breaking away from the waves in strong winds.

In shipbuilding the size of $R_{aerodynamic}$ denotes the resistance of the protruding parts of the hull. Behind the protruding parts or behind the angular joints of the body parts,

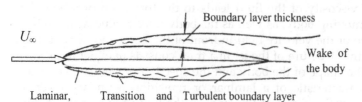

FIGURE 2.2 The development of the boundary layer depending on the Reynolds numbers.

vortices appear that act downstream on the body of the apparatus, which causes an additional vortex resistance of the body R_{vortex}.

Resistance component $R_{non-stationary}$ is one of the most significant among the other components of the resistance. This component of confusion arises when moving on the wave surface of the water or when the body is maneuvering, and is determined by the additional mass added. The visualization is performed by tinted color jets (Kreplin AIAA, 1988). The ellipsoid is located at the angle of attack. The experiment was performed in the hydrodynamic channel of the Institute of Aerodynamics (DFVLR, Gottingen) at low Reynolds numbers. This experiment was performed at a stationary flow.

At the Institute of Hydromechanics of the National Academy of Sciences of Ukraine (NASU), Kiev, numerous experiments on nonstationary flow past canonical bodies have been carried out. The features of the flow structure during acceleration and deceleration of bodies were revealed [86,88,125,206]. The character of flow around the bodies differs significantly depending on the acceleration or deceleration of the bodies. The results obtained are particularly important in assessing the magnitude of the resistance depending on the size of the water surface wave.

It is known that during nonstationary motion along the direction of motion or when maneuvering along a trajectory, there are areas along the path in which the body moves at different angles of attack and with acceleration or deceleration. During acceleration, conditionally as it were, a certain mass of fluid is added to the body, which is called the added mass. When braking occurs, negative acceleration (deceleration) occurs. As a result, the added mass of the fluid also appears and is directed along the trajectory of motion. In this case, the negative attached mass seems to catch up with the body and create additional thrust. Theoretically, it is possible to calculate such a law of nonstationary of motion, which will consist of a section of acceleration and deceleration. If the acceleration is short term, as compared with a longer braking section, the energy spent on accelerating the added mass will be somewhat or completely compensated for at the braking section.

In the simplest case, the equation for the motion of a body with variable speed is:

$$m\frac{dv}{dt} = -\lambda_{11}\frac{dv}{dt} - R_{x-non-stat.} \pm P_e, \qquad (2.16)$$

where m is the mass, dv/dt is the acceleration, λ_{11} is the coefficient of the added mass, $R_{x-non-stationary}$ is the longitudinal nonstationary component of resistance, and P_e is the thrust force. The first component of the right side is the hydrodynamic force of inertial nature, which is calculated taking into account the added mass. Work on overcoming this force is spent on changing the kinetic energy of the fluid surrounding the body as it moves with acceleration. Since the magnitude of the kinetic energy and the process of its change significantly depend on the presence and properties of the fluid boundaries, the value of the added mass will be different when the same body moves at a great depth or on the free surface, and also depends on the nature of the free surface deformations, that is, from the Froude number. In Ref. [157] reference data are given for the calculation of the values of the added mass for bodies of various shapes under various conditions of motion. The second term on the right-hand side of the equation is the resistance force calculated taking into account the effect of acceleration on its magnitude. The method of calculating this component is given in Ref. [297].

2.3 Geometric parameters of the bodies of aquatic animals

When hydrobionts move at a depth equal to the length of their body, most of the components of Eq. (2.11) are insignificant or zero. The most significant are the components of form resistance, viscosity, and nonstationary. In Ref. [175], calculations of the swimming characteristics of fish are given in accordance with the methodology for a rigid body moving stationary. It is argued that the resistance of the fish depends on the shape and Reynolds and Strouhal numbers. The dependences of the shape factor on the ratios of the main body sizes L/B and B/H are given, where L is the body length, and B and H are the width and height of the body, respectively, as well as the dependence of the friction resistance coefficient on the position of the maximum section along the body length and the Reynolds number. An expression is obtained for determining the swimming speed of fish for a given duration of maintaining swimming speed.

Fig. 2.3A shows the distribution of the maximum values of the dimensionless longitudinal pulsation velocity $\sqrt{(u')^2}/U_\infty$ along the longitudinal coordinate x, measured by different authors. Fig. 2.3B is diagram of the formation of CVS in the boundary layer with different flow regimes in the boundary layer [68,70]. Roman numerals denote the boundaries of characteristic types of CVS, and Reynolds numbers correspond to Re_{ls}—the critical number of stability losses, $Re_{kr.1}$—the critical number of the beginning of the transition boundary layer, and $Re_{kr.2}$—the critical number of the

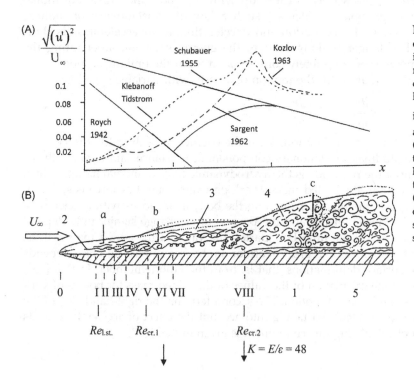

FIGURE 2.3 The scheme of the development of CVS in the boundary layer of a rigid plate: (A) the dependence of the dimensionless longitudinal pulsation velocity $\sqrt{(u')^2}/U_\infty$ of the boundary layer on the x coordinate; (B) formation of CVS along the plate: (1) rigid plate; (2) laminar; (3) transition; (4) turbulent boundary layer (TBL); (5) viscous sublayer of TBL; (a, b, c) locations of sensors for measuring pulsation speeds.

onset of the TBL. Solid arrows show two energy jumps at $Re_{kr.1}$ and $Re_{kr.2}$, defined by the formula:

$$\overline{K} = E/\varepsilon, \qquad (2.17)$$

where $E = \sqrt{(u')^2}/U_\infty$ is the maximum value of the root-mean-square (rms) value of the longitudinal pulsation velocity through the thickness of the boundary layer, and ε is the degree of turbulence of the unperturbed flow.

Usually, the value of ε is determined by the sum of the dimensionless rms values of the components of the pulsation velocity. The most important is the longitudinal pulsation speed. Disturbances of the main flow, penetrating into the boundary layer, interact with the CVS of the boundary layer and cause the development of CVS along the length of the streamlined body. Experimental studies Ref. [70] have shown that in the boundary layer the longitudinal pulsating components increase and their maximum values are located at a certain distance from the streamlined surface, depending on the stage of the CVS development, indicated in Fig. 2.3 in roman numerals.

It was found that at stages VI (the beginning of the transition boundary layer) and VIII (the beginning of the TBL) changes in the structure of the CVS are accompanied by sharp jumps in the pulsating energy calculated by formula (2.17), for example, at $\varepsilon < 0.05\%$ and $E = 1.6\%$, the jump in energy is $\overline{K} = 48$. More details of the diagrams shown in Figs. 2.3 and 2.4 are explained in the following sections.

Fig. 2.4 is a diagram explaining the distribution of various types of loads acting on a rigid axisymmetric body or on a wing profile and in the case when the body surface has a layer of elastic coating. Dashed vertical lines indicate the boundaries of sections of an unregulated uniform elastic surface, lines in the form of dots indicate modified boundaries of the distribution of load parameters and new boundaries of sections after the selection of elastomers with optimal characteristics.

Roman numerals denote the numbers of the corresponding sections.

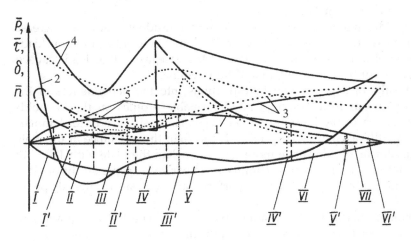

FIGURE 2.4 Standard load distributions along the wing profile: (1) outer contours of the wing profile; (2) pressure $\overline{p}$ distribution along the x axis; (3) boundary layer thickness distribution δ; (4) tangential stress $\overline{\tau}$ distribution; (5) distribution frequency range $\overline{n}$ distribution in laminar and TBL.

The boundaries of the sections are determined by the following requirements:

- Within each section, the gradients of the parameters of the loads must be constant or varying linearly.
- In the area where the sections are connected, the parameters should not contain jumps of values.

Fig. 2.4 illustrates patterns showing that the elastic surface along the wing chord is subjected to various static and dynamic loads. At a constant flow rate in section I, the pressure gradient $\bar{p}$ is maximal, that is, the pressure here drastically decreases, but, remaining positive in sign, compresses the elastomer. The shear stress gradient $\bar{\tau}$ is also maximal, which implies a rapid change in the shear load in the elastomer. In the same section is the high-frequency region of intense laminar velocity pulsations. Due to the maximum gradient δ and its minimum value, the heat transfer gradient is at the maximum. Therefore, in section I, the elastic surface must be sufficiently rigid to be resistant to high loads and their gradients. In this case, the requirement for effective damping of high-frequency velocity and pressure pulsations should also be satisfied. The positive effect may be enhanced by dividing section I into a series of subsections.

In section II, the above values and their gradients change less intensively and do not change sign; the frequency range and the frequencies of unstable oscillations themselves are reduced in this section. Therefore the mechanical properties of the elastomer should differ from those in section I, for example, the surface damping should increase slightly. The magnitude $\bar{p}$, in contrast to in section I, acts here on separation.

In section III, the gradient $\bar{p}$ is positive, and the gradient $\bar{\tau}$ changes sign, that is, the load to break away decreases. The range and frequency oscillations of velocity in the boundary layer also tend to a minimum.

In section IV, the gradient $\bar{p}$ practically becomes zero, that is, the elastomer is subjected to mainly shear stress $\bar{\tau}$. The shear stress $\bar{\tau}$ rises to a maximum and its gradient changes sign. In addition, due to the transition of the boundary layer from laminar to turbulent, the range of magnitude $\bar{n}$ expands dramatically, and the law of growth δ and the coefficient of thermal conductivity also change. In section V, all values change only slightly and can be considered approximately constant.

Sections VI and VII are distinguished by an increase in magnitude $\bar{p}$, that is, an increase in the load on the tear away. In addition to these loads, you should add vibration loads caused by nonstationarity of movement and vibration of the base, as well as loads caused by changes in the properties of the environment and flow regime.

Fig. 2.5 shows the results of experimental investigations into the dependences of the resistance coefficients of various forms of rigid nondeformable profiles on the Reynolds number, including the suction of the boundary layer [256]. The resistance of all laminarized and normal (NACA 0012) axisymmetric profiles is less than the resistance of the plate at TBL. Normalized curvilinear profiles have more resistance than the plate with TBL, but they have the greatest lift coefficients. The accumulated experience in the study of rigid axisymmetric bodies and profiles allows us to perform an analysis of the shape of hydrobionts.

Figs. 2.6−2.8 are photographs of fast-swimming hydrobionts. The fastest aquatic animals are the *Xiphias* fish: swordfish, sailfish, marlin, and tuna. The Reynolds numbers of their swimming can reach $Re_L \approx 10^8$. In Fig. 2.8 are photographs of birds that spend some time underwater when diving. Olushi is a small family of birds from the order of

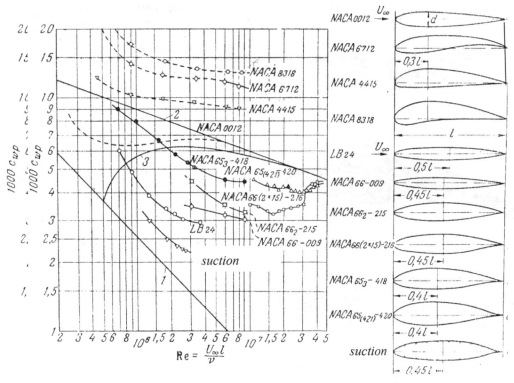

FIGURE 2.5 The dependence of the resistance of various shapes of wing profiles on the Re number: (1) the resistance of a rigid flat plate with a laminar, (2) transition, and (3) TBL [256].

pelican-like (copepods). Olushi are related to birds such as cormorants, pelicans, and phaetons, and there are a total of nine known species of these birds. Olushi are large, with a mass of 3–3.5 kg and a wingspan that can reach 1.3–2 m. The bodies of these birds are streamlined, the neck is of medium length, and the tail is short. Under the skin of the booby in the forehead area are special areas that serve as airbags, that is, they act as shock absorbers when diving. Having folded its long wings along its body, at a speed of about 97 km/h, this bird rapidly dives into water to depths of more than 25 m. The height from which they begin their descent can be 10–30.5 m.

The geometric parameters of hydrobionts should provide minimal hydrodynamic resistance. It is also necessary to take into account the distribution of the load parameters along the body (Figs. 2.3 and 2.4). At the top of Fig. 2.9 is a laminarized profile of the series B TsAGI aircraft wing. On such a profile, the boundary layer remains laminar for 80%–90% of the wetted surface, that is, such a profile has minimal friction resistance and minimal thickness of the boundary layer and the wake behind the profile. Below the profile of the wing are the vertical projections of cetacean bodies: II, orca; III, cetacean dolphin; IV, short-beaked common dolphin; V, sei whale; VI, gray whale; VII, sperm whale; and VIII, smooth whale. It is seen that all bodies have a laminar form. Moreover, the geometry of

FIGURE 2.6 Photos of fast-swimming hydrobionts: (1) short-beaked common dolphin (A) side view, (B) rear view; (2) orca whale; (3) sperm whale; (4) *Carcharodon megalodon*; (5) Adelie penguin (*Pygoscelis adeliae*).

the profile of the body is such that the resulting lifting force of the profile of the body in some cetaceans is directed upwards and in others downward.

It is associated with a lifestyle and corresponds to diving and surfacing. On the one hand, cetaceans can change the shape of their body, unlike solid bodies, and change the direction and location of the application of the resulting hydrodynamic forces acting on their body. On the other hand, cetaceans have almost zero buoyancy and can move in fluids using different body positions, which contributes to their effective maneuverability.

In Fig. 2.10 a comparison of the contours of the bodies of the sperm whale, sei whale, and dolphin [270] is illustrated, and in Fig. 2.11 the laminarized form of an orca (*Orcinus orca*) is illustrated in Ref. [231]. Ref. [175] provides a comparison of the contours of the profiles of the NACA-63, NACA-66, and NACA-67 series. The NACA-63 profile has the

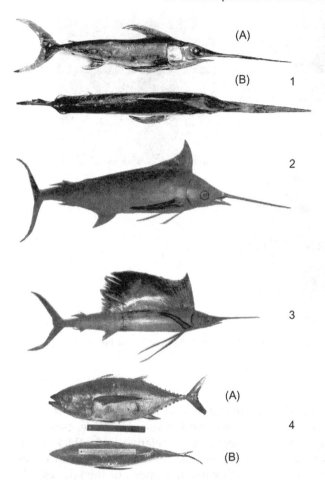

FIGURE 2.7 Photos of some species of fast-swimming *Xiphias* fish: (1) swordfish (A) side view, (B) top view; (2) marlin; (3) sailfish; (4) tuna (a) side view, (b) top view [70].

shape of a trout body, and the wing profiles of the NACA-66 and -67 series are similar in shape to dolphin and tuna bodies. A comparison of the body profiles of fast-swimming sharks—the Australian mako, the white shark, and the thresher shark—is also given in Ref. [231]. The profile of the white shark (upper half of its body) is very similar to the profile of NACA-65, and the profiles of mako and thresher sharks are comparable with the NACA-67 profile and very similar to the tuna profile.

H. Hertel [126,127] reviewed various aspects of hydrobionic studies, in particular, the body shape of various hydrobionts. It is argued that, compared with the shape of modern aircraft, the bodies of large, fast-swimming marine animals have an ideal shape, which allows them to maintain a laminar boundary layer over a large body length and to minimize resistance during fast swimming, with the large relative thickness of the body shape allowing a large mass of locomotor muscles and subcutaneous fat to be accommodated.

These sketches give the approximate profiles, as they show more presumptive "streamlined shapes" than real "laminar shapes." At the same time, they correctly reflect the

FIGURE 2.8 Photographs of high-speed birds when flying in the air: (A) blue-footed booby (*Sula nebouxii*); (B) petrel; (C) blue-footed booby when diving into the water.

relative thickness of the body d/l, where d is the thickness and l is the length. From this scheme, it can be seen that tuna with $d/l = 0.28$ have the highest d/l value. Swordfish have $d/l = 0.24$, and whales and dolphins (porpoises) have a relative body thickness of $0.21-0.25$. The relative thickness values are approximates, as there are significant differences between the species of the family and the individual specimens of the species. The general principle is that the bodies of these marine animals are very thick and that they swim very fast and continuously.

The pointed dolphin head has an advantage over the wing profile. The tuna body also has a laminarized form. An estimate of the body shape of one of the tuna species is also given in Ref. [215]. The profile of the tuna body (Fig. 2.6) corresponds to the profile of NACA-67-021. The boundary layer of up to $x/l = 0.7$ can remain laminar. The oscillation of the tail of the tuna allows the use of a laminarized form in such a way that there is no separation in the tail of the body. Comparison of the tuna and dolphin forms indicates how, in the wealth of animals and fish inhabiting the sea, the body of a laminarized spindle type is best adapted for fast, long-term swimming [127].

The velocity and pressure values at potential flow, thickness of extrusion, and loss of impulse, normal, and tangential components of velocity in a laminar boundary layer, tangential stresses on the wall, several series of rotation bodies with different shapes of meridional section were calculated.

Calculations of laminar flow around 47 models of rotation bodies have been performed. The calculations were tested on a model—the standard in a low-turbulent hydrodynamic tray of the IG SB AS USSR. On this model, the pressure value at the body boundary, its total

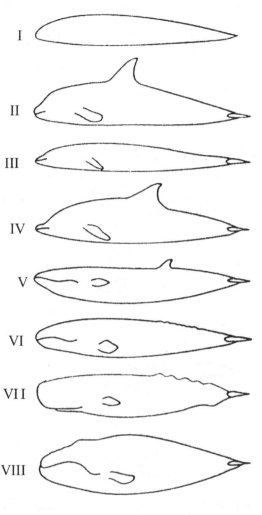

FIGURE 2.9 Profiles of series B TsAGI and some species of cetaceans: (I) wing profile; (II) orca; (III) cetacean dolphin; (IV) white-necked dolphin; (V) sei whale; (VI) gray whale; (VII) sperm whale; (VIII) smooth whale.

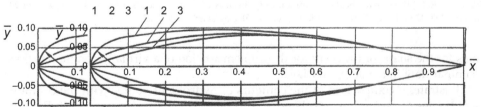

FIGURE 2.10 Comparison of the contours of the body of revolution, equivalent to the shape of the body of the sperm whale (1) Ref. [192], sei whale (2) Ref. [184], and dolphin (3) Ref. [233].

resistance, the profiles of the tangential velocity component in the boundary layer, and the separation point were determined experimentally. The calculated material obtained made it possible to select the body of revolution with minimal resistance and determine the

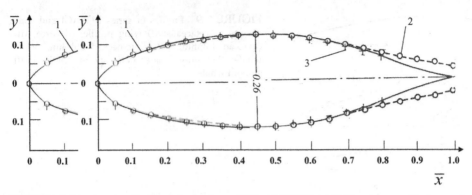

FIGURE 2.11 Laminarized orca body shape [231]: (1) NACA-66-026 laminar profile for equivalent body of revolution, (2) reduced to symmetrical lateral, and (3) symmetrical abdominal (+) orca profiles.

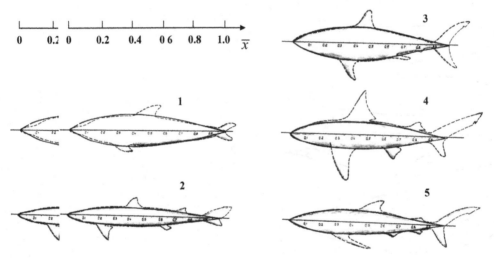

FIGURE 2.12 Comparison of the contours of hydrobionts (dashed lines) and models of rotation bodies (solid lines), calculated in Ref. [9]: (1) short-beaked common dolphin, model 32; (2) Black Sea shark Katran, model 22; (3) blue shark, models 32 and 25; (4) mako shark, model 25; (5) gray shark, model 23.

requirements for the shape of the body for the implementation of a virtually continuous flow around a laminar flow in the boundary layer.

Fig. 2.12 illustrates a comparison of the contours of hydrobiont bodies with the contours of models of bodies of revolution calculated according to the method described in Ref. [9]. The contours of the bodies of short-beaked common dolphin, the Black Sea shark of the qurans, $l = 1.05$ m; $U_\infty = 5$ m/s; blue shark (*Prionace glauca*), $l = 1.5$ m; $U_\infty = 10$ m/s; mako shark (*Isurus oxyrinchus*), $l = 1.55$ m; $U_\infty = 20$ m/s, and gray shark (*Hexanchus griseus*), $l = 1.1$ m; $U_\infty = 10$ m/s. Fig. 2.12 shows separately the longitudinal axis of the body and the magnitude of the dimensionless coordinate. In Refs. [231,280] theoretical drawings of Azovka dolphins, white-necked and bottlenose dolphins, as well as tables with relative geometrical sizes of cetaceans are presented.

In Refs. [72,73], the influence of the shape of the nasal contours of an axisymmetric body and the Reynolds number on the body resistance were studied, and in Ref. [40] the impact of the vortex structures in the flow on the characteristics of the boundary layer were studied.

2.4 Geometric parameters of the caudal fin

The shape of cetacean fins has an optimal geometry, corresponding to the best aerodynamic profiles. Fins of cetaceans have the ability to change their geometry, and the tail fin changes its span and shape cyclically during the process of oscillation. An analysis of the shape of the longitudinal sections of the caudal fin of the Dall's porpoise with a span of $l_z = 0.49$ m was performed in Ref. [231]. With distance from the longitudinal axis of the fin, the maximum profile thickness decreases from 19.6% to 15.7% of the chord length of this fin, and the maximum section distance from the profile tip increases from 15.7% to 19.6% of the chord length. The contours of the tail fin profiles were also compared with the contours of the Zhukovsky Nezh-1 and NACA 66_3 aviation profiles, and a graph comparing the profiles of the caudal and dorsal fins of various dolphin profiles with the Nezh-1 profile is also presented. In Refs. [98,100] it is stated that the indicated profiles of the caudal and dorsal fins of fast dolphin species also coincide in form with the little-known special airborne profiles of Wortman FX-05-19 and Eppler EA-6, designed for the range $Re_b = (0.5-3.0) \cdot 10^6$. The profiles of the longitudinal sections of the caudal, dorsal, and pectoral fins of the Black Sea bottlenose dolphin are given in Ref. [166]. It is argued that they almost coincide with the symmetric NACA profile with a large maximum thickness of 18%−26%, 25%−35% separated from the front edge of the profile. The profiles of the cetacean fin sections are similar to the contours of the shape of their bodies in the horizontal plane in accordance with the correction for the Reynolds number.

Fig. 2.13C shows a photograph of the caudal fin of the fin whale [270] and Fig. 2.13D shows the short-beaked common dolphin caudal fin [13], taken, respectively, on the deck of a whaling ship and on lodgment after the dolphin was lifted from a pool. Fig. 2.13A shows the shape of the caudal fin of a slow sperm whale and a fast-swimming fin whale, reduced to the same span [270]. The form of the sperm whale is almost triangular and has a maximum wetted surface, and its relative elongation is 4.18, and the relative elongation of the fin is 6.7 with the same sweep. The shape of the fin of fast hydrobionts reduces the inductive resistance and friction resistance of the fin surface, which increases the efficiency of the fin movements. Shown in Fig. 2.13B are profiles of middle sections of whale tails in comparison with the Zhukovsky profile which shows that the profile of the sperm whale fin has a maximum thickness $\bar{x}_{max} = 0.25$ and is closest to the Zhukovsky profile, while the maximum thickness of the more high-speed fin whale tail is $\bar{x}_{max} = 0.3$ [270].

Fig. 2.14 shows photographs of the caudal fin of the short-beaked common dolphin (A) when entering the water after a jump, the prepared caudal fin of the swordfish (B) and its middle part (C) (the longitudinal axis of this fin is shown by a solid line) and the vertical fin of the short-beaked common dolphin (D) on which a ruler is located. The pectoral and dorsal fins have laminarized profiles and are arrow-shaped with low profile and inductive resistances. In the area of the ruler in Fig. 2.14D on the body of the dolphin there is a smooth trough, which helps to reduce the harmful effects of large vortices forming in the

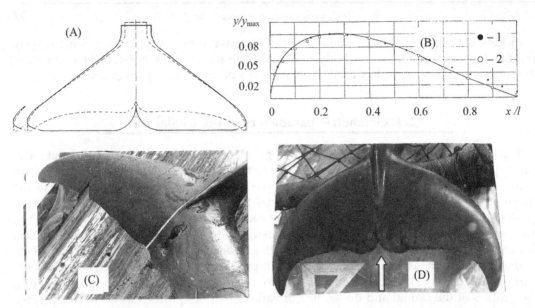

FIGURE 2.13 Comparison of the contours of the caudal fins of various cetaceans: sperm whales (solid curve) and fin whale (dashed curve) (A); (B) sperm whale (1), fin whale (2) and Zhukovsky's profile (solid curve); (C) photograph of the caudal fin of the fin whale; (D) photograph of the caudal fin of the short-beaked common dolphin.

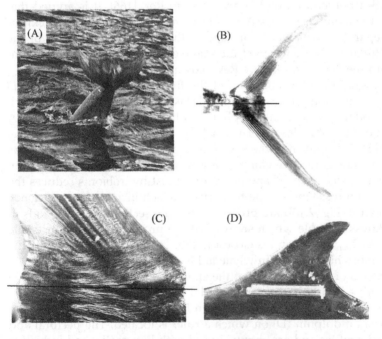

FIGURE 2.14 Photos of the tail fin of the short-beaked common dolphin (A), swordfish (B and C) and the vertical fin of the short-beaked common dolphin (D).

area of the junction of the hull and the vertical fin. In aviation, there is a rule for smoothly changing the magnitude of transverse areas along the longitudinal length of an aircraft. The dorsal fin, which creates additional resistance, serves as a stabilizer. The lateral fins are mobile and play a large role in the stabilization and control of movement, and also serve as horizontal rudders in the dolphin. It can rotate them, take them aside and press to the body at high speeds. It is known that experimental measurements showed a significant dependence of the pressure distribution on the side fins on the angles of attack. Therefore, the lift coefficient increases 30 times when increasing the angle of attack from 0 to 6 degrees and decreases 2.5 times when changing from 0 to −6 degrees relative to the value at a zero angle of attack [183,184]. The V-shaped arrangement of the side fins increases the stability of the movement of the dolphin.

The main value of the tail stem of a dolphin is in the oscillatory movement of the tail fin. The stem consists of articulated vertebrae and muscles that provide it with a bending motion. The muscles terminate in tendons extending to the vertebrae of the stem, which has lateral thickening and which helps to reduce the resistance of the stem during its oscillatory movements in the vertical plane and to increase the lateral surface, which ensures the implementation of complex maneuvers in the horizontal plane.

The tail fin, which is the main method of propulsion for the dolphin, performs flexural-oscillatory movements with the help of dorsal and abdominal muscles in the same plane as the stem. It is divided into two lobes, elongated in the transverse direction and connected to the stem by means of a movable articulation, which allows the fin to perform oscillatory movements relative to the stem and, consequently, to control the flow of the caudal fin. Four tendon strands attached to the dorsal muscles are connected to the caudal fin, while the abdominal muscles are connected to the fin by only three tendon strands. The cross sections and dimensions of these strands are the same. Alternating tension of the upper and lower tendons causes upward and downward movement of the stern fin. A more detailed study of the structure of the motor muscles of the tail section of a dolphin is available in Refs. [1,4]. The trailing edge and end portions of the fins are more elastic than the fin itself. Under the action of hydrodynamic forces, the lateral ends of the fins are bent in the direction opposite to the vertical movement of the fin (Fig. 2.14A). This bending contributes to the rapid formation of end vortex chords, which significantly reduce the inductive resistance of the caudal fin as a small elongation wing and increases its efficiency. The tail fin possesses, in addition, another important structural feature: its rear edges in the area of the longitudinal axis of symmetry of the body have smooth curves (in Fig. 2.13D indicated by an arrow). These curves of the right and left parts of the caudal fin overlap with each other. The morphological structure of the caudal fin allows the span and area of the caudal fin to change at different speeds, which also significantly affects the inductive resistance of the fin. In addition, these axial curves make it possible, when the tail fin is repositioned, to flow a certain amount of fluid in the region of the axis of symmetry from one side of the fin to the other. It also affects the inductive resistance and continuity of the flow past the caudal fin, in the wake of the body and the thrust-weight of the dolphin. The noted deformation features of the dolphin's caudal fin are clearly visible in Fig. 2.14A. It can be seen how the fin's shape deforms upon entering the water, which reduces its resistance.

Fig. 2.14B and C shows the design of the tail fin of a swordfish, which also allows the span of the fin to be changed. The body structure of a swordfish is given in described in Ref. [127]. In Fig. 2.14C specific longitudinal ordering of the outer fin cover is clearly visible, which allows longitudinal vortex structures to be formed that reduce fin resistance.

In Refs. [225,231], data on the structure of the caudal fins of various hydrobionts are given. In particular, the patterns of the features of the structure of the caudal fins are shown in

Fig. 2.15. The geometrical similarity of the shape of the caudal fins in various dolphin species is confirmed by the general linear dependence of the fin span λ on the body length L:

$$\lambda = 0.24\,L \qquad (2.18)$$

The body length of the investigated dolphins was within $1\,\text{m} \leq L \leq (3–5)$ m. For larger orcas, the lengths were $5\,\text{m} \leq L \leq 9$ m:

$$\lambda = 0.24 + 0.48(L - 5) \qquad (2.19)$$

Based on the analysis of the connection of the tail fin area S with the dolphin body length L, the dependence for $1\,\text{m} \leq L \leq (3–5)$ m was obtained (Fig. 2.15B):

$$S = 0.015\,L^2 \qquad (2.20)$$

Comparing the graphs in Fig. 2.15A, and their dependencies Eqs. (2.18)–(2.20), which determine them, we obtain an expression for the relative span λ of the dolphin's tail fins for $l \leq L \leq 3$:

$$\lambda = \frac{l^2}{S} = 3.85 \qquad (2.21)$$

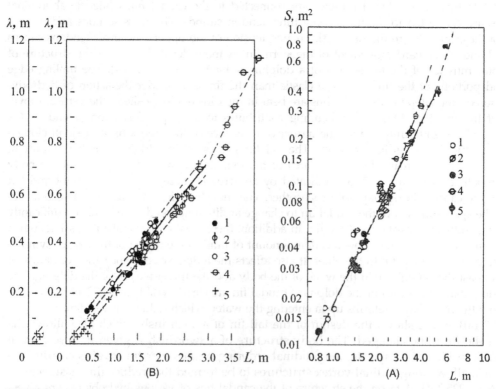

FIGURE 2.15   Dependences of the swing (A) and area (B) of the caudal fin on the body length. For (A): 1, ordinary dolphin; 2, bottle-nosed dolphin; 3, porpoise; 4, Atlantic beluga; 5, large whale sei whale embryos; for (B): 1, porpoise; 2, bottle-nosed dolphin; 3, common dolphin; 4, Atlantic beluga; 5, rock whale [231].

Thus, $\lambda \approx 4$ is the optimal value of the relative span of the tail fin of small dolphins. For a large orca with $L = 6$ m, $\lambda = 5.75$.

Fig. 2.16 shows the dependence of the area of the caudal fin S on the body length l or its mass P, as well as the dependence of the body length on the mass [43].

Based on the publications in the Bionica collections [22,31] and their own measurements [43], the indicated regularities were obtained (Fig. 2.16). Some of these data were extrapolated when the value of P was calculated from measurements of the volume of the body, taking into account neutral buoyancy. These patterns allow us to make three main conclusions:

1. The form of patterns is uniform for all types of hydrobionts. From this we can conclude that the mechanisms of their adaptations are associated with hydrodynamic characteristics of motion.
2. The most universal dependence is $S(l)$. Hydrodynamic resistance depends on the magnitude of the elongation and the length of the body. To overcome the resistance, hydrobionts need to develop a propulsion force created by the caudal fin. This dependence determines the ratio of S to the length of the body in the oscillatory mode of motion.

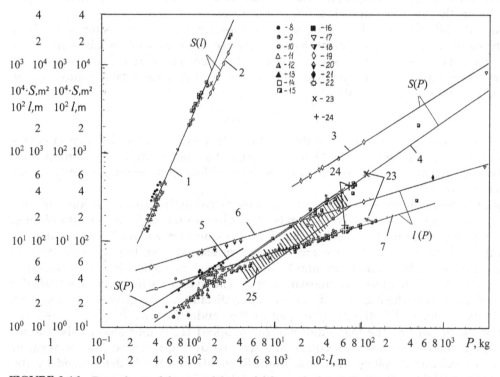

FIGURE 2.16 Dependence of the area of the caudal fin on the length (1, 2) and mass (3, 4, 5) of the bodies of hydrobionts, as well as the dependence of their length on the mass (6, 7) of pelamida (8), tunny tuna (9), yellowfin tuna (10), mackerel (11), striped (12), longfin (13), spotted (14), big-eyed (15); and in sand sharks (16), herring (white) (17), mako (18), blue (19), marten (20), six-tapped (21), katran (22); Azovka dolphins (23), short-beaked common dolphins (24); dependence of wing area on bird mass (25).

3. The difference in the laws of $S(P)$ and $l(P)$ is determined by the range of velocities of movement of hydrobionts. As in Ref. [231], all species of hydrobionts are classified by the value of U. The body mass determines the muscular energy or energy reserve of the body. However, in hydrobionts, the same P value can be contained in different volumes of the body, that is, the body has a different wetted surface and different hydrodynamic resistance. The speed of movement is also influenced by the degree of perfection of the adaptive functions of the outer covers of hydrobionts. All this determines the peculiarities of the dependences $S(P)$ and $l(P)$ for various types of hydrobionts.

2.5 Swimming speeds of aquatic animals

It is necessary to know the real swimming speeds of hydrobionts in order to assess the compliance of the adaptations of the body systems with the hydrodynamic laws of flow around the body. There are three characteristic swimming speeds: slow navigation with unlimited travel time, cruising speeds that animals can withstand for 15–20 minutes, and forced speeds that animals can maintain for a few seconds.

In Refs. [71,109,169,175,231] there are sources of foreign and domestic studies in which these types of hydrobiont movement speeds were recorded, as well as tables in which data on movement speeds for the specified types of movement are given. It was proposed to normalize the averaged maximum swimming speed of cetaceans, taking into account the main determining factors [231]:

$$U_{max} = U(F, L, T, B),$$

where F is the form of the body, L, is the length of the body, T, is the duration of swimming at a given speed, and B is the biological state of the animal, which is determined by the impact of enemies and herds, environmental conditions (temperature, salinity, water transparency, etc.).

Table 2.1 shows the short-term impulse maximum swimming speeds of some species of cetaceans, averaged over several seconds. According to the table, for small dolphins $T = 3$ s, for large dolphins (orca, beluga whale) $T = 5$ s, and for whales $T = 10$ s. Table 2.1 shows the biological length of cetaceans, which is measured from the tip of the nose to the fork of the caudal fin. The maximum L_m and average L zoological length of each species are averaged over numerous measurements. The average values are: $\overline{L} = L/H$, the relative elongation of the body; $\overline{x} = x/l$, the relative distance of the greatest cross section from the body tip; $\overline{l} = l/L$, the relative span of the caudal fin; and $\overline{h} = h/l$, the relative height of the dorsal fin. The maximum navigation speed U_m is indicated as short-term, impulse, maintained for several seconds, and reached during instantaneous removal in the state of maximum voltage. The maximum Reynolds number is determined by the values of L and U_m. These are approximate data on swimming speeds, they were obtained by observing wild animals at sea and partly during experiments in a specially made channel as a result of long training sessions.

The relative span of the tail fin decreases with an increase in the Reynolds number in the range $\overline{l} = 0.22 - 0.24$. The exception is the high-speed orca and low-speed whales—gray and sperm whales—with $\overline{l} = 0.26 - 0.29$. The tail fin stands for the size

TABLE 2.1 Kinematic and dynamic parameters of hydrobionts.

Cetacean species	Length of body (m)		Relative values		Fin sizes		Maximum speeds		Reynolds number
	L_{max}	L_{midl}	L/H	x/L	l/L	h/L	U_{max}	U_m/L	$10^{-6}Re$
Dolphins									
Short-beaked common dolphin	2.6	1.6	5.3	0.42	0.24	0.1	13.2	8.2	16
Striped dolphin	2.4	1.8	4.6	0.42	0.23	0.08			
Pantropical spotted dolphin	2.1	1.9	5.7	0.4	0.22	0.1			
Northern right whale dolphin	2.5		6.5	0.45	0.19	0			
Common bottlenose dolphin	3.6	2.3	5.1	0.4	0.23	0.09	12.5	5.4	22
Pacific shortheaded dolphin	2.5	2.2	6	0.38	0.25	1			
Pilot whale	6	4	5.2	0.38	0.23	0.07	11.3	2.8	35
Orca	10	6			0.28	0.16	13.4	2.2	62
Black finless porpoise	1.6	1.2	4.5	0.4	0.26		6.8	5.2	
Harbor porpoise	1.8	1.4	4.8	0.38	0.24	0.06	8.6	6.1	6.3
									9.2
Dall's porpoise	2.1	1.7	5	0.4	0.24	0.08	10.4	6.1	13
Beluga whale	6	4.5	4.3	0.4	0.24		6.1	1.35	21
Whales									
Sei whale	18	15	5.8	0.38	0.25	0.035	1	1.02	180
Fin whale	24	20	6.2	0.41	0.21	0.025	13	0.65	200
Blue whale	30	24	6.3	0.43	0.22	0.013	15	0.63	280
Humpback whale	18	12.5	4.5	–	–	0.022	7.5	0.6	72
Gray whale	15	12	4.9	0.32	0.26	0	7	0.58	65
Sperm whale	20	15	5.2	0.31	0.28	0.025	7.2	0.48	83

of the body. The average relative height of the dorsal fin is different in dolphins and whales, as well as in dolphins of various species. The orca has the highest mean relative height of the dorsal fin: $\bar{h} = 0.1$ (females) and $\bar{h} = 0.16$ (males), with the maximum value for males $\bar{h} = 0.22$. In whales, the average relative height of the dorsal fin is small.

In Table 2.1 classification is given to the following species of cetaceans. Dolphins: short-beaked common dolphin (*Delphinus delphis*), striped dolphin (*Stenella coerulealbus*), pantropical spotted dolphin (*Stenella attenuata*), northern right whale dolphin (*Lissodelphis borealis*), common bottlenose dolphin (*Tursiops truncatus*), Pacific shortheaded dolphin (*Lagenorhynchus obliquidens*), pilot whale (*Globicephalus melas*), orca (*Orcinus orsa*), black finless porpoise (*Neomeris phocaenoides*), harbor porpoise (*Phocaena phocaena*), Dall's porpoise (*Phocoenoides dalli*), beluga whale (*Delphinapterus leucas*). *Whales*—sei whale (seyda-noy) (*Balaenoptera borealis*), Fin whale (*Balaenoptera physalus*), blue whale (*Balaenoptera mysculus*), humpback whale (*Megaptera nodosa*), gray whale (*Eschrichtius gibbosus*), and sperm whale (*Physeter catodon*).

Fig. 2.17 shows the dependence of the maximum relative swimming speed U_m/L of 18 species of cetaceans on the length of the body L. The graph does not show the dynamic similarity of the system, but the relative speed of cetacean fast (*I*) and slow (*II*) species. At the same dimension, the ratios for lines *I* and *II* are written as [personal communication]:

$$\frac{U_{m.}}{L} \approx k_s L^{-0.915} \tag{2.22}$$

$$U_{m.} \approx k_s L^{0.085}, \tag{2.23}$$

where the dimensional coefficient is:

$$k_s = \frac{U_m}{L}\big|_{L=1} = 5.5 \div 12. \tag{2.24}$$

This coefficient corresponds to the "potential" rapidity of embryos of length $L = 1$ m for various cetacean species. In Ref. [231] it was shown that one of the embryonic forms of small dolphins and large whales (with an embryo length of 1 m) formed the shape of the body and fin. They are programmed as a result of evolution for various species of cetaceans, which with further growth are born in water and begin to actively swim immediately. It can be approximately obtained from the dependencies shown in Fig. 2.17 that the comparative potential speed of various types of cetaceans k_s is:

- short-beaked common dolphin (*D. delphis*), sei whale: 12;
- common bottle-nose dolphin, orca: 11.4;

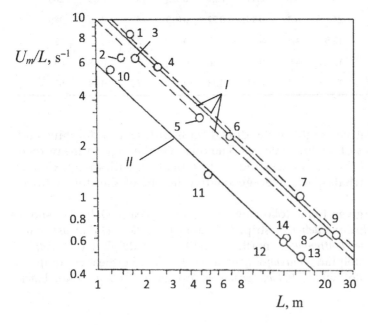

FIGURE 2.17 Dependence of the relative maximum swimming speed of fast (*I*) and slow (*II*) cetacean species on body length: (1) short-beaked common dolphin, (2) harbor porpoise, (3) Dall's porpoise, (4) common bottle-nose dolphin, (5) pilot whale, (6) orca, (7) sei whale, (8) fin whale, (9) blue whale, (10) unbred porpoise, (11) beluga whale, (12) gray whale, (13) sperm whale, (14) humpback whale [231].

- pilot whale, fin whale: 10;
- harbor porpoise: 8.2; and
- beluga whale, gray whale, sperm whale: 5.5.

From relations (2.22) and (2.23) it can be seen that the length of the body has little effect on the absolute maximum swimming speed of cetaceans. This is of biological significance for the swimming together of cetacean families with small calves.

An important characteristic of hydrobionts is the duration of the characteristic swimming speed. In Ref. [231], the maximum swimming speeds of hydrobionts were systematized depending on the load-bearing time (Fig. 2.18). The following types of swimming duration T are distinguished: 10−29 min, a frightened animal; 2−4 h, pursued by a vessel; 1 day, migrating.

It is shown that with an increase in the duration of swimming, the speed decreases. The maximum swimming speed of trained dolphins is significantly lower than dolphins swimming in natural conditions. Thus, the speed of movement is determined by the time during which the hydrobiont can withstand the load. In addition, the speed of movement is determined by the size of the musculature, that is, the size or length of the animal.

In Ref. [231], a comparison of the hydrodynamic parameters of various hydrobionts and manmade objects [294], whose sea trials were carried out under laboratory and full-scale conditions, was presented in the form of a four-parameter graph $L - U_m$, $Re_L - \overline{U}_m$ (Fig. 2.19).

On the graph, each point for cetaceans corresponds to the average size and maximum swimming speed the species. For manmade objects, the size in length and speed are shown

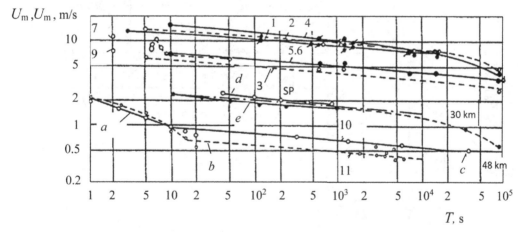

FIGURE 2.18 Dependences of the absolute maximum speed of swimming of hydrobionts on the duration of the load bearing: (1) short-beaked common dolphin; (2) orca; (3) beluga whale; (4) blue whale; (5) gray whale; (6) sperm whale; and for those trained in captivity; (7) pantropical spotted dolphin; (8) Pacific bottlenose dolphin; (9) Pacific short-beaked common dolphin. Experiments with fish in hydrodynamic pipes: (a) impulse, (b) long, (c) migratory mode. Sports swimming (SP) champions: (d) with flippers, (e) freestyle. (10) Salmon 0.18 m long; (11) rainbow trout 0.24 m long [231].

according to the source data. At present, the speeds of manmade objects have increased, their speed can be refined, for example, by referring to Ref. [294]. Hydrobionts have characteristic parameter and criteria values. Fish have a wide range of absolute maximum swimming speeds. Cetaceans are large, but the range of maximum swim speeds is limited. In Fig. 2.19, arrow A and the oblique dashed lines show the range of swimming speeds of fish. Swordfish and other xiphoids have maximum swimming speeds. The horizontal arrow B and the horizontal dashed lines show the cetacean swimming speed range. Solid lines and letters CP and TT on the graph indicate the region of the parameters of hydrodynamic models, tested, respectively, in cavitation pipes and towing tanks. The majority of high-speed hydrobionts in their individual motion parameters have the same characteristics as modern high-speed manmade means of moving in an aquatic environment. Some hydrobionts have speeds higher than those of manmade objects. This is observed at the maximum swimming speeds of hydrobionts, which can be maintained only for a short time. At the same time, the power supply of hydrobionts is substantially less than that of manmade objects.

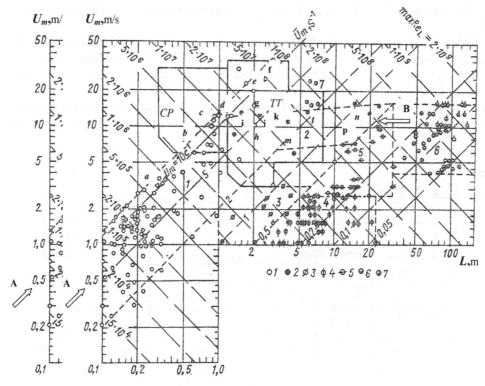

FIGURE 2.19 Hydrodynamic parameters of hydrobionts and manmade objects. Four-parameter chart. (1) Fish: (a) trout, (b) salmon, (c) wahoo, (d) barracuda, (e) tuna, (f) swordfish, (g) mako shark, (h) blue aul; (2) dolphins: (i) short-beaked common dolphin, (j) bottle-nosed dolphin, (k) pilot whale, (l) orca, (m) beluga; whales: (n) striped whale, sei whale, fin whale, blue whale, (p) gray, hunchbacked, sperm whales; (3) transporters of light divers; (4) self-propelled manned vehicles; (5) ultra-small submarines; (6, 7) foreign submarines and torpedoes. CP and TT are the regions of the parameters of hydrodynamic models tested in cavitation pipes and towing tank [231].

2.6 Equipment for investigation of the laws of swimming aquatic animals

To understand the features of economical functioning of hydrobionts, it is necessary to conduct experimental studies in natural conditions, however, it is not possible to measure this with sufficient accuracy. Therefore, in experimental hydrobionics, as in classical hydromechanics, it is necessary to carry out studies of hydrobionts and their analogues with reversed motion. When developing hydrobiological installations, devices, etc. it should be borne in mind that the test conditions of a hydrobiont should either differ be as close as possible to its natural conditions. Due to the great complexity and laboriousness of creating such facilities, studies were first conducted in a basin ($15 \times 7 \times 2$ m^3) located at the Biological Station of the Karadag branch of the Institute of Biology of the Southern Seas (Crimea) and in a network enclosure ($40 \times 8 \times 5$ m^3) located near the coast [243]. On the body of trained dolphins, sensors were installed in front of a vertical fin on suckers or in the region of a vertical fin on a belt or fixed by other methods [6,253,265,266]. Some of these results will be presented in the future. The existing experimental facilities and small specially made facilities for preliminary hydrobionical studies were used [166,175,222]. Over time, the plants were upgraded for specific studies with hydrobionts. Additional equipment and devices were developed in order to obtain reliable measurement results. As not all analogues can be investigated in standard hydrodynamic or existing hydrobionic installations, equipment and devices have been developed for studying analogues of hydrobionts [70,115,125,145,156,166,170, etc.].

The Institute of Hydromechanics of the NASU has designed and manufactured several new experimental facilities for the study of hydrobionts and their analogues. Previously, an annular biohydrodynamic installation was made for studying the kinematics of swimming fish, which was improved and a new modified installation constructed [244]. In both installations, the principle of studying fish swimming was carried out on the basis of the use of visual-motor reflexes (opto-motor responses) of the objects under study during reversed movement. In this case, the fish make swimming movements, but remain in the same place. This is convenient for filming. The difference from the known similar ring installations, which have a number of drawbacks, was that the water and the walls of the channel were stationary. From the outside of the transparent walls of the channel, a screen was moved, formed by alternating black and white vertical stripes. The opto-motoric reaction consisted of the fact that the fish under study remained in place, but made movements with different parameters determined by the speed of movement of the screen. The annular channel (2) has a cross section of 0.25×0.25 m^2, and its outer wall is made of organic glass. The inner diameter of the channel is 1 m, and the outer diameter is 1.5 m. The installation is mounted on the lower frame (4), which is an octagon shape, with a circumference of 2.18 m. The annular channel is mounted on the upper octagonal frame (7), which, with the help of eight vertical posts (5), is attached to the lower base (4). Under the channel there is a rotating wheel (3) with a painted white vertical wall (1), on which black vertical metal strips of different width are installed with an arbitrary pitch being fixed. Wheel (3) is mounted on a vertical shaft (6), which through a two-stage V-belt transmission (9) with a gear ratio $n = 25$ is driven by an electric motor into rotational motion.

For an experimental study of the kinematics of swimming in high-speed fish, a biohydrodynamic tube was constructed, in which the opto-motor response of the fish under investigation was used due to the black grid in the work area at the bottom [139]. The biohydrodynamic installation consists of a biohydrodynamic pipe installed on the base frame, a DC electric motor, and measuring and recording equipment. The water in the pipe is driven by a three-bladed screw with a diameter of 0.4 m. The flow velocity varies within 0.5−8 m/s. To obtain speeds less than 0.5 m/s, bypass device (7) is used. The honeycomb cell size is 25×24 mm^2, the nozzle has a fivefold preload and is made in the Vitosinsky form. The working part of the rectangular cross section has dimensions of $0.4 \times 0.4 \times 1.8$ m^3. Rectangular well (26) has a cross section of 0.25×0.35 m^2 and is designed to load the studied hydrobionts. Mirror (10) with a size of 0.6×1.5 m^2 is installed at an angle of 45 degrees to the side wall of the work area and allows pictures of the movement of hydrobionts to be taken simultaneously in two planes.

To study the kinematics of dolphin swimming, V.P. Kayan designed a canal that was manufactured at the Institute of Hydromechanics of the NASU and assembled in the area of the Cossack Bay (Crimea). Fig. 2.20 shows the experimental setup. Channel (1) is made of a metal frame installed on a solid foundation, ensuring the preservation of the horizontal structure and straightness of its longitudinal axis.

The length of the channel is 65 m and with a cross section of 2×2 m^2 it provides the necessary length of working area in which the animal moves at a uniform speed. The transverse dimensions of the channel allow the hydrobiont to move in "comfortable" conditions. In the middle of the channel the side wall is glazed for filming. On the opposite wall of the glazed area (2) there is a grid (3) with a cell size of 0.5×0.5 m^2. For catching an animal there is a penstock (4). The animal is loaded into the canal by means of a hoist crane (5) and a specially designed cage (6) in which the animal is placed from the basin of

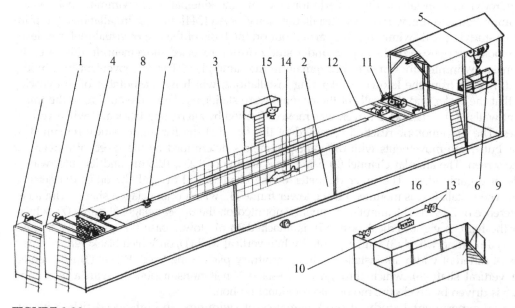

FIGURE 2.20 Diagram of a biohydrodynamic device for studying the kinematics of hydrobiont swimming: (1) channel, (2) glazed section, (3) coordinate grid, (4) pen grid, (5) hoist crane, (6) dolphin box, (7) small trolley, (8) monorail, (9) platform for equipment placement, (10) control panel, (11, 12) electric drives, (13, 15) cine cameras, (14) test animal, (16) electric stop [175,244,245].

the aquarium. To train an animal for swimming speed, a trolley (7) with a lure is provided, moving along a monorail (8) installed above the free surface of the water along the longitudinal axis of symmetry of the channel. Opposite the glazed part of the channel, a platform (9) is installed to accommodate the necessary equipment when conducting experiments. The remote controller (10) allows you to turn on the electric drive (12), which sets in motion the trolley (7) and the electric drive (11), which sets the driving grid (4). In the field of view of camera (13), there is an electric stopwatch (16), which is turned on simultaneously with the cine camera (13).

2.7 Swimming features of aquatic animals

The swimming of aquatic organisms in natural conditions is always unsteady. The characteristic features of the swimming of aquatic organisms are described in detail in Refs. [61,175,231]. The filming of dolphins swimming under natural conditions was first performed by Jacques-Yves Cousteau. The systematization of free-swimming frames was performed in [231] (Fig. 2.21).

Fig. 2.22 shows the results of processing a kinogram of swimming dolphins [231]. The trajectories of movement of a short-beaked common dolphin with a length of 1.32 m and a harbor porpoise with a length of 1.33 m have been investigated. The top dorsal and extreme points of the pectoral fins, which are close to the CG of the body, move almost along a straight line trajectory, while the fork of the tail fin moves along a sinusoidal trajectory. Both animals are the same size, but the high-speed short-beaked common dolphin oscillates with smaller amplitude. The oscillation wavelength λ has the largest span A at the fork of the caudal fin, the smallest span of oscillations at the upper point of the dorsal fin and at the extreme point of the pectoral fin, and the intermediate value of the span at the extreme point of the nose. According to Fig. 2.22 the relative spans A/L of the extreme points of the caudal fin, the dolphin's nose and the CG of the body are approximately in proportion: AHP: An: ACT = 1.0: 0.4: 0.05 [231].

Systematic data of dolphin swimming kinematics obtained in a network aviary are given in Ref. [243]. With the help of high-speed filming, the swimming of a harbor porpoise with a length of 1 m was investigated. At a swimming speed of 1.02 m/s and an oscillation frequency of the tail fin of 0.73 Hz, trajectories of the fork of the tail fin were plotted, with the dependencies f (V/L), A/L (V/L), St (A/L), and St (V/L).

Despite the newly obtained results, it became obvious that further research should be carried out in stationary hydrobionic plants. In Ref. [144], the obtained results of dolphin swimming kinematics when moving in a channel are given (Section 2.6, Fig. 2.20). Studies were conducted with six bottlenose dolphins with lengths of $L = 2.35 - 2.65$ m. On the abscissa axis, the trajectory of the rectilinear motion of the root chord of the caudal fin was postponed, the frame numbers on the film and the time t (s). The vertical amplitude of A_o of the caudal fin is plotted along the ordinate axis. The speeds U and V are determined according to the kinograms based on time derivatives:

$$U = \frac{\partial s}{\partial t}, \; V = \frac{\partial A_0}{\partial t}, \tag{2.25}$$

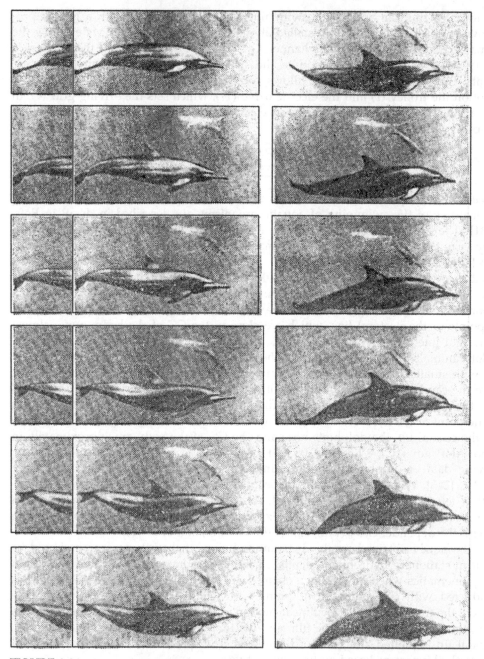

FIGURE 2.21 Kinogram of one cycle of underwater swimming in the sea of an ordinary dolphin (from top to bottom). The oscillation frequency of the caudal fin is $n = 2$ Hz, the Strouhal number is $Sh_A = 0.37$ [231].

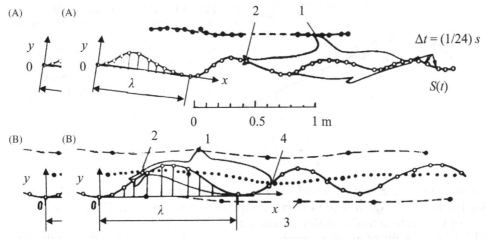

FIGURE 2.22   The trajectory of the characteristic points of the body of the dolphin of the short-beaked common dolphin (A) and harbor porpoise (B): (1) the upper point of the dorsal fin, (2) fork of the caudal fin, (3) extreme point of the lateral fins to the extreme point of the nose [231].

The instant speed of movement of the center of pressure of the caudal fin is determined by the formula:

$$W_i = \sqrt{U^2 + V^2} \qquad (2.26)$$

In this case, the angles β are determined graphically, directly from Fig. 2.23A, angles γ are obtained from the ratio of the component velocities:

$$tg\gamma = -\frac{V}{U}, \qquad (2.27)$$

angles of attack are found as the difference of two angles $\alpha = \gamma - \beta$.

From the above results it follows that the translational velocity $U(t)$ of the movement of dolphins is uneven. During this period, the caudal fin twice crosses an almost straight path of translational motion and is twice, in extreme positions, shifted through a zero angle of attack. Therefore the speed of translational movement in one period, twice changes from maximum to minimum and, conversely, from minimum to maximum. It is greater when the tail fin moves down and smaller when the fin is lifted. By reversing the fin in the extreme upper and lower positions, the constancy of the positive thrust force is ensured. The greatest accelerations are observed during the passage of the caudal fin with the angle of attack of the median aligned position of the dolphin's body. When swimming with a small average speed $U_0 = 2.3$ m/s, the instantaneous values of speed differ by $\pm$ 3%, and at high speeds the ripple speeds can reach $\pm$ 15%. The curve of variation of the transverse velocity, $V(t)$, naturally has the same character as the curve of variation of the fin oscillation amplitude, but with a phase shift of $\pi/2$. The extremes of the $V(t)$ curve occur at the moments when the tail fin has the middle, aligned dolphin position. The ratio of the maximum value of

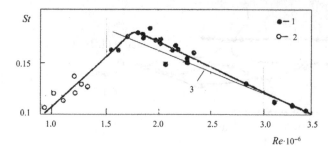

FIGURE 2.23 The dependence of the Strouhal number on the Reynolds number of dolphin swimming kinematics: (1) experiments with harbor porpoise and short-beaked common dolphin in a circular channel in 1963−1965 [264]; (2) experiments with harbor porpoise in an aviary, published in 1968 [243].

V/U_0 is about 0.87. The angle of attack $\alpha(t)$ during each half-period of oscillation of the caudal fin has a maximum area corresponding to the position V_{max}. The phase angle shear of the vertical and angular oscillations is $\delta = \pi/2$.

In Ref. [264], the results of an experimental study of the dependence of the oscillations of the caudal fin of the Black Sea harbor porpoise (*P. phocaena*) on the speed of movement, made in 1963−1965, were published. The studies were carried out in a large circular glazed basin of the Marine Hydrophysical Institute of the Ukrainian Academy of Sciences [271]. The main advantages of this basin when conducting research on the movement of dolphins were the circular shape of the pool, with a large radius of curvature and, accordingly, a small curvature with sufficient width, the presence of a glazed part for observing and recording the movement elements of the animals, and the possibility of frequent changes of sea water, the purity of which is necessary for conducting cinematic and biological measurements. Nine animals participated in the experiments (porpoises and short-beaked common dolphin), whose body lengths ranged from 1.1 to 1.6 m, and the swimming speed was 0.3−0.5 m/s.

The movement of animals was recorded in the horizontal and vertical planes of high-speed and conventional movie cameras using synchronous filming. To obtain quantitative data on the dolphin's trajectory movements, a grid with a side of a square equal to 0.2 m was plotted on the inner opaque part of the wall and the bottom of the channel. The time was measured by the interval between frames. Kinogram areas where the dolphin movement was uniform and the trajectory of the CG in the vertical plane were almost straight were processed. The curvature of the trajectory of movement of dolphins in the horizontal plane was negligible and not taken into account in the calculations. These investigations are supplemented by the results of experiments conducted in 1968 in a marine enclosure [243]. The error in the first measurements did not exceed 5%, and in the second, 10%.

The body of dolphins can be divided into three main parts: the nasal (to the cervical vertebra), the middle (from the cervical vertebra to the posterior edge of the dorsal fin) and the aft areas, consisting of the stem and tail fin. The nose part of the dolphin's body performs oscillatory movements in the vertical plane. The frequency of these oscillatory movements is consistent with the frequency of flexural-oscillatory movements of the aft part of the hull, but their amplitude is incomparably smaller than that of the caudal fin. The magnitude of the amplitude of the dolphin's nose decreases significantly with increasing speed. At the maximum speeds of the dolphin, the trajectory of its nose approaches a straight line.

The analysis of the dolphin body motion was performed in an inertial coordinate system. In this case, the x axis was directed along the dolphin body, the y axis was perpendicular to the x axis and in a vertical plane. The origin of coordinates was located at the beginning of the rostrum. In this coordinate system, the oscillation of this type can be described by the equation:

$$y = A_0 e^{\beta x} \cos\{\alpha(t - x/c)\}, \tag{2.28}$$

where y is the vertical deviation of any point of the hull, x is the distance along the body, t is the current time, c is the wave propagation velocity along the body, $\alpha = 2\pi/T$, where T is the oscillation period, A_o is the maximum amplitude of transverse oscillations of the nasal tip of the rostrum, β is a factor that characterizes the rate of increase of the amplitude of transverse oscillations along the body from head to tail, and is a function of the speed of movement of the dolphin. In the first approximation, it is possible to take an inversely proportional dependence of β on the dolphin's speed $\beta = k/u_o$, where k is the proportionality coefficient and u_o is the velocity of the dolphin's CG.

The oscillatory movements of the nose part of the dolphin's body in the vertical plane can be considered as the result of the rotation of a quasi-solid body about some transverse instantaneous axis. At equal angles of deviation from the abscissa axis, the total pressure impulse in the bow during the oscillation period is almost zero, therefore the effect of such movements on the creation of the dolphin thrust force can be neglected. However, it should be noted that in determining the moments of hydrodynamic forces acting on a dolphin, the movement of the bow of the hull cannot be ruled out, since the resulting pressure difference on the forward part of the hull creates a restoring hydrodynamic moment that ensures the dolphin returns to its original position. Considering the shape of the nose contour, it can be expected that the nature of the pressure distribution along the dolphin's body in the nonstationary flow regime will be significantly affected by the change in the angle of attack of the flow past the bow section.

In the middle part of the dolphin's body, between the pectoral and dorsal fins, its CG is located, the translational movement of which, averaged over the period, can be assumed to be uniform, despite the periodic cyclical nature of the dolphin's oscillations. This gives reason to consider the movement of his body in a quasistationary approximation. Analysis of numerous films showed that the bending-oscillatory movements of the tail part of the dolphin's body are made in a vertical plane up and down from its longitudinal axis of symmetry. At the same time, the amplitudes of the transverse deviations of different points of the tail section of the hull along the longitudinal axis are not the same: they increase towards the stern, substantially unevenly. With the indicated movements, the stem of the caudal fin can also create additional thrust.

The curve described by the dolphin propulsion blade in the vertical plane is close to a sinusoid. When a dolphin thruster is operating, the tail fin is established at a certain angle of attack to the incident flow, which leads to the emergence of a lifting force. The projections of the lifting forces on the direction of motion from each blade as they move both up and down will be turned in the same direction and in the sum make up the propulsion thrust.

Thus, using flexural-oscillatory movements of the caudal fin the dolphin gives kinetic energy to the surrounding fluid. In the flowing stream, this energy is converted into the potential energy of the hydrodynamic pressure field, and hydrodynamic pressures occur in the aft part, by their magnitude significantly exceeding the hydrodynamic pressures acting in the bow section. As a result, the main vector of hydrodynamic pressures acting on the surface of the body and the dolphin's tail fin, at any moment of uniform or uniformly accelerated motion, will be directed toward the translational movement of the dolphin. The main vector of hydrodynamic pressures is the force that ensures the movement of the dolphin in the water or, in other words, is a thrust created by the dolphin propulsion. However, the dolphin thrust force during a single period of oscillations of the caudal fin changes quite dramatically: in the extreme positions of the caudal fin the transfer of the kinetic energy of the surrounding fluid stops, since the angles of attack are close to zero and the main vector of hydrodynamic pressures vanishes when reaching the intersection of the tail fin of the x axis, when the speed of the caudal fin is greatest. Therefore the forward movement of the dolphin during each period is nonstationary.

The rear edge and end parts of the fins are more elastic than the fins themselves (Section 2.3, Fig. 2.6B and Section 2.4, Fig. 2.16G). Under the action of hydrodynamic forces, the ends of the fins are bent in the direction opposite to the vertical movement of the fin. The analysis of the motion pictures showed that when the dolphin moves in the tail area of the hull there are no return currents, neither were any concentrated vortex chords of considerable tension observed. Apparently, the flow around the dolphin's body is continuous. It should also be noted that the flow of fluid adjacent to the body of the dolphin does not deviate from the direction of the outlines of the body and the reason for this, apparently, is the specified bend of the "free" ends of the caudal fin. It must be assumed that the thrust force at the free ends is not created due to this bending.

As the swimming speed increases, the amplitudes of oscillations of the dolphin's nose and tail parts significantly decrease. When the tail part of the body deviates upward on the lower surface of the body, in the tail part there are prerequisites for the appearance of positive pressure gradients and separated flow. However, the sucking forces and the specific structure of the skin of dolphins prevent this. At this moment, negative pressure gradients appear on the upper surface of the tail part of the body. Due to the specific structure of the caudal stem flattened in a vertical plane, there is a flow of fluid on the stem from its upper surface to the lower one, which increases the amount of suction force. When the tail part of the body is deflected downward, the flow around the upper and lower parts of the body becomes mirrored. As a result, over the period of oscillation, a redistribution of pressure occurs along the dolphin body, so that the negative pressure gradient must shift to the tail part of the body.

Under the action of the increased hydrodynamic pressure, the *span of the caudal fin decreases* and the amount of fluid flowing from one side of the caudal fin to the other in the region of the axial curvatures of the rear edge increases. A change in pressure in a specific circulatory system of the caudal fin leads to a change in its elasticity, which also affects the continuity of the flow around the dolphin's body [231].

The obtained films of an experimental investigation of the kinematic characteristics of the movement of dolphins are presented in the form of a graphical dependence of the Strouhal number $St = Af/u_0$ on the Reynolds number (Fig. 2.23), f and A are the frequency and amplitude of oscillation of the caudal fin. At low Reynolds numbers, the product Af increases in proportion to the swimming speed of the dolphin. With the increase in the Reynolds number, this product grows at a much slower rate than the speed.

This is explained by the fact that the speed of a dolphin is directly proportional to the number of oscillations of the tail fin per unit of time. In turn, the oscillation amplitude of the caudal fin is a function of frequency. Moreover, at small oscillation frequencies, the amplitude of the caudal fin increases and reaches a maximum, and then decreases with increasing oscillation frequency, approaching its optimal value. With an increase in the amplitude of oscillations of the caudal fin, the total thrust force during the period of oscillation also increases.

However, an increase in the amplitude of oscillations of the caudal fin also leads to an increase in the total resistance to the movement of the dolphin. As a consequence, starting from a certain moment, the dolphin uses another opportunity to increase the swimming speed; namely, it increases the frequency of oscillation of the caudal fin. Consequently, the factor determining the thrust of a dolphin thruster is the speed of the transverse movement of its caudal fin, the average value of which during the oscillation period is equal to the product of $2Af$.

In Fig. 2.23, the left part of the curve (point 2) was obtained from the results of experiments with a harbor porpoise in an aviary, therefore, this pattern should be considered taking into account the large scatter of measurement points. In an open-air cage, a wild animal requires a lot of training time and the arbitrariness of the results obtained using moving strips of the grid applied to the moving cage of the open-air cage should be taken into account. If you do not take into account the left side of the curve shown in Fig. 2.24, then the averaging of points (1) without taking into account points (2) will change the slope of the right side, and the regularity will be expressed by the straight line (3).

The relative lateral speed of oscillation of the dolphin's tail fin can be represented as:

$$\bar{u}_k = \frac{\bar{v}_y}{\sqrt{gl}} \frac{L}{D}, \tag{2.29}$$

where $\bar{v}_y$ is the speed of oscillation of the caudal fin in the vertical plane, l is the span of the fin, L is the length of the dolphin from the start of the rostrum to the fork of the caudal fin, D is the reduced diameter of the dolphin cross section in the midsection, and g is acceleration due to gravity.

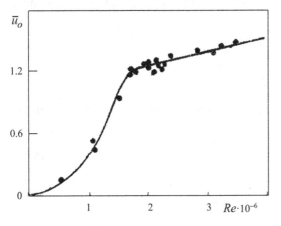

FIGURE 2.24 Relative transverse velocity of the movements of dolphin's caudal fins, the short-beaked common dolphin and harbor porpoise, depending on the Reynolds number [263,264].

The relative transverse velocity (2.29) can be interpreted as the composition of the Froude number for the caudal fin on relative elongation of the dolphin, that is,

$$\bar{u}_k = Fr\frac{L}{D}, \qquad (2.30)$$

where
$$Fr = \frac{\bar{v}_y}{\sqrt{gl}}.$$

Dependence of the relative transverse velocity on the Reynolds number is presented in Fig. 2.24.

The experimental results shown in Figs. 2.23 and 2.24, indicate the presence of two dolphin swimming modes.

In the first mode, the tail fin moves with large amplitudes and low oscillation frequency, and in the second mode it moves with low amplitudes and high oscillation frequency. Similar conclusions were obtained in Ref. [231] when analyzing the experimental data of N.P. Semenov and A.F. Kudryashov, as well as when processing the kinograms obtained during the movement of dolphins in a rectangular channel [245] with uniform motion and depending on the magnitude of the acceleration of motion [142]. Thus, from all this we can draw the following conclusions:

- The forward movement of the dolphin, in contrast to the steady-state motion of solids, is characterized by nonstationarity of motion, continuous flow, and the presence of an additional force of thrust, resulting from a specific oscillatory motion.
- The amplitude-frequency characteristic of the thruster has two modes of movement of the dolphin—in the first mode, the movement of the whole body is sinusoidal, in the second the whole body moves almost stationary, and the tail part is like a flapping wing mounted on a flexible holder.

The generalized kinematic characteristics of various types of hydrobionts, obtained in the facilities outlined in Section 2.6 (Fig. 2.24), are considered in Ref. [175].

Experimental results can be divided into three types of patterns that characterize the laws of swimming aquatic organisms, depending on the their movement methods. The averaged regularity of points (1) characterizes representatives with eel-like swimming and can be represented by the equations:

$$f = \frac{2U}{L} + 1, [1/c], \qquad (2.31)$$

$$\frac{U}{L} = 0.5 \ (f-1), \qquad (2.32)$$

For fish with the combined swimming technique (points 2–5), a straight line can be drawn, described by the equations:

$$f = 1.33\frac{U}{L} + 1, [1/c], \qquad (2.33)$$

$$\frac{U}{L} = 0.75 \ (f-1), [1/c]. \qquad (2.34)$$

Line A, obtained by Bainbridge and line B, obtained by Magnuson and Presscot, are expressed, respectively, by Eqs. (2.35) and (2.36):

$$\frac{U}{L} = 0.75 \ (f - 1.33), [1/c]. \tag{2.35}$$

$$\frac{U}{L} = 0.64 \ (f - 0.76), [1/c]. \tag{2.36}$$

In the case when the dolphin moves almost evenly, for values $0.5 < U/L < 2$, the set of points (8) can be expressed using the equations:

$$f = 1.05 \frac{U}{L} + 0.25, [1/c], \tag{2.37}$$

$$\frac{U}{L} = 0.95 \ (f - 0.25), [1/c]. \tag{2.38}$$

Points (6) and (7), characterizing the swimming of dolphins with significant acceleration, are located higher, and points (9) (deceleration) are below these lines.

In addition to the oscillation frequency, as indicated in Refs. [263,264], an important characteristic for creating thrust by the tail fin mover is the amplitude of oscillation of the caudal fin A or the amplitude of oscillation of the trailing edge of the fin $2A$.

The approximating straight line is described by analytic dependence:

$$\frac{A}{L} = 0.51 (lg \ Re)^{-0.9} \tag{2.39}$$

In the shaded area adjacent to the straight A and bounded by equidistant dashed lines differing from Eq. (2.39) by $\pm$ 12%, most of the experimental A/L values of the aquatic organisms studied are found.

Through all the experimentally obtained points, one can draw an averaged approximating curve, whose equation has the form [175]:

$$lg \ St = (5.16 - 1.15 \ lg \ Re - 4.7)^{-1} \tag{2.40}$$

The maximum deviation of the experimental points from the average curve does not exceed 20%. Eq. (2.40) allows, for a given range of Reynolds numbers, to calculate the required amplitude-frequency characteristic of a waving fin mover.

In Fig. 2.23, we can see various laws of the dependence of the Strouhal number on the Reynolds number obtained in experiments with dolphins in a network enclosure (speed) and in a circular basin [264]. Experiments with fish were carried out in the circular channel and in the biohydrodynamic pipe (Section 2.6, Fig. 2.24), and experiments with dolphins were performed in the biohydrodynamic channel (Section 2.6, Fig. 2.20). To obtain reliable results, it is necessary to carry out experiments with hydrobionts in accordance with the methods given in Chapter 1. When comparing these patterns obtained in an experimental study in a circular pool (Fig. 2.23) [264] and in a straight channel, it was found that the patterns on the slope of the curve for dolphins are the same, and the magnitude of the Strouhal numbers differs significantly. This is explained by the fact that in the circular channel the dolphins during the experiments caught the centrifugal force and moved steadily with low speed. In addition, the dolphins were wary, as swimming conditions differed from the conditions of movement in their natural habitat. At the same time, the dolphins, which were investigated in the straight channel, were well trained and moved along the usual trajectory, and the difference was that in the straight channel the dolphins moved with different accelerations. Straight lines (8) and (9) average the experimental points as the dolphins move, with low acceleration and deceleration, respectively, almost steadily. Straight lines (6) and (7) average the experimental points during the movement of dolphins with high and medium acceleration, respectively. All this must be considered in the practical implementation of the results.

Theoretical hydrodynamic models for dolphin swimming were proposed in Refs. [168,214].

Bioenergy of swimming aquatic animals

3.1 Basic regularities of bioenergy

The basic regularities of bioenergy are discussed in detail in Refs. [94,175,231]. The bioenergy of swimming is of great importance, as it allows, in the first approximation, to calculate the active power of hydrobionts and their hydrodynamic resistance. Biologists and physiologists investigate the energy expenditure of animals mainly through the metabolism. There are basic, active, and intensive metabolism, and also the level of exchange—the energy spent by animals (Joule)—and the intensity of exchange—specific energy per unit of hydrobiont weight (J/W). The level of energy exchange of a hydrobiont, referred to a unit of time, is power (N, watt), and the intensity of exchange per unit of time is the specific power q (w/N). The dimension of 1 w/N = 1 m/s.

The power of the main metabolic rate, N_o, expresses the average minimum power of the body in relative rest under certain external conditions of rest, which is spent on the "inner workings" of organs and tissues (the work of the heart, lungs, digestive system, etc.). Along with the main exchange, a more rigorous concept of "standard exchange" (standard metabolism) is used, which is determined by extrapolation, as the level of "zero activity" in the thermo-neutral zone, in the temperature range at which energy consumption does not change. The numerical values of the power of the main and standard exchanges are usually close. The power of active metabolism of N_a expresses the average additional energy costs of a hydrobiont during the muscular load of active swimming. Active exchange is several times greater than main metabolism (at high swimming speeds up to the order and more times). The total exchange power, N, is the total average power of active swimming of a hydrobiont. It is conditionally accepted that when swimming, the main metabolism of a hydrobiont is the same as at rest. Therefore the power of energy metabolism can be expressed as $N = N_o + N_a$ ($N_a \gg N_o$). Accordingly, for the specific power of total q, basic q_o and active q_a exchange of hydrobiont of mass m, the following relations are valid: $q = N/m$, $q_o = N_0/m$, $q_a = N_a/m$.

The effective power of N_e for horizontal swimming of a hydrobiont with an average speed U_o = constant is expressed through the thrust force $\overline{T}$ or its equal average resistance force R:

$$N_e = \overline{T}\, U_o = \overline{R}\, U_o.$$

The relationship between N_a and N_e can be determined by the formula:

$$Ne = \eta_{id}\eta_m N_a = \eta_{id}N_m,$$

where η_{id} is the ideal average hydrodynamic efficiency at a given mode of swimming of a hydrobiont, taking into account the loss of power for kinetic energy in its wake, η_m is the efficiency of muscles of a hydrobiont engine, which takes into account the efficiency of converting the biochemical energy of an animal into mechanical work with muscle power N_m. On average, $\eta_m \approx 0.2$, however, the value of η_m varies greatly [231].

The main factor in the exchange of warm-blooded animals is total body weight. Their body temperature is maintained approximately constant: $t \approx 36°C–37°C$ regardless of the ambient temperature. In many species of birds, as a result of modern processing of all previous studies, formulas for the average standard exchange [231] for non-sparrow birds have been established:

$$N_0 = 3.8\, m^{0.724},\ q_o = 3.8\, m^{-0.276},\quad 3 \cdot 10^{-3} \le m \le 10^2, \tag{3.1}$$

for passerines:

$$N_0 = 6.25\, m^{0.723},\quad q_o = 6.25\, m^{-0.277}, \tag{3.2}$$

where m is the mass, kg; N_0 is power, Watt; and q_o is the specific power of the main exchange, W/kg.

As a result of processing experimental data, similar mean dependences of basal metabolism on body weight were obtained for various mammals [231]:

$$N_0 = 3.4\, m^{0.74},\quad q_o = 3.4\, m^{-0.26},\quad 20 \cdot 10^{-3} \le m \le 3.7 \cdot 10^3. \tag{3.3}$$

These dependencies are satisfied with the data on the main exchange of humans weighing 70–80 kg and dolphins weighing 50–200 kg (the basic exchange patterns are the same for mammals).

The main factors for the main metabolism in fish are the temperature of the environment and body weight. The curve of the dependence of the main exchange of fish on temperature is determined by the exponential curve [231]:

$$q_o = 0.07\, e^{0.1t},\ 5° \le t \le 27°, \tag{3.4}$$

where t is the water temperature, °C. When the water temperature rises, oxygen consumption by fish per unit of time per unit mass increases dramatically in the main exchange: from 5°C to 25°C. In Ref. [231], the dependence of the main exchange of fish on body weight is also given:

$$N_0 = 0.44\, m^{0.8},\quad q_o = 0.44\, m^{-0.2},\ 10^{-5} \le m \le 10 \tag{3.5}$$

The data are given for one temperature ($t = 20°C$). The nature of the dependence of the main metabolism on body mass is parabolic for power, and hyperbolic for specific power, that is, with increasing fish mass, N_0 decreases. With an increase in the mass of fish by one order of magnitude, the specific power of the general exchange decreases on average by 15%, and to reduce the specific power by one order of magnitude, the weight of the fish must be 10^5 times greater. For all hydrobionts, identical power dependencies are valid for the main exchange power N_0 and the specific power of the main exchange q_o on the body mass of the type $N_0 = am^n$, $q_o = am^{n-1}$, where the dimensional coefficient a determines the corresponding values of the main exchange for the animal mass $m = 1$, and the dimensionless exponent $n < 1$, which varies for certain types, characterizes the intensity of the decrease in these values with an increase in body weight $m > 1$ kg. In general, the values of this indicator vary in the range of $0.67 \leq n \leq 1$. Here, the lower value $n = 0.67$ is limited to the particular case of proportionality of the main exchange of the body surface area of a certain animal (since $S \approx m^{2/3}$), and the upper value $n = 1$ to the particular case of proportionality the main exchange of the body mass itself ($N \approx m$), which occasionally occurs in some species. The above stable values of the exponents $n \approx 0.75$ in expressions (3.1)–(3.5) are average statistics for the main metabolism of the corresponding groups of animals, intermediate between the values of the dependence of metabolism both on the surface and on body weight [231].

The average power of the general exchange N is proportional to the average power of the main exchange N_o:

$$N = k N_o = k\, am^n. \tag{3.6}$$

The dimensionless coefficient of multiplicity k is determined by the nature of the work performed by the animal, its activity, and duration. At the same time, the active exchange of the same mammal can exceed the basal metabolism at $k = 1.6$; 4.8; 25 times (Fig. 3.1). The main metabolism of bottlenose dolphins corresponds to the basal human metabolism (direct 1). In the porpoise and bottlenose dolphins, in the excited state, the main metabolism exceeds the norm by more than two times (direct 1′).

The coefficient of multiplicity k, according to experimental data for land mammals and birds, does not depend on their mass. The exponent n in the formulas for the sought dependence on the power of active metabolism on the mass of animals retains the same magnitude as for the main metabolism: $n \approx 0.75$. For hydrobionts, the multiplier $(k - 1)$ is expressed in terms of the average swimming speed U_o in each mode of the corresponding duration of swimming T to fatigue, therefore the multiplier can be:

$$(k - 1) = \varphi\,(U_o) = \varphi^{-1}(T). \tag{3.7}$$

The power of the active exchange of warm-blooded hydrobionts can be represented as follows:

$$N_a = \varphi\,(U_o)\,\psi\,(m) = \varphi^{-1}(T)\psi(m). \tag{3.8}$$

The exponent $\psi\,(m)$ is known $n = 0.75$, and the speed function $\varphi\,(U_o)$ depends on the flow regime in the boundary layer of the hydrobiont and should be a power function.

In Ref. [231], the recalculated results of experimental data obtained by Matyuhin [199] and Brett [94] in the study mainly of small fish are given.

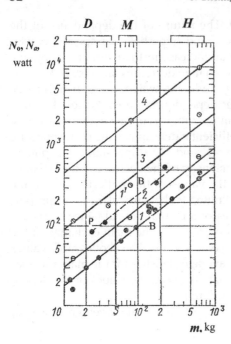

FIGURE 3.1 The power of the main N_o (1, 1' excited state) and active metabolism of N_a (2–4), depending on the total mass of warm-blooded animals and humans: D, dog; M, man; H, horse; P; porpoise; B, bottlenose dolphin [231].

3.2 Experimental investigations into the energy expenses of dolphins in a state of relative rest

The results of calculations are known from experimental data of the main and active metabolisms in fish. To obtain similar results for high-speed hydrobionts is difficult. In 1967, at the Institute of Hydromechanics of the National Academy of Sciences of Ukraine, Kiyv, a method was developed for carrying out such measurements on dolphins. Special masks with two suckers were made, which were put on the head of the dolphin in the area of the blowhole and nose. The technique of fixing suckers to the head was previously investigated (Fig. 3.2). The water in pool 1 was at a depth of 1 m, in which the dolphin could swim slowly. Were trained with dolphins, which were wearing masks with two suckers with valves. After the dolphins became used to breathing through a mask, measurements were carried out according to the method described in Refs. [138,175].

In Ref. [138], experiments carried out in the summer of 1967 in the Crimea to determine the basic parameters of respiration and calculate the energy consumption in the state of relative rest (standard exchange) in common bottlenose dolphins (*Tursiops truncatus*) and harbor porpoises (*Phocaena phocaena*), the results of which are presented here. Energy metabolism was determined by the consumption of oxygen and the release of carbon dioxide from animals placed in special water baths. Parameters of external respiration were calculated using standard techniques. Hydrobionts breathed through a specially made mask with wider bends with low-inertia valves. Exhaled air was taken into a spirometer bag made of thin elastic rubber, the scale of which was graduated by gas watches. The composition of exhaled air was determined using the Ors-Fisher gas analyzer.

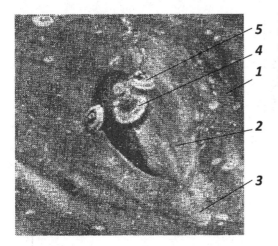

FIGURE 3.2 Photo of a common set of suction cups for mounting the sensor of the pulmonary ventilation flow meter on the dolphin's head: (1) swimming pool, (2) dolphin head, (3) dolphin nose, (4) central suction pad with valve, (5) side suction cups with valves (top view).

Five common bottlenose dolphins (19 experiments) and five harbor porpoises (11 experiments), having a weight of 29.5–40 kg, participated in the experiments. The state of the animals during the experiments was controlled by the frequency of respiration and pulse. We took into account data that did not differ significantly before, during, and after the experiments. The results are shown in Table 3.1. In the last column of Table 3.1, the energy consumption is expressed as a percentage of the basic human exchange with the dolphin in terms of dolphin characteristics. It was assumed that the energy metabolism increases in proportion to the weight of the animal in the 0.75 degree [149,222,223,241,296].

Experiments were carried out with common bottlenose dolphins and harbor porpoises at water temperatures of 21°C–24.5°C and 17.3°C–22.5°C, respectively. To determine the average breathing rate in each experiment, 30–60 counts were performed and to determine the expiratory volume there were 15–30 measurements. The oxycaloric coefficient was taken in accordance with the average value of the resulting respiratory coefficient. The common bottlenose dolphins which participated in the experiments had been conditioned to work with the experimenters and they were used to the experimental conditions. In the course of the experiments, these animals willingly took food from their hands and even tried to play with toys thrown into the water. The harbor porpoises did not respond well to training, and were restlessly, which affected the experiment results.

In the common bottlenose dolphins, the respiratory rate was stable and averaged 1.87 ± 0.036 breaths/minute. The average expiratory volume on different days differed for the same animal by 1.5–2 times. The value of the respiratory coefficient (0.69) indicates that the substrate of respiration is fat. In the harbor porpoises, the respiratory rate often changed and depended on the behavior of the animal during the experiment. The mean respiratory rate changed 2.2–10.9 times. However, respiratory volumes of harbor porpoises are more stable than those of common bottlenose dolphins. Although their extreme values in individual experiments differed 1.5–2 times, the average values were very close, averaging 1.17 L with a standard deviation of ± 0.05 L from the mean (the respiratory volumes are given in the BTPS system, i.e., at temperature, pressure, and humidity of exhaled air).

TABLE 3.1 Study of energy consumption in a state of relative rest.

Dolphin	Weight (kg)	Amount of experiences	Respiratory volume (BTPS) (liters)	Frequency of breathing (min)	Pulse	Composition of exhaled air (%)	
						O_2	CO_2
Bottlenose	215	5	5.55	1.81	77	11.3	6.4
	155	3	3.27	1.72	94	10.8	6.8
	150	3	2.86	1.99	94	11.0	6.8
	145	4	3.32	1.93	73	11.1	6.8
	170	4	2.91	1.91	68	10.8	6.7
Bottlenose (average)	170	–	–	1.87 ± 0.031	–	–	–
Porpoise (average)	34.6	11	1.17 ± 0.05	6.24 ± 0.806	105	14.3	3.6

Dolphin	Respiratory coefficient	Consumption oxygen		Energy costs		
		l/min	mL/min kg	cal/min kg	kcal/day	to energy costs human (%)
Bottlenose	0.7	0.808	3.71	17.54	5428	137
	0.69	0.486	3.14	14.63	3431	110
	0.69	0.484	3.23	15.07	3255	107
	0.7	0.453	3.66	18.23	3807	128
	0.69	0.498	2.91	12.28	3344	100
Bottlenose (average)	0.69 ± 0.005	–	3.47 ± 0.123	15.76 ± 0.599	3937 ± 196	116
Porpoise (average)	0.6 ± 0.024	0.333 ± 0.0213	9.97 ± 0.755	45.92 ± 3.44	2246 ± 125	125

In the common bottlenose dolphins, the amount of energy exchange was close to the main or standard exchange. The data obtained on the standard exchange of common bottlenose dolphins coincided almost entirely with the well-known classical Benedict curve for terrestrial warm-blooded animals [112]. Data for the harbor porpoises were located above the specified curve. In a state of relative rest, the common bottlenose dolphins, with an average weight of 170 kg, consume about 4000 kcal/day. With a weight of 90 kg, this would equate to 2500 kcal/day, which corresponds to an "input" power of 117.6 W (0.16 hp).

In hydrodynamic calculations, it should be taken into account that in dolphins the muscles of the torso and tail are combined into a single functional system. This musculature makes up 80%−90% of the muscle mass of the animal, and 40%−50% of its total weight. The musculature of a person is also 40% of his total weight, but the number of muscles providing movement to a person is much lower. It is also necessary to take into account the value of the Archimedean force when dolphins are swimming. The main energy consumption during the movement of a dolphin is due to hydrodynamic resistance.

Cetaceans, compared with terrestrial mammals, have large oxygen reserves. When breathing, they achieve a high degree of oxygen utilization—the oxygen capacity of the blood is 20%−21%. Muscles of cetaceans contain a huge amount of myoglobin. The content of myoglobin in the muscles of the female Black Sea bottlenose dolphins reaches 3600 mg%, which is 15 times more than in the heart muscle of a rabbit or guinea pig.

3.3 Indirect estimates of the energy expenses of dolphins

In Refs. [175,231], a method for determining the active power of dolphins by analyzing the results of filming dolphins jumping in natural swimming conditions is presented. The dependences of the relative height of the dolphin's jump on the angle of its body's exit from the water and the height of the jumps on the body length are given. In Ref. [231] there is a table of the impulse specific recoil of dolphin power as a result of the processing of kinogram jumps.

Some species of marine mammals are distinguished not only by high swimming speed, but also by the ability for deep-sea diving. The duration of diving is due to the ability to shut off external breathing, with oxygen provided by the body's oxygen reserves accumulated in muscle myoglobin, hemoglobin of the blood, and the lungs. The most important sources are oxygen stores in the muscles: in diving mammals, the concentration of myoglobin muscles is 5−10 times higher than in nondiving animals.

There is a difference in the structures of the motor red and white muscles in high-speed fish and high-speed dolphins. In high-speed fish, red muscles that are saturated with myoglobin and work for a long time are allocated at separate areas in the lateral parts of the body and are located among the white muscles that work for a short time. In high-speed dolphins, the fibers of the red and white muscles are distributed evenly and diffusely in different functional muscles. The motor musculature of the tail region of dolphins is characterized by hypertrophic development of two pairs of muscle complexes located, respectively, above and below the spinal column of the longest muscle—the lifter—and the hypaxial muscle—the lowerer—with the weight of the first being 1.5–2 times greater than the second. The masses of the muscles performing lateral bends of the tail are 15–20 times lower than the masses of the muscles moving the tail in a vertical plane.

In Ref. [231], on the basis of well-known methods for assessing bioenergy data in sports, a calculated graph of the dependence of the maximum specific output of power on external mechanical work q_M on the duration of swimming of aquatic organisms T (Fig. 3.3) is given. Experimental data are approximated by curve (1):

$$q_M = 1.73 \left[lg(1+T) \right]^{-1.25} \tag{3.9}$$

where T (s); q_M (m/s). The specific output of power is taken per 1 N of the total weight of the athlete, with an average mass of $m = 75$ kg. For warm-blooded hydrobionts, significantly different from humans in terms of total body mass and relative muscle mass, formula (3.9) takes the following form with a correction factor k:

$$q'_M = kq_M, \quad k = \left(\frac{m}{m_0} \frac{c_1}{c_0} \right)^{-0.25}, \tag{3.10}$$

where m is the total body mass of the hydrobiont being calculated, and m_M/m is its relative mass of muscle.

The generalized power yield curves are equally consistent with the data from various sports. They are also consistent with the calculated data for dolphin and salmon jumps for pulsed maximum values of specific recoil power of short duration, which exceeds the daily power values by about 20 times. At the same time, the specific output of the power of well-trained athletes is less than that of athletic champions by about 30%. It can be argued that the data in Fig. 3.3 can be used to determine the power output of warm-blooded hydrobionts to the mechanical work of swimming.

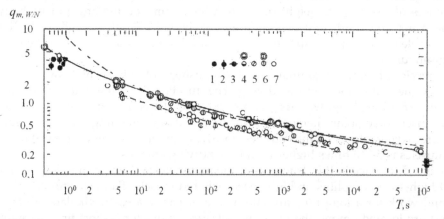

FIGURE 3.3   Dependence of the maximum specific power output q_M of hydrobionts on the duration of swimming T. Generalized curves: (1) calculated approximation by the formula (3.9); (2 and 3) physiological assessment for champion athletes and well-trained athletes. Experimental points for hydrobionts: (1) jumping dolphins, (2) jumping salmon, (3) spawning migration of chum salmon; athletes: (4) rowing, (5) running along a track and uphill, (6) exercises on a cycle ergometer [231].

3.4 Quasistationary bioenergetics calculations of dolphin swimming

Based on the above data, Ref. [231] considered a quasistationary calculation of the swimming dynamics of hydrobionts, taking into account the available biological and physiological data. In this case, the corresponding average values of the swimming speed and the specific power of the active exchange are used for the same duration of swimming, and the inertial forces and moments are not taken into account.

The main vector of external forces acting on hydrobionts consists of two almost balanced vertical forces, weight G and buoyancy Q, and two horizontal oppositely directed and equal forces, thrust P and medium resistance R, which are determined during steady motion based on the maximum average speed U_o for a given duration of swimming T.

The average thrust force P is expressed through the appropriate power of N_e by its muscular system:

$$P = \frac{N_e}{U_0} = \eta \frac{q'_M G}{U_0} = \eta \frac{k q_M \rho_T g c_3 L^3}{U_0}, \qquad (3.11)$$

where η is the propulsion hydrodynamic coefficient taking into account the loss of power during the operation of the propulsion complex "body–tail fin" in water, q'_M is the specific output of muscle power per unit of body weight for q_M and k values by formulas (3.9) and (3.10), $G = \rho_T g c_3 L^3$ is the body weight of a hydrobiont in air, where ρ_T is the density of the body, $g = 9.81$ m/s^2 is the acceleration due to gravity, and c_3 is the coefficient of similarity of masses with the length of hydrobiont L, which is calculated from the tip of the nose to the fork of the caudal fin.

In Ref. [231], the dependences of the mass m kg on the length L m of various hydrobionts are given, which are approximated by straight lines and indicate similarity for all the considered hydrobionts. The mass of a hydrobiont is calculated by the formula $m = c_3 L^3$, where $c_3 =$ constant, is determined by the indicated dependencies. For example, for the Black Sea harbor porpoises $c_3 = 14.6$ kg/m, for orcas $c_3 = 12.05$ kg/m, and for Mediterranean common tuna $c_3 = 15.75$ kg/m. Similar calculations for fish were performed in Ref. [117].

Under the above assumptions, the hydrodynamic resistance R of a hydrobiont consists of three main components: the friction resistance of the smooth body surface F_r, the resistance of the shape of R_F, and the inductive resistance of the fins R_i:

$$R = F_r + P_\phi + P_\pi = C_x \Omega \rho \, U_0^2/2 = C_x \, c_2 \, L_2 \, \rho \, U_0^2/2 \qquad (3.12)$$

where C_x is the dimensionless coefficient of force of hydrodynamic resistance, referred to the total area Ω of the wetted surface of the hydrobiont, including the surface area of the fins, ρ is the density of water, c_2 is the dimensionless shape factor, so that $\Omega = c_2 L_2$. In Ref. [231], $c_2 = 0.443$ for the harbor porpoise and $c_2 = 0.412$ for the common bottlenose dolphin were determined.

The C_x coefficient for well-streamlined solids is expressed in terms of the dimensionless friction coefficient of a smooth flat plate C_f: $C_x = nC_f$. Then from Eq. (3.12) we get:

$$C_f = \frac{2\eta k q_M \rho_T g c_3 L}{n c_2 \rho U_0^3} = \frac{2\eta k q_M G}{n \rho \Omega U_0^3}. \qquad (3.13)$$

In practice this can be taken for hydrobionts as $\rho_T = \rho$. Then, using the well-known Blasius formula for the laminar flow regime in the boundary layer of a hydrobiont, from expression (3.13) we obtain the dependence of the velocity U_o, which is given in Ref. [231]. This allows us to analyze the influence of the components of Eq. (3.13) on the hydrodynamic characteristics of hydrobionts.

Fig. 3.4 shows the results of the dependences of C_x and C_f on the Reynolds number, determined as a result of bioenergy calculations for Black Sea bottlenose dolphins, harbor porpoises, short-beaked common dolphins, common bottlenose dolphins, and oceanic belugas and orcas [231]. The results of the calculations are compared with experimental studies of the corresponding models of $A_{B.d}$ and D and field experiments [140–143], processed according to the theory [192–194] for bottlenose dolphins (2). C_f (Re) curves for all dolphin species are arranged according to their biological characteristics, which determine the flow regime. The smallest hydrodynamic resistance is possessed by the fastest dolphin, the short-beaked common dolphin, and the largest with the same Reynolds numbers, the large slow-swimming belugas. In addition to the small short-beaked common dolphin, the bottlenose dolphins and orcas are efficient swimmers. The mode of their boundary layer is

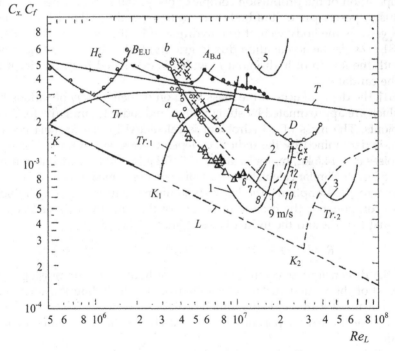

FIGURE 3.4 The dependence of the drag coefficient of dolphins on the Reynolds number. Bioenergy calculation (C_f) [231]: (1)short-beaked common dolphin $L = 1.6$ m, (2) common bottle-nosed dolphin $L = 2.3$ m, (3) orca $L = 6$ m, (4) harbor porpoise $L = 1.4$ m, (5) beluga, $L = 4.5$ m. The natural experiment (C_x) [142]: common bottlenose dolphins $L = 2.35$ m $\div$ 2.65 m; $B_{E.U}$, almost uniform; a, accelerated (active); b, slow (passive) swimming. Towing in experimental basins: $A_{B.d}$, rigid bottlenose dolphin model $L = 2.01$ m; H_c, carcass of a harbor porpoise $L = 1$ m; D, popup hard model "dolphin" of laminarized form $L = 1.6$ m [118].

transitional, and the coefficient C_f is lower than for a smooth plate in the turbulent mode T (Fig. 3.4) by almost sixfold. In Fig. 3.4, the law of resistance of a smooth rigid plate with a laminar boundary layer is denoted by the letter L, and in the transitional flow regime in the boundary layer by the letters Tr. The letter K indicates the critical Reynolds numbers corresponding to the beginning of the transition flow regime in the boundary layer. Compared with the $A_{B.d}$, towed rigid model of the bottlenose dolphin and the emerging laminar body D, the hydrodynamic resistance of dolphins during their movement is substantially less. A decrease in the resistance of fast dolphins occurs at relatively high Reynolds numbers $Re = 5 \cdot 10^6 - 7 \cdot 10^7$.

Fig. 3.5 shows similar dependences of the resistance coefficient on the Reynolds number [104]. Dependencies (12) and (13) are plotted on the basis of experimental investigations of various axisymmetric rigid bodies in comparison with dependencies (10) and (11), obtained experimentally for flat smooth rigid plates. The experimental data are of interest: (1−6) in Fig. 3.5 [140−142], recalculated according to Ref. [231]. When braking during active swimming (1) and moving by inertia (6), the drag coefficient of dolphins has a maximum value in the process of measuring in the hydrodynamic channel (Section 2.6, Fig. 2.27). During braking, dolphins increase the negative angles of attack of the lateral and caudal fins, which leads to a significant increase in their resistance, and, consequently,

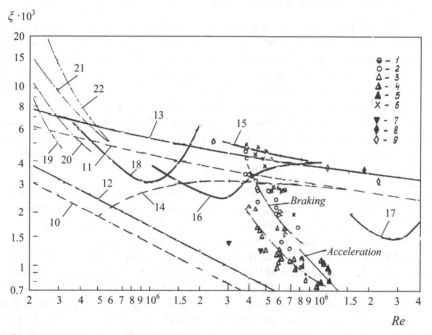

FIGURE 3.5 Dependence of resistance coefficients of hydrobionts and their models on the Reynolds number: common bottlenose dolphins, movement with accelerations: (1) (−0.1−0.3) m/s², (2) (−0.2−0.08) m/s², (3) (0.1−0.3) m/s², (4) (0.35−0.7) m/s², (5) (0.75−1.5) m/s², (6) acceleration during inertial swimming; (7−9) data from Ref. [233]; (10, 11) laminar and turbulent friction of the plate; (12, 13) laminar and turbulent viscous resistance of axisymmetric models; (14) transient friction for the plate; (15−18) towing resistance of a dolphin model; (19−22) resistance of various fish types [61,66].

to an increase in resistance of the whole body. Under extreme braking, the dolphins unfold the body across the stream, which leads to a flow separation, like a cylinder located across the flow. The "braking" curve averages points (2) as the magnitude of the magnitude negative acceleration decreases, and the drag coefficients decrease as compared to (1) and (6).

The dash-dotted line averages experimental points in the process of increasing acceleration (3, 4). At the same time, with increasing Reynolds numbers, the drag coefficients decrease and approach line (12), to laminar flow around rigid axisymmetric bodies. The straight line "Acceleration" averaged the drag coefficients at the maximum recorded values of dolphin swimming acceleration (5). The same values of the resistance coefficient were obtained as for (3, 4), but with large Reynolds number values. Most of the values of the drag coefficients for various dolphin swimming regimes are in the region of transitional flow around known solid axisymmetric bodies. This allows dolphins to swim with various combinations of acceleration/braking. For example, with a sharp acceleration in some way the dolphins drag reduction, but at the same time consume more power. Then the dolphins rest and move with a slowdown. Their resistance increases, but remains below the turbulent viscous flow around axisymmetric models (13).

3.5 Comparison of human and dolphin bioenergies

In Refs. [52,312], a hypothesis is proposed to explain some aspects of the energy of dolphins. Estimation of the power of the main N_o and active exchange of N_a in warm-blooded animals and humans is carried out, as described in Refs. [89,195,231], based on the use of oxygen consumption QO_2 by the whole organism or reduced to unit weight. On the basis of such an estimate, the values of $N_o \approx 100$ W and $N_a \approx 2.2$ kW were obtained in Ref. [231] for humans and dolphins. Comparison of estimates of exchanges obtained from the use of oxygen consumption with the previously considered capacities of only one component of energy loss (heat loss to the environment) shows that the estimates of capacity based on the use of oxygen consumption are clearly underestimated. Such a comparison result cannot be described as unexpected, considering the following:

- The process of the main energy conversion of the body, the oxidative cycle (the tricarboxylic acid cycle, the Krebs cycle), takes place in warm-blooded animals without oxygen. The latter reacts only at the final stages of phosphorylation [195].
- The hypothesis has been described that the thermal energy released during the oxidation of hydrogen in oxygen in the final stage of phosphorylation is spent on maintaining the thermal regime of the body. In this part we pay attention to the approximate correspondence of the previously mentioned value of the active exchange power of 2.2 kW, obtained on the basis of the use of oxygen consumption, and the measured average value of the heat loss power of the swimmer, equal to 2.1 kW.
- QO_2 does not characterize the intensity of oxygen metabolism in individual organs and tissues of the same organism. Table 3.1 presents data from Ref. [195] on the oxygen supply to human tissues.

The data given in Table 3.2 show that the muscles making up the largest relative body weight are characterized by relatively low specific oxygen consumption.

TABLE 3.2 Oxygen consumption.

Components	Skeletal muscle	Skin	Digestion	Liver	Brain	Heart	Kidney	Lungs
Weight of organ (% of total weight)	40	10	4	20	4	0,4	0,4	1,4
O_2 consumption at rest (% of QO_2)	25	2	19	20	15	9	5	4

It follows from the above that the use of oxygen consumption as a quantitative assessment of the bioenergy of the whole organism is incorrect. Estimates of power for oxygen consumption should be used only to compare individual bioenergy processes.

For a comparative analysis of microvibrations on the surface of human skin and a dolphin performed in Refs. [52,312], their velocity was determined as the multiplication of the circular frequency on amplitude from records of the dependences of the instantaneous microvibration values on time, which are given in Ref. [231]. The following average values of speeds were obtained: humans, 0.35 mm/s; common bottlenose dolphins in the air, 1.13 mm/s; and common bottlenose dolphins in water, 1.88 mm/s. Compared to a human, the speed of vibration of a common bottlenose dolphin is 3.2 times higher in air than in humans, and 5.4 times higher in water. Since the average power of oscillatory processes is proportional to the square of the velocities, the corresponding excess in power of the common bottlenose dolphin looks very impressive: 10.2 times more in air and 29.2 times greater in water. In the common bottlenose dolphin, the excess considered above is achieved due to the specific structure of the skin surface. It is also significant that the microvibrations on the surface of the common bottlenose dolphins in water as compared with air have a higher frequency filling and slightly faster speed. From the point of view of mechanics, this is paradoxical: the transfer of an oscillating surface from a medium with a lower viscosity to a medium with a high viscosity expanded the frequency spectrum of oscillations upward. It can be considered that the explanation for the phenomenon under consideration should be sought in the plane of adaptive electromagnetic control of the dolphin's skin.

The results of measurements in the air of the distribution of potentials on the skin surface of dolphins, in particular, common bottlenose dolphins, are given in Ref. [211]. Based on this distribution, the average potential was determined, which is 170 mV for the common bottlenose dolphin. In humans, when measured by an identical method, the average potentials are equal: 18 mV at normal points and 57 mV at biologically active points. The apparent excess of 9.4 times the average potential of a common bottlenose dolphin compared to a human has a morphological explanation: dolphins have a more powerful branched nervous system. It is known that the dolphin is very well heat insulated with a fat layer. From this point of view, the skin should be depleted of thermal energy, and then control with the help of the nervous system becomes problematic due to the known sharp decrease in propagation speeds of the nerve impulse during thermal depletion.

In order to confirm or refute the presence of thermal depletion, an analysis of the heat transfer from the skin of a dolphin and that of a swimmer has been performed in Refs. [52,312]. The investigation of heat loss was based on the use of a cylindrical model of a heat source. Using the Searle method for such a model for the media interface, the following was obtained:

$$t_p^0 - t_e^0 = t_c^0 - \frac{p_l}{p(\lambda_1 + \lambda_2)} \ln \frac{D_p}{D_c},$$ (3.14)

where t_p^0 and D_p are, respectively, the temperature and diameter of the media interface, t_c^0 and D_c are, respectively, the temperature and diameter of the heat-radiating cylinder, t_c^0 is the ambient temperature, p_l is the linear density of thermal power, and λ_1 and λ_2 are the specific thermal conductivities, respectively, of the environment with a heat source and the environment.

Known thermal conductivity of dolphin skin is $\lambda_{11} \approx 0.209$ W/m/K [221]. The thermal conductivity of nonvascularized human skin as well as adipose tissue is also equal to $\lambda_{12} \approx 0.209$ W/m/K. At the same time, muscle tissue with normal blood flow is $\lambda_{13} \approx 0.532$ W/m/K. We draw attention to the fact that the specific thermal conductivity of fatty tissue is only 2.5 times less than that of muscle tissue, and is equal to the thermal conductivity of the skin. Therefore the well-known assumption that the dolphin is well insulated with fatty tissue is not correct. For D_c, corresponding to the boundary of subcutaneous adipose tissue with $D_p = 0.3$ m, $t_p^0 - t_c^0 = 0.1°C$ and $t_c^0 = 37°C$ from Eq. (3.14) we have $p_l = 246$ W/m², which corresponds to the heat flux density for the cylindrical model, approximately equal to 390 W/m², which refutes the presence of thermal depletion of the skin of the dolphin.

We now compare the power of heat transfer of a dolphin with a similar power of heat transfer to that of a swimmer. To do this, we turn to the well-known results of studies of human heat transfer. In Ref. [11], the results of studies of the radiative and convective components of heat exchange in various regions of the human body at rest are given. Averaging the results given in this work for different areas of the body, we find that the average density of the body heat flux is 62 W/m², which is approximately 250 W in terms of power over the entire surface area of the skin. Ref. [113] presents the results of the study of heat transfer from the skin of a swimmer. These results show that under unsteady cooling mode the heat loss of a swimmer at water temperatures of 15°C–27°C is proportional to the temperature difference between the surface of the skin and water, and in some cases exceeds 10 kW. The output to the stationary heat transfer mode occurs within 7–10 min, and the heat flux density in the stationary mode is in the range of 250–800 W/m², which in terms of power over the entire skin surface area will be from 1–3.2 kW. The average value of this power is 2.1 kW. Thus, the previously calculated density of the heat flux of the dolphin corresponds to the range of the heat flux density of the swimmer's skin.

The frequency of microvibrations for experimental investigations is in the range from 11 to 16 Hz. At the same time, the well-known results of the dependence of the change in the force of muscle contraction on the frequency of stimuli during its electrical stimulation Ref. [235] show that in this frequency range there is a rather high growth rate of the muscle contraction force taking place with increasing frequency of stimulation. With an increase in the frequency of electrostimulation of more than 20 Hz, the growth of the contraction force practically stops, and muscle fatigue rapidly increases. Thus, the frequency range of microvibrations corresponds to the optimal in terms of the energy efficiency of muscle tissue.

Analysis of the electromagnetic interaction of the skin surface of a dolphin and the surrounding aquatic environment was based on the use of the Maxwell–Tamm equation for inhomogeneous media [268,290]. In accordance with this equation, the density vector of the bulk electrical forces f_o is equal to:

$$f_0 = \rho E_c - \frac{1}{2} E_c^2 grad \ \varepsilon + \frac{1}{2} grad \left(E_c^2 \delta_c \frac{\partial \varepsilon}{\partial \delta_c} \right), \tag{3.15}$$

where ρ is the bulk density of charges, E_c is the electric field strength, ε is the dielectric permeability, and δ_c is the density.

Consider, first of all, the contribution of the second term Eq. (3.15) to the studied values of the density of electrical forces acting on the skin surface of the dolphin. To estimate this contribution, we first determine the order of the values of the dielectric permeability gradient. It is of interest to consider the gradient at the interface of the dolphin's skin−water environment. We use the approximate ratio in the direction perpendicular to the interface between the media:

$$|grad\varepsilon| = \frac{\varepsilon_0(\varepsilon_2 - \varepsilon_1)}{l_{12}}, \tag{3.16}$$

where ε_0 is the dielectric constant, ε_2 and ε_1 are, respectively, the absolute dielectric permeability of the skin of the dolphin and the aquatic environment, and l_{12} is the width of the transition zone.

The results of experimental investigations of the dependence of the absolute dielectric constant of muscle tissue, skin, and other tissues are given in Refs. [8,272]. The approximation of the results of these works for the frequency range up to 10 kHz allowed us to obtain the following frequency dependence:

$$\varepsilon_2 \approx 7.91 \cdot 10^6 / f^{1/2}. \tag{3.17}$$

In the low-frequency range, the extremely large values of ε_2 in Eq. (3.17) have a biophysical explanation [8,272]. Moreover, substances with such values of ε_2 are called energetically saturated media. At the same time, in artificial materials there is still a long technological path to such values of dielectric permeability. For the aquatic environment in the low-frequency range $\varepsilon_1 \approx 80$, that is, $\varepsilon_2 >> \varepsilon_1$.

The choice of the value of l_{12} in Eq. (3.16) is justified on the basis of two points: first, the outer layer of the epidermis prevents the passage of water molecules, and, secondly, for the formation of such an indicator of the environment as ε_1, at least one layer of water molecules is necessary. This determines the choice of the value of l_{12} equal to the maximum size of the water molecule, that is, 0.138 mm [105,147]. In areas of high hydrodynamic resistance, the surface field strength can reach $E = 1$ mV/mm. For $f = 10$ Hz, for the modulus of the second term Eq. (3.15), $|f_o| = 80$ kHz/m^3 was obtained, and for $f = 20$ Hz, $|f_o| = 57$ kHz/m^3. The considered force action leads to a *change in the structural construction of the molecules of the liquid* in the layer adjacent to the skin of the dolphin, which, in turn, leads to a decrease in hydrodynamic resistance.

When a dolphin is moved from a weakly conductive air environment to the marine environment there is a smoothing of zones with different biopotentials on the surface of the skin, and, on the other hand, with the appearance of folds at their borders, a local increase in the surface field strength occurs. This explains the previously mentioned paradox in relation to the upward expansion of the microvibration spectrum of the skin surface.

4

The morphology of the body systems of some aquatic animals

4.1 The structure of the body systems of dolphins

The research results presented in the earlier chapters testify to the properties of high-speed hydrobionts, which allow them to save energy and achieve high hydrodynamic qualities when moving. Below some studies of the morphology of velocity hydrobionts, necessary for understanding the features of their hydrodynamic characteristics, are described.

The skeleton of whales is spongy, and tubular bones are missing. The bone brain, which in other mammals is located in the tubular bones and performs the function of blood formation, in whales fills the pores of all bones of the skeleton, and is much larger. Accordingly, more red blood cells are released. The spine has four sections: cervical, thoracic, lumbar, and caudal. The thoracic region bears 10—17 pairs of ribs, of which only the first 2—8 pairs are articulated with the sternum. Intervertebral discs give the spine, especially the tail, greater mobility. The absence of the hindlimbs, sacral spine, and pelvis increases the freedom of movement of the caudal stem.

It is now known that, to a greater or lesser degree, network (7) is well developed in all mammals leading an aquatic life. One of the functions of this structure is the regulation of pulsating pressure in the circulatory system, which develops as a result of breath-holding in cetaceans. Intravital angiography has shown that the dolphin brain is supplied with blood not directly from the carotid arteries (as usually happens in mammals), but only through the paravertebral network. In addition, a fascinating network ensures the functioning of the air bags and the entire dolphin echolocation system.

In the area of each vertebra, there are autonomous circulatory, lymphatic, and innervation systems that control narrow annular skin. Each section is served by a separate vessel. With active swimming, a wavelike flexural oscillation of the body occurs, and a wave moves along the body.

It can be assumed that the system of self-regulation of skin damping is activated sequentially in sections with a shift in time and phase. Thus, in each section there is an alternation of the phases of the greatest activity (work) and relative rest (recovery) of the self-regulation system, the local action of which is optimized.

One of the main features of the circulatory system of cetaceans is a high muscular hemoglobin content. As shown for a typical dolphin, individual muscle groups vary significantly in hemoglobin concentration: the heart and head muscles have a minimum hemoglobin concentration of 1200 mg/%, which is still several times higher than that of land mammals, and the maximum in the dorsal and abdominal muscles (up to 3500 mg/%). In dolphins, the muscles account for about the same proportion of hemoglobin, as that contained in the blood, meaning that these cetaceans significantly differ from terrestrial mammals, in which muscular hemoglobin accounts for 15%–25% of the total hemoglobin.

Fig. 4.1 shows the pattern of interweaving of the circulatory and lymphatic systems in the skin of mammals. The lymphatic system is a part of the vascular system that complements the circulatory system. It plays an important role in the metabolism and cleansing the cells and tissues of the body. Unlike the circulatory system, the mammalian lymphatic system is open and does not have a central pump. The lymph circulating through it moves slowly and under slight pressure. As a result of filtration of plasma in the blood capillaries, the liquid enters the extracellular space, where water and electrolytes are partially associated with colloidal and fibrous structures, and partially form the aqueous phase. This forms a tissue fluid, part of which is reabsorbed back into the blood, and part of it enters the lymphatic capillaries, forming the lymph. Thus, the lymph is the space of the internal environment of the body, formed from the intercellular fluid.

The formation and outflow of lymph from the extracellular space are subject to the forces of hydrostatic and osmotic pressure, and occur rhythmically. The lymph node is a peripheral organ of the lymphatic system that performs the function of a biological filter through which lymph flows from organs and other parts of the body. In Fig. 4.1 it can be seen how the supplying arterial, venous, and lymphatic vessels are closely intertwined. Of particular

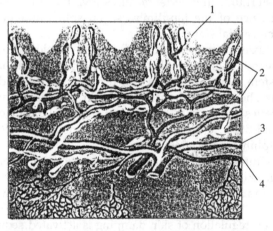

FIGURE 4.1 Diagram of the circulatory and lymphatic systems of vessels in the skin of dolphins (according to V.N. Nadezhdin): (1) dermal papillae, (2) vessels of the lymphatic system, (3) vessels of the venous system, (4) vessels of the arterial system, (5) fragment of the vessel of the lymphatic system with internal valves.

importance is the compact arrangement of the capillaries of these vessels in the area of the dermal papillae (1).

The epaxial musculature is innervated by the dorsal branches of the spinal nerves. In the cervical region, the dorsal branches form the dorsal cervical plexus, from which six to seven branches extend in fanlike form in the dorsal, dorsal-cranial, and dorsal-caudal directions between the semianistoma and the multiparticle muscles.

Of these, the latter branch is especially developed, the branches of which end at the level of 10—11 thoracic segments. The branches of the cervical plexus innervate the spinous, semidiscontinuous, partitioned, and under the partitioned muscles in the region of the anterior thoracic segments, the belt muscle and the cranial part of the longest muscle, separate branches pass into the skin. In the thoracic region, the dorsal branches separate and form three main branches: the medial, intermediate, and lateral. The medial branch passes to the mastoid processes of the subsequent vertebrae in the dorso-lateral direction, innervating caudally the mytoma of the multiparticle muscle and the dorso-medial part of the longest muscle. The terminal branches (in most cases, two) pass in the dorso-lateral direction into the skin. According to Ref. [275], the medial branches in the innervation of the longest muscle do not participate, moreover, the so-called skin branches of the medial branches described in this study innervate not only the skin, but also the medial, dorsal parts of the multiparticle muscle. The medial branches of the first and sixth lumbar neurotomes have some peculiarities, each forming two independent branches, or innervating two myotomes (5) and (6), respectively.

Vascular capillaries originate from arterioles, which are located in the bed of dermal ridges. A large arteriole passing in the platen sends one or several capillaries to each papillary. A powerful capillary network, the so-called vascular plexuses (Fig. 4.2), is located in the dermal ridges and the adjacent part of the underpapillary layer. They join the nerve bundles and fibers, forming a neurovascular plexus. Based on the gas exchange study, it was shown in Refs. [150,151] that the respiratory system cannot provide the energy material with the power

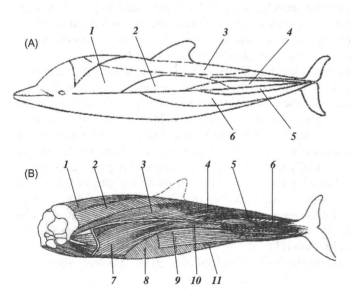

FIGURE 4.2 The scheme of the musculoskeletal muscles of the short-beaked common dolphin. (A) Ref. [2]: (1) the longest muscle (longissimus), (2) lumbar rib (iliocostalis), (3) m. spinalis, (4) m. intertransversari supl, (5) m. hypaxialis, (6) m. intertransversari inf. (B) Ref. [275]: (1—6) raising the tail stem, (1) spinalis, (2) transverse spinalis, (3) longissimus, (7—11) ventral, lowering the tail stem, (8) obliquus abdominis externus, (9) hypoxialis, (10) iliocostalis lateralis [4,287].

developed by the dolphin muscles to achieve known swimming speeds. Ref. [1] provides a detailed review of well-known studies of the motor muscles of dolphins and their innervation.

Fig. 4.2A shows the scheme of the dolphin's musculature, studied in Ref. [1], which largely coincides with the description in Ref. [210], as well as the scheme of the dolphin's musculature, shown in Ref. [275] (Fig. 4.2B).

The longest muscle is divided into superficial and deep, while the tendons from the outer portion, penetrating the inner muscle, are gradually attached to the spinous processes of the caudal vertebrae. Part of the tendons located deeper, reaches the last caudal vertebrae, penetrating to the lobes of the caudal fin. Each tendon, in addition to the portion attached to the vertebrae, has a portion that forms a stretch around the reaching tendons, as well as participating in the formation of a powerful aponeurosis around the muscle. The tendons, reaching and entering the lobes of the caudal fin, diverge in it, causing tufts on both sides. Reaching the edge of the blade, the fibers are bent downwards and are perpendicular to the upper layer. Thus, the skeleton of the blade is formed by the fibers of all suitable and their tendons. The function of this powerful muscle is to unbend and bend the upper part of the body, and when its parts stretch, tendon-giving to the vertebrae of the caudal blades, the angle between the blades and the tail stem during its movement from top to bottom. A similar picture is observed from the ventral side. Investigation of the attachment of tendons penetrating the blades shows that the angle formed between the blades and the stem of the dolphin's body section is a continuation of the trunk musculature. Almost all muscle tissue goes into the tendons around the middle of the tail section. Then, only tendons are directed caudally, the mass of which gradually decreases, as they are partially attached to the spinal column and partially participate in the formation of aponeuroses.

All the individual and common aponeuroses are formed by part of the tendon fibers of the muscles ending in a given place. In addition, fibers that come from the opposite side and are located at a certain angle of the fibers of this side take part in the formation of aponeuroses. The aponeurotic case covering muscles of a tendon is especially expressed in the area of a stalk of a tail section. The most powerful muscles of both the epaxial and hypoxial muscles [275] end in processes (ostitsyh, transverse), which can be represented as the long arms of the levers. This greatly increases the effectiveness of their actions. This is also evidenced by the appearance and development in the process of evolution of the hemal bones, as a result of which the place of attachment of the hypoxial muscle tendons is removed from the longitudinal axis of the body. The epaxial musculature in the thoracic region is innervated by the branches of the thoracic nerves, in the middle and tail parts of the body, with the help of the dorsal nerve trunks that are there, reaching caudally to the tail fin blades. The hypoxial muscles are innervated by nerve branches extending from the ventral trunks. In total, four nerve trunks are formed by branches, which are located dorsally and ventrally from the transverse processes. These trunks are directed caudally to the blades of the caudal fin, passing along the path of the corresponding segmental branches to the above muscles. In the area of the caudal lobes, they form thick nerve ramifications. A large number of different nervous structures in the forms of nerve plexuses, terminal endings, receptors (sensors and motor plaques) and elements of automatic direct and feedback with the central nervous system are also found in the area of the musculature of the caudal region [1].

The location of the internal organs in the body of dolphins is given at www.cetacea.ru/gen2.htm.

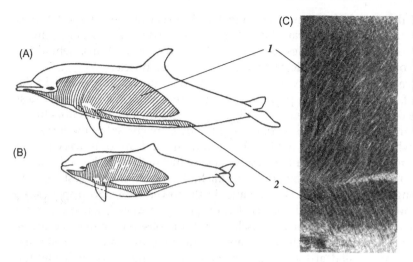

FIGURE 4.3 The direction of the fibers of the skin muscle in the short-beaked common dolphin (A), porpoise (B), and a photograph of the lateral (1) and ventral (2) areas of the skin muscle of the bottlenose dolphin (C) [287].

R.M. Surkina studied in detail the structure of the skin of the Black Sea short-beaked common dolphins (*Delphinus delphis ponticus* B.), bottlenose dolphin (*Tursiops tursio* Fabr.) and harbor porpoise (*Phocaena phocaena* L.) [286]. For the first time, the dermal muscle of dolphins were discovered and studied [287]. Fig. 4.3 shows a scheme of the skin muscle on the body of the short-beaked common dolphin (A) and harbor porpoise (B). The dolphin's skin muscle (musculus cutaneus) lies deep in the skin between the dermis and subcutaneous fatty tissue; it surrounds the neck, chest, belly and back of the animal. Muscle portion of the muscle has a complex configuration and is divided into separate sections. It is located almost along the entire body of the dolphin along its lateral and abdominal surface, starting from the angle of the lower jaw and continuing to the anus and a little further. Skin Muscle — Flat muscle consisting of striated muscle fibers. Muscle thickness varies in the bottlenose dolphin from 2 to 20 mm, in the short-beaked common dolphin from 1 to 18 mm, and in the harbor porpoise from 1 to 12 mm in different parts of the body.

In the muscle, there are several separate sections, each of which is characterized by a different direction and size of muscle bundles and fibers that make it up. Separate lateral (1, lateral) and abdominal (2, ventral) areas of skin muscles are distinguished on each side of the dolphin's body. In the anterior part of the body, the lateral areas of the skin muscle travel down and from both sides of the head reach the throat and converge together under the lower jaw, forming one unpaired mandibular area. In the mandibular portion of the lateral part of the dermal muscle, muscle bundles are directed, generally perpendicular to the longitudinal axis of the animal's body, when changing to the lateral areas they change direction and are set obliquely from the top upwards from head to tail.

The mandibular portion is the smallest area of skin muscle. The muscles here are located on the ventral surface of the lower jaw in the form of a triangle, the sharp top of which is located at a distance 4—6 cm from the end of the lower jaw in the short-beaked common dolphin and 2.5—3.5 cm in the harbor porpoise. The base of the triangle faces the chest. Muscle bundles and fibers are fixed in the skin and, as if stretched between the angle of the lower jaw, are perpendicular to the longitudinal axis of the body. As the distance from the end of the lower jaw increases, the width of the muscle layer increases and the direction of the muscle bundles changes somewhat. The unpaired muscle flap is divided and gives rise to two separate, independent bundles that stretch along both sides of the neck toward the pectoral fins. Behind the eyes, these bundles are connected to the widest lateral part of the skin muscle.

The lateral area is located for the most part at the lateral surface of the body and is in the form of a lanceolate leaf. Starting from the posterior corner of the eye, it reaches its greatest width at the level of the pectoral fin and then gradually narrows in the caudal direction. In the front part of the lateral area, individual muscle fibers and bundles are curved in the form of a gentle arc, the apex of which faces the eye. Behind the pectoral fin and up to the very end of the segment in the direction of the muscle fibers, their angle of inclination on the body of the animal is different. The most dense and powerful (the thickest, widest, and longest) muscle bundles are located along the edges of the lateral leaflet in the lower, side, and upper parts, and in the center the muscle is sparse and constructed of loosely arranged, narrow muscle bundles. A particularly dense arrangement of muscle bundles is observed along the lower edge of the lateral region, where they lie tight to one another in the form of a tape. Most muscle bundles of the lateral region end at the middle of the adjacent one above or below the underlying muscle bundle and lie down on top of one another, parallel but at an angle to the surface of the skin. All muscle bundles continue into bundles of connective tissue fibers, which are attached to the skin, mainly in its dermal layer, or form connective tissue fascia and aponeuroses, located on the border of the dermis and subcutaneous fat.

The ventral of the narrowest and longest part of the skin muscle begins in the region of the pectoral fin, where large isolated spindle-shaped muscle bundles lie. They are attached to the humerus of the pectoral fin, and along the midline of the chest between the pectoral fins they are connected using very short and sometimes barely perceivable tendons or with the same powerful muscle bundles on the opposite side of the body. Along the dolphin's body, the ventral portion stretches as a ribbon from the pectoral fin to the tail and ends at the anus. The structure and length of the posterior region of the male and female are different. Muscle tape for almost the entire length consists of powerful, almost square or laterally compressed muscle bundles, which are located in the muscle layer very tightly, and that lie parallel to each other. All muscle bundles begin with tendons and stretch, without interruption, from the top edge of the tape to the bottom edge. Located on the edge of the muscle plates, muscle bundles and fibers of all parts of the skin muscle are transformed into thick bundles of connective tissue fibers, forming extensive flat connective tissue plates—aponeuroses in the skin. Like muscle, aponeuroses are located in the thickness of the skin of a dolphin at the border of the dermis and subcutaneous fat, and have a different thickness in different parts of the body. The dorsal muscle bundles of the lateral area becomes a dense, wide, and long aponeurosis, which stretches from the muscle along the lateral and dorsal surface of the dolphin's body and merges with the tendon plate of the same muscle of the other side of the body along the midline of the back.

Along the lower edge of the lateral region, the muscle bundles continue into short connective tissue bundles, also forming a short and wide tendon plate, and connecting with a tendon suture with the same bundles extending from the lower ventral portion of the skin muscle. Close to the pectoral fin, the tendon suture passes into the muscular suture, along which the muscle bundles of the lateral and ventral sections join (Fig. 4.3C). Down from the abdominal area of the skin muscle, a short tendinous plastic begins, which with its connective tissue bundles forms a suture along the midline of the abdomen, connecting with the same tendon plate of the other side of the body and interlacing in the fasciae of the skeletal muscles.

The aponeurosis, which is a continuation of the dermal muscle, encloses the entire body of the animal from the skull to the end of the caudal stem, where it thins and fits tightly to the tendon cords from the packages of skeletal muscles of the body in the tail fin. With the help of connective tissue strands and plates, the skin muscles are connected not only with the pectoral and caudal fins, but also with the dorsal fin of the dolphin.

All muscle fibers and bundles are connected with each other by thin muscular fibers, passing from one bundle to another, or with the help of connective tissue (Fig. 4.3C). Thin and thick collagen and elastic fibers and bundles form a lining around each of the muscle bundles and bundle small bundles into larger ones. Each of these muscle bundles continues into connective tissue bundles and fibers connecting the muscle with the skin. In addition, the dermal muscle is pierced obliquely by bundles of connective tissue coming from the dermal layer of the skin into the subcutaneous tissue. Due to this structure of the connective tissue, the skin muscle is very firmly connected with the dense and thick layer of skin located above it—the dermis—and much less firmly with the subcutaneous fatty tissue [286].

Subcutaneous adipose tissue, which lies under the skin muscle, has a very loose, soft, and supple formation, with great mobility in all directions and the ability to deform. This creates the prerequisites for the mobility of skin with respect to the skeletal muscle and skeleton. The presence of abundant innervation in the skin muscle suggests that the muscle bundles of the skin muscle can contract in various combinations, from contraction of individual sections to contraction of the entire muscle. It is possible that the contractions alternate and move along the skin muscle. In addition, it can be assumed that the function of the muscles in different parts of the body is different.

4.2 Dolphin fin morphology

In Sections 2.3, 2.4, and Refs. [175,231], data on the geometric structure of the fins of aquatic organisms are given. Various aspects of dolphin fin morphology are given in Ref. [4]. In Refs. [217−220], the shape and partial structure were investigated, and a mathematical modeling of the shape of the vertical fin of dolphins was proposed. In Ref. [295], the morphology of the pectoral fin muscles of harbor porpoises was studied in detail. The pectoral fin has the most developed shoulder girdle, where the bulk of bone and muscle tissue are concentrated. The shoulder girdle is represented by a flattened scapula, devoid of significant depressions on the lateral and medial surfaces. The scapular cartilage is minimal. Acromial and coracoid processes are well developed, and the clavicle is missing. The shoulder is a shortened cylindrical bone with pronounced elevations at the proximal end. The bones of the forearms are flattened, and of almost identical length. The diamond-shaped muscle (1) occupies the area between the fascia, which covers the dorsal muscles, and the cranial half of the vertebral margin of the scapula. This poorly developed flat muscle starts from the fascia and is attached to the outer surface of the spinal edge of the scapula. The diamond-shaped muscle (1) pulls the vertebral edge of the scapula up and forward.

The shoulder—atlas muscle (2) begins in the region of the first cervical vertebra, continues in the direction of the outer surface of the scapula, partially covering the dorsal half of the deltoid muscle (3), and passes into a distinct aponeurosis, growing together with the fascia, which covers the superficial muscle group of the scapula. This aponeurosis is attached along the spinal edge of the scapula and partially in the upper corner of the caudal edge of the scapula.

The large-toothed muscle (6) is subdivided in porpoise into three parts of unequal size:

- The cranial part is a small weak bundle of muscles, located in the cranial upper corner of the scapula (on its inner surface), which goes from the area of the upper third of the first rib;
- The middle part is an underdeveloped flat muscle bundle located in the central inner surface of the scapula. It starts from the upper third of the first and second ribs, goes into the aponeuroses, covering the muscles, nerves, and vessels of the subscapularis area;
- The caudal part is most developed and is formed by two powerful teeth, the first of which is a ribbon-like muscle, which goes from the central part of the third to fifth ribs to the upper back angle of the scapula, where it forms a common aponeurosis with the middle part; the second flattened prong also goes from the upper third of the second to fourth rib to the medial surface of the scapula.

The broadest muscle of the back (8) is a well-pronounced lamellar muscle, originating with thin bundles of tendon fibers at the level of the fourth—sixth ribs, at the center of the chest height. The deltoid muscle (3) is the most powerful muscle of the outer surface of the scapula, occupying its greater half, covering the prostate and, in part, the major muscles (4). It starts from the cranial process and is attached to the front upper part of the neck of the shoulder. The deltoid muscle provides lifting of the fin and keeps it in a perpendicular position in relation to the body. The abdominal muscle (4) is located in the central part of the outer surface of the scapula with a slight backward shift. The abdominal muscle begins at the spinal edge of the scapula and is attached with tendon cords to the anterior upper edge of the neck of the shoulder. By contracting, the muscle lifts the limb and slightly turns the humerus backward. The large round muscle (5) forms the caudal edge of the muscle group of the scapula, begins at the caudal edge of the scapula, and is located partly along the scape-side of the scapula and on its outer and inner surfaces. The large round muscle is fixed mainly to the inner surface of the shoulder and raises the humerus, causing at the same time raising of the front edge of the fin. The triceps muscle of the shoulder (9) is the only representative of the muscles of the free limb; it is located in the form of a small but clearly defined cord along the caudal edge of the shoulder joint. It begins immediately above the articular cavity of the scapula and is attached to the body of the humerus.

In Refs. [175,231], data on the morphological features of dolphin fins are presented. The tail fin tissues consist of five layers: (1) the epidermis, (2) the papillary dermis, (3) the layer of tendon tissues, (4) the reticular layer of the dermis, and (5) hypodermal formations. The hypodermis forms the core of the fin blade. All the other layers mentioned are located ventrally and dorsally from it.

The epidermal layer of the skin of the caudal fin is thick, but reaches especially great power in areas that are exposed to high-pressure ambient flow. In the short-beaked common dolphin, this layer is 1—1.5 mm thick and 2—2.5 times thicker on the front side than on the dorsal or ventral surface of the fin. The structure of the papillary layer is the same as on the surface of the dolphin's skin. Dermal papillae have an oblique direction. The smallest angle of inclination of the dermal papillae is in those parts that have the greatest hydrodynamic pressure. In the papillary layer, bundles of collagen fibers are located throughout the entire thickness, and intertwine with each other and create a continuous coating that serves as a support for the epidermal layer.

A layer of tendon strands with diameters of up to 1−1.5 mm on the dorsal and ventral sides of the blades are parallel to each other, oriented along the fin lobes. This layer, in the form of a kind of frame, densely covers the layers of the dermis and hypodermis located below. From the blades, tendon yarns are directed to the tail stem. Most of these cords end at the caudal vertebrae, fascia, and ligaments, while a smaller proportion continues as tendon trunk muscles.

Compared to the skin of the body, the dermal layer is thicker, and has a more complex interweaving of support elements from collagen and elastic fibers. Unlike the papillary, the reticular layer consists of collagen fibers that are larger and assembled into powerful bundles that are interwoven with elastic elements. The net layer of the dermis is oriented differently compared to the same layer in other parts of the body.

A hypodermal layer of lobes of caudal fin is indistinctly separated from the adjacent layer of dermis. This is due to the fact that in the hypodermis of the fin there are no huge accumulations of fat cells, which is seen in other parts of the body. The collagen fiber bundles of the hypodermis are therefore tightly pressed together. In terms of the location of the supporting elements and the direction of the bundles, their hypodermis resembles the dermis, which is connected to the function of these tissues in providing fin rigidity. Most branches of the trunk muscles end in the caudal stem, on the spinous and transverse processes of the caudal vertebrae. The muscles that lower and raise the caudal fin generally end with their tendon ends on the bodies of the caudal vertebrae, also without going beyond the limits of the tail stem. Only a small number of the tendons of these two large muscles pass into the caudal blade.

One of the specific features of the vascular system of cetaceans is the numerous plexuses of arteries and veins. In Refs. [292,293] a description of the detected type of plexuses is given, called complex vessels. Such vessels consist of one thick-walled muscular artery, closely surrounded by 6−20 thin-walled veins that braid the central arterial vessel (Fig. 4.4). In the lobes of the caudal fins and in the caudal stem there are an overwhelming number of such complex vessels.

There are also single veins. The blood supply to the tissues of the caudal fin is made by one unpaired caudal artery, which travels from the abdominal aorta to the caudal region. Of all the branches of the caudal arteries, only the medial caudal artery is developed in dolphins (Fig. 4.5). It is necessary for effective regulation of pressure and blood flow velocity. In the studied cetaceans, the medial caudal artery is located under the spinal column in the cavity formed by the hemicial arches of the vertebrae, which provides reliable protection

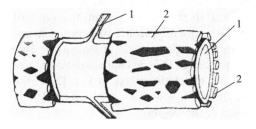

FIGURE 4.4 A complex dolphin blood vessel diagram: (1) artery, (2) veins [231].

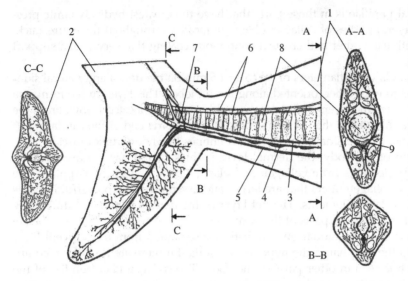

FIGURE 4.5 The circulatory system of the caudal stem and fin of the short beaked common dolphin: (1) caudal stem, (2) caudal fin, (3) caudal arterial vessels (a. caudal medialis), (4) caudal venous vessels (v. caudal medialis), (5) lateral venous vessels (v. caudal lateralis), (6) superior venous vessels (v. caudal lateralis superior), (7) arterial node, (8) caudal vertebrae, (9) hemal arches. A–A and B–B - cross sections in the region of the caudal stem, and C–C in the region of the caudal fin. Accordingly, these letters indicate the location of the cross sections [175,231].

for the fin's sole blood supply source. In Fig. 4.5, the fin is conditionally rotated 90 degrees for clarity of the structure of the circulatory system.

The tail artery, even before branching in the caudal stem, has a large internal diameter exceeding the lumen of the arteries of the heart and liver. In this case, the relative masses are 1.3%, 1.9%, and 1.8%, respectively. In the caudal fin of cetaceans, there are no own muscles or other active organs; therefore, the abundant blood supply to the caudal fin is connected with the fulfillment of the important function of an effective organ for creating an emphasis. The effectiveness of the fin propulsion is enhanced by reducing the friction resistance and the hydroelastic effect of the tail fin tissues.

In the cavity of the hemal canal of the spine (Fig. 4.5), the medial caudal artery is closely surrounded by a bundle of 3–5 large veins, which anastomose with each other, and are essentially the median medial vein. They take blood from the numerous venous branches of the caudal stem and are involved in the removal of blood from the lobes of the caudal fin along with the paired branches of the lateral caudal vein.

Large paired branches of the lateral caudal vein in the caudal stem, immediately after exiting the fin lobes in the form of four independent arterial lines, are located both on the dorsal and ventral sides of the spine. Thus, dolphins have two types of veins in the caudal stalk and partly in the lobes of the fin. The first are strongly anastomosed, connected in complexes with arteries, and the second are single and thick-walled. In the caudal region, the veins of the complex arteriovenous vessels are most common. Consequently, in the presence of a single arterial trunk in the tail region, blood outflow from this region occurs at least along five large main veins. A similar ratio of arterial and venous vessels (1:5) is characteristic only of cetaceans. In connection with the adjustable hydroelastic effect described in Ref. [226], the walls of single veins in dolphins are thick, more like the thick walls of arteries, as in single veins the blood flow is regulated by changing the lumen of the vessels.

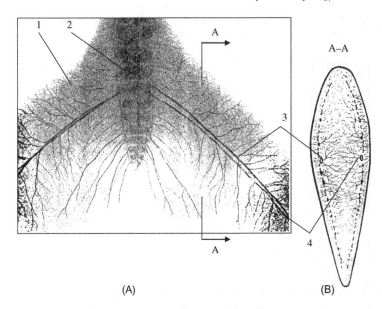

FIGURE 4.6 X-ray photograph of the caudal fin of a dolphin: (A) top view, (B) longitudinal section of the blade, (1) tail fin, (2) caudal vertebrae, (3) upper arterial complex vessels, (4) lower arterial complex vessels. A–A - cross sections in the region of the caudal fin, letters A indicate the location of the cross sections [231].

The most effective regulation of blood flow occurs in complex vessels with their thin-walled veins (Fig. 4.4). With an increase in blood flow, the walls of the arteries of the complex vessel expand, which leads to a narrowing of the lumen of the veins. They are compressed and reduce the capacity of the veins, which causes a more significant filling of the network located in the fin tissues. The paired ventral and dorsal blood branches split off from the main complex vessel in the hemiacal canal from the artery and vein in the region of each caudal vertebra (Figs. 4.5 and 4.6). In addition, a powerful paired branch extends from this vessel to each lobe of the fin, which runs along the middle part of the blade to its very end. This branch can be considered the main distribution branch (Fig. 4.6). The cross section of this branch is approximately equal to the cross section of the main caudal artery. The paired distribution arteries from above and below each lobe (four in total) depart in the same place from the tail line almost at a right angle, forming a kind of arterial node (7 in Fig. 4.5), which is absent in other mammals. A similar venous node is present in the complex veins surrounding the arteries. In addition, this node is formed not only by the medial tail vein, but also by single paired branches of the lateral tail vein. The common node of the complex vessels in the short-beaked common dolphin is located in the region of 6−7 caudal vertebrae, and in the Dall's porpoise at the 10th caudal vertebra, counting from the caudal end of the spine (Fig. 4.6). Thus, the arterial node provides the same and simultaneous distribution of blood flow and its velocity in both lobes of the caudal fin. Outgoing from node (7), the right and left ventral branches of the complex vessels reach the surface layers of the ventral side of the lobes and are located directly above the ventral layer of tendon strands.

Dorsal branches are located near the surface under the dorsal tendon. The main distribution network of complex vessels, vascularizing more than three-fourths of each lobe, is located in the dermal layer directly under the continuous tendon sheath. Subsequent large

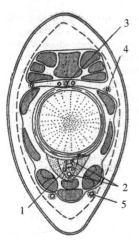

FIGURE 4.7 Scheme of the main arteries and veins of the tail stem of the sei whale: (1) caudal arterial vessels (a. caudal medialis), (2) caudal venous vessels (v. caudal medialis), (3) lateral venous vessels (v. caudal lateralis), (4) upper venous vessels (v. caudal lateralis superior), (5) inferior venous vessels (v. caudal lateralis inferior) [176,231].

and medium branches from the distribution complex vessel are located in the same plane under the tendon layer. They form a dense, strongly anastomosed surface network of large vessels (Fig. 4.6). Anastomoses of this network are especially clearly visible at the edges of the blade, where connections are observed not only between adjacent branches, but also between the vessels of the ventral and dorsal sides.

Figs. 4.5 and 4.7 show the distribution of the considered vessels of the circulatory system of the caudal fin in the dolphin's tail stem (cross section).

The main complex vessel passing in the hemal canal at the base of the blades forms a complex distribution node. It separates the five main complex vessels, of which two upper and lower vessels go to the left and right lobes of the fin, and one is located caudally along the longitudinal axis of the body. The venous node is more complex: both veins of the main complex vessel and single (lateral) veins are involved. All in the area of the common distribution node form numerous venous anastomoses. As a result of the interaction of these two nodes, five complex distribution vessels and single veins of blades are formed.

A comparison of the circulatory systems of the fins of dolphins and whales is given in Refs. [175,231]. There are three types of branching of the circulatory vascular system in the fins of cetaceans: trunk, loose, and semiloose. Radiographs of all three types of branching are given in Refs. [175,231], and a detailed description of their structure is given. Trunk type branching is discussed in detail above. It is characteristic of high-speed species of cetaceans, mainly dolphins.

The loose type is characterized by the absence of a distribution node. On the main complex vessel there is a series of consecutive nodes, from which more than 10 independent vascular branches extend, reaching remote parts of the blades, each of which serves a narrow strip of tissue. Typical of this type of branching is the sperm whale tail fin, which dives deep into cold water, and its complex vessels also perform another function—the body's thermal control.

The semiloose type of branching is intermediate between the other two types of branching. This type of branching is found in the relatively slow beluga and the high-speed sei whale. It can be argued that the speed characteristics of cetaceans are characterized by the

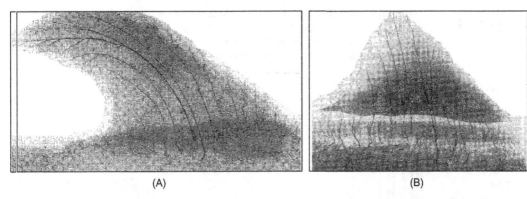

(A)   (B)

FIGURE 4.8 X-ray photograph of the arteries of fragmented complex vessels in the median plane of the dorsal fins of pantropical spotted dolphin (*Stenella attenuata*) embryos (A) and a sectioned arterial network in the back under the fin of long-winged dolphins (B) [231].

simultaneous functioning of the interconnected adaptive functions of various body systems, in particular, those considered in the previous sections.

Fig. 4.8 shows X-ray photographs of the circulatory system of cetacean vertical fins. The circulatory system has its differences in the three types of cetacean fins in accordance with their functional purpose. All three types of fin have a sufficiently developed vascular system. Paired pectoral fins (rudders for depth and maneuvering) have a specific bone system and a powerful muscular system, which provides fins with sufficient rigidity. Their circulatory system is less developed compared to other fins. The dorsal fin (Fig. 4.8), performing the function of a passive motion stabilizer, does not have bone formations, but does contain a more extensive blood network of complex vessels. In the dorsal fin, separate complex vessels pass only in one row in the median plane of the fin thickness, which coincides with the longitudinal plane of the dolphin body. The vessels do not have a common distribution node (a loose type of branching of the circulatory vascular system). Another feature of this type is that in the dorsal fins complex arteriovenous vessels are separated from each other. These are single large vessels that do not constitute a special network of an extensive vascular system of the dorsal fin. The arteries of the complex vessels in the dorsal fin are a continuation of the sectioned arterial vessels located in the body of cetaceans under the dorsal fin. The arteries in the dorsal fin are surrounded by veins and turn into complex arteriovenous vessels.

4.3 Features of the structure of dolphin skin

In Ref. [231], data on the development of tail fin skin in embryos of the fast whales are presented (Fig. 4.9). With a total cetacean embryonic development period of about 11 months, the dermal ridges are laid in the embryo at the 6th month, and the dermal papillae only in the 8th month. In accordance with the principle of embryogenesis, this indicates that the above symptoms are special (uncommon). In view of the very late embryonic appearance of the dermal

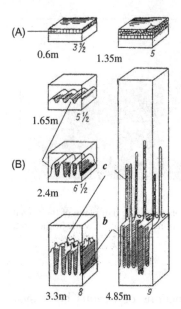

FIGURE 4.9 Diagram of the development of the upper layers of the fintail skin in whale embryos: (A) epidermis, (B) dermal ridges, (C) dermal papillae; the numbers on the left are the body length (meters) and the numbers on the right are the embryo age (months) [231].

ridges and papillae, their functions are associated with the subsequent new life conditions of the newborn cetacean, that is, with active swimming immediately after birth.

Fig. 4.10 shows the structure of the skin of dolphins according to V.E. Sokolov [278,279]. Longitudinal epidermal septa are directed along the body (indicated by an arrow).

In connection with the principles and peculiarities of aquatic organism swimming (Chapters 1 and 2), it is necessary to more thoroughly examine the structure of the skin of dolphins.

In 1967–1968, when conducting research on live dolphins, captured dolphins were placed in special boxes containing seawater. On the raised parts of the animal body, the skin looked hydraulically smooth, as it was covered with a thin film of water. However, if this film was blown off or removed with gauze, then clearly observed microfolds were observed, running mostly in the transverse direction relative to the longitudinal axis of the body (Fig. 4.11). In different species of dolphins, the sizes of these microfolds vary. Of the three dolphin species studied (the bottlenose dolphin, the short-beaked common dolphin, and the harbor porpoise), the minimum size of microfolds was found in the harbor porpoise, which also has the smallest body size. The average sizes of microfolds for short-beaked common dolphins are given on fixed material (Fig. 4.12). According to Ref. [121], microfolds are as follows: a sine wavelength is 608–1360 μ, and the wave amplitude is 230–340 μ. At the microfold sizes shown in Fig. 4.12, the dolphin's skin is hydraulically smooth up to a speed of 8 m/s. Given that microfolds are smaller in live dolphins, then, it would appear that, even at high swimming speeds, the dolphin's skin remains hydraulically smooth.

Observations show that these microfolds are clearly visible when bending on the concave side of the body and are completely smoothed on the convex side. Currently, the functional purpose of microfolds is not clear. It can be assumed that in view of the mobility of the upper layer of the epidermis, microfolds on the dolphin body alternately appear and disappear, as if

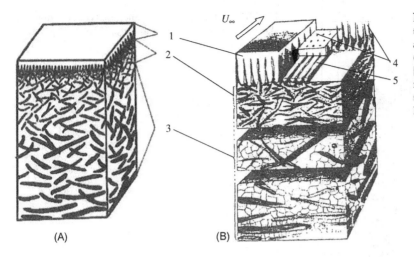

FIGURE 4.10 The structure of the dolphin skin according to V.E. Sokolov (A) Ref. [277] and (B) Ref. [231]: (1) epidermis, (2) dermis, (3) subcutaneous fatty tissue, (4) dermal papillae, (5) longitudinal epidermal septa.

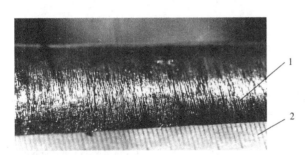

FIGURE 4.11 Microfolds on the skin surface of a short-beaked common dolphin: (1) microfolds, (2) ruler [33,50,70].

moving along the dolphin body due to oscillatory body movement and skin muscle work. In contrast to Fig. 4.10, according to our measurements [14], the dermal papillae are located along the dermal ridges at a set inclination angle. The measurements showed that the angles of inclination of the short-beaked common dolphin (*D. delphis*) dermal papillae over the body change dramatically (Fig. 4.13).

The angles of inclination were counted on histological specimens counterclockwise in the plane of the correct cut between the axis of the dermal papilla and the tangent to the lower surface of the epidemic. The correct cut was considered to be such a cut where the contours of the dermal papilla had the correct shape, that is, when the cutting plane ran along the dermal roller. The results of the measurements showed that the smallest slope of the dermal papillae ($\alpha = 10-25$ degrees) were found in the front part of the body, where the greatest dynamic pressure occurs, in the middle part of the body, where the negative pressure distribution occurs, in the dorsal and ventral parts of the tail stem, where there are large pressure gradients during oscillatory motion. The greatest slope ($\alpha = 55-80$ degrees) is observed in the skin of the beak, in the areas in front of the eyes, and also in the lateral part of the stem, that is, in the areas with low pressure gradients, below the lateral fins and in the section behind the dorsal fin, where the hydrodynamic shadow occurs.

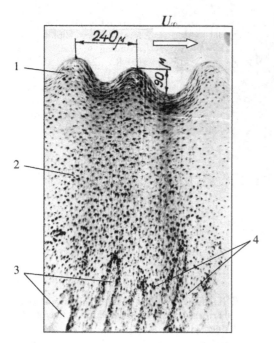

FIGURE 4.12 Transverse microfolds on the dolphin skin: (1) skin surface, (2) epidermis, (3) dermal papillae, (4) epidermal papillae [14].

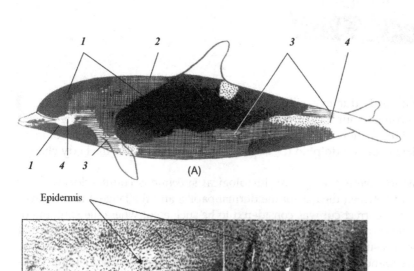

(A)

Epidermis

Dermal papillaes

(B)

FIGURE 4.13 The angles of the dermal papillae in the skin of dolphins: (A) distribution of the angles of inclination along the body, (B) angles of inclination in different areas of the skin, (1) $\alpha = 10-25$ degrees, (2) $25-40$ degrees, (3) $40-55$ degrees, (4) $55-80$ degrees [14,70].

The distributions of loads along the body are shown in Figs. 2.4, 2.5, and 2.15 (Section 2.3). In particular, Fig. 2.15 shows the pressure distribution along a rigid axisymmetric body of revolution, flowing around a uniform flow. This kind of load determines the form resistance (Section 2.2, Fig. 2.1). Between $x/l = 0.1$ and $x/l = 0.7$ is a region with negative pressure. Dolphins move both evenly without oscillation of the body (by inertia), and when the body oscillates, while, as can be seen in Figs. 2.29 and 2.30 (Section 2.7), the central part of the body moves relatively uniformly. Unlike solids, in dolphins, the central part of the body has moving ribs and a deformable body (Section 4.1). Mechanisms must be developed to compensate for the effect of negative pressure on the dolphin's body in the above range of body lengths. As can be seen in (Section 4.1), it is in the area of negative pressure that the main musculature and skin muscle are located along the body, which, along with the lungs filled with air, make it possible to maintain and regulate a well-streamlined body shape, depending on the speed of movement. In this case, the dermal muscle also allows regulation of the thickness of the skin and the angles of inclination of the dermal papillae ($\alpha = 10-25$ degrees) throughout the distribution of negative pressure along the body. In addition, other loading parameters (Section 2.3, Figs. 2.4, 2.5, and 2.15) affect the body in this area, in particular, shear loads of the boundary layer and various pulsating loads of the boundary layer affecting friction resistance (Section 2.2, Fig. 2.1). The skin muscle in this case has another purpose—to regulate the parameters of the structures of the skin, which also affect the friction resistance. Thus, the indicated features of the structure of the dolphin body systems correspond to the principles given in Section 1.3.

In Ref. [242], photographs of the location of the dermal ridges in the skin of various dolphin species are shown. It is argued that an oblique wrap around the dolphin body leads to a decrease in the effective Re number, and to an increase in the length of the laminar boundary layer due to a decrease in the path length of the fluid particles around the dolphin body.

According to our data [14,70,288], the location of the dermal rollers is somewhat different from the location specified in Ref. [242]. The direction of the dermal ridges corresponds to the flow lines of the fluid around the dolphin's body, taking into account the kinematics of the oscillation of the body during swimming. Fig. 4.14 shows the location of the dermal ridges of the short-beaked common dolphin. They are directed mainly along the longitudinal axis of the body, deviating in the middle part of the body from the midline downwards and upwards. Such an arrangement is more justified in connection with the oscillatory movements of the tail part of the body. In addition, the difference in density of connective tissue rollers and epidermal outgrowths may be important.

The thickness of the epidermis in short-beaked common dolphins is 1–2 mm, while in the epidermis there are dermal papillae, starting on the surface of the rows of dermal rollers (Figs. 4.11 and 4.12). Studies into the direction of dermal rollers across the entire surface of the body and the fins of the short-beaked common dolphin have shown that the dermal rollers in the dolphin's skin on the entire surface of the body are directed mainly along the body of the dolphin and differ in some parts of the body [288]. Fig. 4.15 shows the directions of dermal rollers along different parts of the dolphin's body. From Fig. 4.15 (1, 2) it can be seen that, in the tail part of the body, dermal rollers deviate on the dorsal part upwards from both sides of the body and interlock with each other in the upper (dorsal) axial part of the body and stem, and on the underside of the body the dermal rollers deviate downwards and converge in the lower (abdominal) axial part of the body. This direction of the dermal

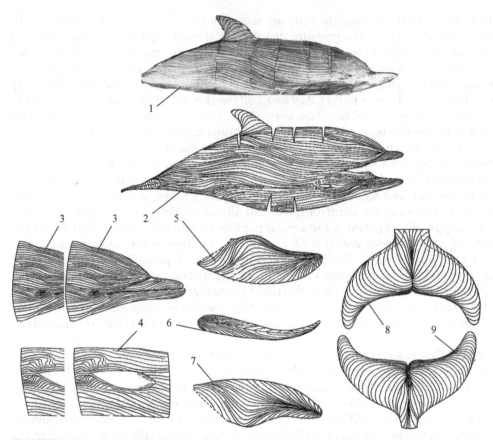

FIGURE 4.14   The location of dermal rollers in the skin of the short-beaked common dolphin: (1) side view, (2) skin sweep, (3) head skin, (4) in the pectoral fin region, (5) upper surface of the pectoral fin, (6) anterior edge of the fin, (7) lower surface of the thoracic fin, (8) ventral part of the caudal fin, (9) dorsal part of the caudal fin. (8) Scanning the surface of the skin of the fin, the dotted line indicates the contour of the fin [14,70,288].

rollers corresponds to the flow lines of the fluid during the oscillatory movement of the tail part of the body during the movement of the dolphin.

The direction of the dermal rollers in the head region (3) corresponds to the spreading flow lines of the fluid in the head of a similar axisymmetric rigid body. A research photograph is shown in Refs. [70,107], which shows a picture of the visualization of vortex structures that form in an angular articulation of a longitudinally streamlined body of rotation and a wing profile. The chaotic structure of dermal rollers in a similar place on the dolphin's body (4) was obtained as a result of evolution under the influence of the vortex pressure field, given in Ref. [107]. The picture of dermal rollers on all fins of the dolphin (1, 2, 5–9) was formed under the influence of the flow pressure field as they flow around in accordance with the functions of the corresponding fins. The structure of the dermal rollers on the vertical fin is determined by the streaming current lines on the elliptical wing of small elongation and the formation of an end vortex on the fin, which determines the inductive resistance of the vertical fin. The structure

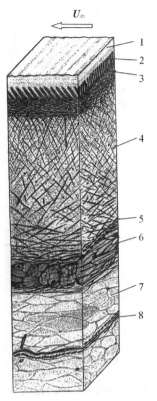

U_∞

FIGURE 4.15 The structure of the skin of dolphins: (1) the epidermis with microcap on the outer surface, (2) the papillary dermis, (3) under the papillary dermis, (4) the mesh layer of the dermis, (5) the connective tissue layer adjacent to the skin muscle, (6) skin muscles, (7) subcutaneous fatty tissue, (8) connective tissue layer, coming from skeletal muscles [16].

of the dermal rollers on the pectoral fins was formed under the influence of the flow of the stream from the end part of the fins, which are functionally mobile. To reduce the resistance, the pectoral fins are so positioned that the stream drains from them mainly from the end part.

The structure of the dermal rollers on the caudal fin is developed under the influence of an alternating dynamic load on the surface of the caudal fin during its oscillation. The lines of dermal rollers indicate that the vortex bundles flowing from the trailing edge of the fin are concentrated in two places—on the extreme areas of the lobes and in the central region of the fin. This is in contrast to the rigid oscillating wing, in which the vortex bundles are concentrated along the edges of the wing, forming inductive resistance (Section 2.2, Fig. 2.1). In Figs. 2.17 and 2.18 (Section 2.4), it can be seen that the tail blades of dolphins in the area of the longitudinal axis of the body are of such a shape that the blades can overlap each other. This form allows an additional pair of vortex bundles to be formed in the middle of the fin. In addition, dermal rollers of the caudal fin testify to the deformation and change in the span of the fin.

The above results enabled V.V. Babenko to develop a scheme for the structure of the skin of dolphins on the basis of the experimental studies by R.M. Surkina (Fig. 4.14). The results of the morphological studies by Surkina were obtained through creative collaboration with Babenko, who focused on the hydrodynamic influence of the structural features of the

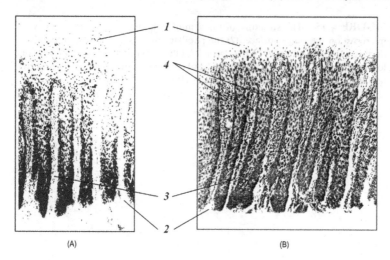

FIGURE 4.16 The form of the dermal papillae with low blood flow (A) and with increased blood flow (B): (1) epidermis, (2) dermal rollers, (3) unfilled dermal papillae, (4) dermal nipples with increased blood flow [14].

morphology of the skin and the functioning of dolphin body systems. The skin of dolphins has no hair or skin glands. Fig. 4.15 details the structure of the skin. The cells of the epidermis (1) are arranged in the form of regular rows of arches. There are no fat cells in the papillae of the dermis (2). The dermal papillae are well innervated [121,129]. The deep layers (4, 5) of the dermis, dermal muscle (6), subcutaneous fatty tissue (7), and connective tissue layer (8) are penetrated by elastic tangles of elastic and bundles of collagen fibers. Their combination with fat cells gives the skin mobility and elasticity.

A significant feature of the upper layers of the skin of dolphins is extensive saturation of an extensive network of small blood vessels. Each dermal papilla has one or two capillary vessels and the capillaries of the lymphatic system (Figs. 4.3). It is known from hemodynamics that the pressure in blood vessels of the capillary vessels reaches seven atmospheres [254]. The energy exchange and circulation of blood in dolphins changes dramatically with a change in swimming speed, so the blood vessels have variable filling.

Fig. 4.16 shows photographs of the dermal papillae with their different blood supplies. With an increase in swimming speed, blood circulation increases, which significantly changes the elastic-viscous properties of the whole body, including the fin elasticity and the elasticity of the upper layers of the dolphin's skin.

This change is explained by both mechanical work and an increase or decrease in temperature of the skin layers, as they regulate the process of thermoregulation of animal tissues [14]. Living tissues are similar to polymeric materials, in which the elastic-viscous properties change dramatically when the temperature changes [14].

The well-defined direction of the dermal longitudinal rollers (transverse microwaves), and especially the distribution of the angles of inclination of the dermal papillae along the body, suggest that this is of great importance for damping coherent vortex structures (CVSs) of the boundary layer of dolphins [70]. As a first approximation, it can be expected that plane disturbances in the form of longitudinal waves such as Tollmien-Schlichting waves in the boundary layer can be successfully damped by the angles of inclination of the dermal papillae, and three-dimensional disturbances can be damped due to the existence of rows of epidermal outgrowths and dermal ridges in the skin, including when due to the different densities of these longitudinal rows.

Figs. 2.4 and 2.5 (Section 2.3) show standard load distributions along the wing profile. Similar load distributions will be on the streamlined body of rotation. When a dolphin swims at different speeds, these distributions change along the body. When swimming at high speeds, in particular, significant pressure drops can occur. In order to preserve the well-streamlined shape of the flexible body and to maintain the optimal elastic-damping characteristics of the skin, the deeper layers of actively working tissues and, above all, the skin muscle are essential. Figs. 4.13 and 4.15 show the location of the outer layers of the skin of dolphins along the body. To determine the patterns of distribution of individual skin layers, the thickness distribution of these layers along the dolphin body was measured. Fig. 4.17A shows the results of measuring the thickness of individual skin layers along the body of a high-speed short-beaked common dolphin (1.45 m) [16] and in Fig. 4.17B along the body of a porpoise (1.27 m) [227]. V.V. Babenko proposed a scheme for the location of points on the body surface for carrying out measurements (Fig. 4.17A), which were carried out on skin samples of short-beaked common dolphin skin taken at the intersection points of the transverse section planes and planes along the body. The section planes in the diagram are denoted by section numbers, and the planes along the body are denoted by solid, dashed, dash-dotted, dashed with crosses, and large dashed lines. Since the measurements were carried out on samples fixed in formalin (c−e), and on preparations when pouring material with celloidin (B), a change to the thickness of the skin layers should be taken into account due to the shrinkage of skin samples caused by the action of the preparations. The measurement results indicate that the accuracy of the measurements at a confidence level of $\alpha = 0.68$ does not exceed 15%.

In the short-beaked common dolphin, the thickness of the epidermis (Fig. 4.17b) varies somewhat in different places along the body, but on average is a constant value of ~ 1.5 mm, which increases to a value of ~ 1.7 mm in the area most dynamically loaded when the tail part of the body oscillates—sections 6−9. The thickness of the suckling layer (Fig. 4.17c) increases on average from 0.2 mm in section 2 to 1.2 mm in section 8, and then remains constant. In the upper (solid) and lower (large dashed) lines, the thickness of this layer increases substantially. When the stem oscillates in this area, the hydrodynamic load and flow slant sharply increase, causing intense three-dimensional perturbations. To stabilize these loads, it is necessary to increase the size of the underpapillary layer. The thickness of the subcutaneous fatty tissue (Fig. 4.17g) has a maximum value in the region of sections 2−4 and decreases at section 6, then at section 8 the thickness increases again, finally decreases significantly at section 9.

The increase in the thickness of this layer is primarily due to the thermal insulation of the main internal organs of the dolphin and the thermal insulation of the tail part of the body, which flows around a more vortex flow, increasing the heat from the dolphin's body. In general, the thickness of the skin of the short-beaked common dolphin (Fig. 4.15), as seen in Fig. 4.17g, along the arcuate line, decreases at the start and in the caudal part of the body, with the exception of the skin area located on the upper part of the head, which experiences the main dynamic flow loads. That is why in the region of sections 1−3 above the head of a dolphin there is a hard strip before the breathing hole. Such a pattern of skin thickness in general resembles the pattern of growth of the thickness of the boundary layer. Fig. 4.17B shows the results of a study of the patterns of distribution of skin thickness along the body of a porpoise [227]. The thickness of the skin of a porpoise was measured in three

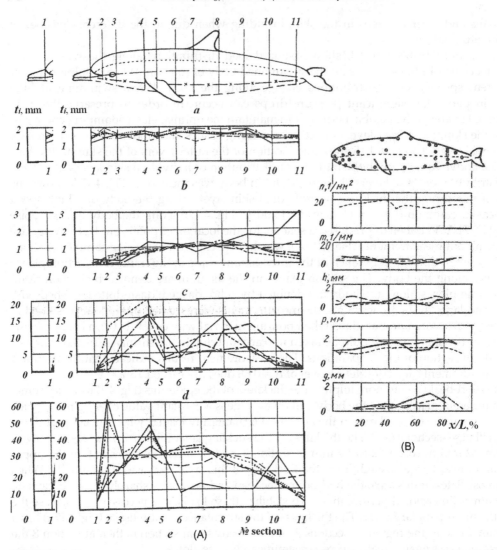

FIGURE 4.17 The change in the thickness of the skin layers of short-beaked common dolphin (A) and porpoise (B) along the body. For the short-beaked common dolphin: (a) is the scheme used to measure; (b–d) are lines of thickness, respectively, of the epidermis, under the papillary layer, subcutaneous fat, generally the skin [16,227]. For the porpoise: the designation of the curves m and n is given in the text, p is the thickness of the epidermis, h is the thickness of the papillary layer, g is the thickness of the papillary dermis [227].

lines along the body [278], shown on the contour of a dolphin. The thickness of the epidermis p in porpoises almost coincides with the thickness of the short-beaked common dolphin epidermis. In the epidermis, there is an outside thin monolithic layer of a membrane type, which below is converted into epidermal septa and epidermal papillae, which are monolithically connected to similar structures of the papillary layer. The thickness of the epidermal and dermal papillae is almost the same (Fig. 4.17). On average, the thickness of epidermal papillae of the porpoise is $h = 0.75$ mm. Consequently, the thickness of the located membrane above layer h is also equal to 0.75 mm. However, the epidermal membrane consists of an outer, continuous cell layer monolithically connected to vertical parallel rows of monolithic cell partitions located between the rows of dermal papillae and parallel epidermal septae by other parallel rows of partitions in the form of rows of epidermal papillae located in the longitudinal rows of the dermal papillae between them. Thus, an important hydrodynamic role is played not only by the layer of dermal papillae, filled with blood and lymphatic capillaries and the smallest nerve endings, but also by the complex epidermal membrane.

The dolphin skin papillary layer (Fig. 4.14) has a strictly ordered microstructure, which is characteristic only of cetacean skin. The dermal papillae are located on the dermal ridges, which are oriented along the stream flowing around the dolphin's body. On a segment 1 mm long, normal to this direction, the average number of $m = 7.5$ mm^{-1} dermal rolls is placed, and the average number of dermal papillae per 1 mm^2 of the body surface of the dolphin is $n = 16.5$ (mm^2)$^{-1}$. Thus, the approximate number of dermal papillae on the entire surface of the dolphin is 3×10^6 (three million units).

The structure of the skin of the vertical fin of dolphins was studied in detail in Refs. [216−220].

Based on the structural features of dolphin skin, technical analogues of damping integuments have been developed [22,23,27].

4.4 The structure of shark body systems

The body shape of the shark is divided into three main groups [114,234,300]. The *first* group is pelagic fish. These are active predators. They have the excellent hydrodynamic forms of fast swimmers: the blue shark (*Prionace glauca* L.) swims at a speed of 10 m/s, salmon at 5 m/s, and tuna at 6 m/s. The *second* group is sharks living in coral reefs and deep lagoons, in the surface zones of the ocean. They have a thinner, slender body with a streamlined shape, the caudal fin is asymmetrical, and the caudal stem is thick. The most typical representative of this group is the gray shark (*Careharhinus menisorrah*). The *third* group is typically bottom-living animals leading a sedentary lifestyle. Among these sharks there is not a single species adapted for fast swimming (Scyliorhynidae). It can be assumed that in the course of a long evolution, fast sharks have developed some hydrodynamic adaptations that enable fast swimming with minimal energy expenditure. These mechanisms are most likely found in the skin.

Research on the structure and location of the placoid scales along the body of sharks has received much attention in various countries. In Ukraine, studies in this area were performed by Zayets [99,100,301−304].

In Ref. [122], data on the structure of the body systems of various species of sharks are given. A comparison of the location of the internal organs of dolphins (www.cetacea.ru/gen2.htm) and sharks shows a significant difference, primarily due to the absence of lung in sharks. A large area in sharks is taken up by the stomach and liver.

The number of gill slits (2, 3) in different species of sharks varies from five to seven pairs.

On the inner surfaces of the gill sacks there are a large number of dark red folds—gill petals. These petals have a significant total surface area, permeated with a huge number of blood capillaries. Blood oxygenation is carried out with flow ventilation during the move-ment of sharks. Water in the gill slits and blood in the capillaries move in the countercur-rent. The gas exchange efficiency of a small shark, *Squalus acanthias*, is estimated by the surface area of the gill lobes as $0.37 \text{ m}^2/\text{kg}$ of shark weight.

In the gill sac water flows around both sides of the slit. In some species of slow-swimming sharks, the gas exchange process takes place in the active mode: water enters the expanding throat (1), through the internal gill slits (2), penetrates to the petals of the gills (4), and further into the cavity in front of the external gill slits (3) closed by water pressure. The exhalation begins with the approach of the gill arches and, consequently, a reduction of the pharyngeal volume. The contiguous petals do not allow water to get into the pharynx again, the water is directed to the outer gill cavities and under pressure comes out, giving oxygen to the gill petals and saturating with carbon dioxide. A passive mode of gas exchange is provided by fast-swimming sharks with flow-through ventilation, while the gills are washed by water entering the mouth and exiting the outer gill slits. Some shark species use both types of breathing. In addition, the effectiveness of respiration is provided by the myoglobin content in the blood, which combines a large number of molecules with oxygen. This carries of $3-5 \text{ g}/100 \text{ mL}$ of blood to the organs of the blue and hammerhead sharks, and $14 \text{ g}/100 \text{ mL}$ of the blood of mackerel, white, and mako sharks [122].

In Refs. [70,130,133], the functioning of the gill apparatus of sharks was modeled. The flow of water around an ellipsoid of rotation was experimentally investigated using a laser ane-mometer. Polymers solutions were injected through an annular gap located in the nose of the model. The distributions along the model of longitudinal averaged and pulsation velocities over the thickness of the boundary layer were measured [70]. In-situ studies of injection of polymer solutions into the boundary layer were performed in Refs. [204,205]. The results obtained are presented in detail in Parts II and III of this monograph.

In Ref. [97,134] various aspects of the circulatory system of the gills of fish, which ensure high efficiency of their work, were investigated. The shape and size of the respiratory plates along the length of the gill lobe varies to ensure the effectiveness of the gill apparatus. This work investigated the optimal process of gas exchange in the gills. The task of studying the hydraulic characteristics of the gill apparatus remains relevant.

In Ref. [304], the distribution of red muscle in the body of sharks was investigated. Through the process of evolution, sharks have been enabled to swim at high speeds and have a number of characteristic features not only of their external structure (well-streamlined body shape, symmetrical tail fin, lateral carinae on the tail stem, etc.), but also a number of anatomical fea-tures of various organ systems. The locomotor function of pelagic fish is enhanced by the development of more powerful lateral muscles of the body, and at the same time the structure of the myotomes is complicated, and in the most best swimming forms, a well-separated, deep-lying red muscle appears in addition to the superficial lateral layer of red muscles.

In fish musculature, the differentiation of muscle fibers into tonic (red) and non-tonic (white) is expressed much more sharply than in other animals. Red muscles, according to Ref. [92], in comparison with whites are a more developed and perfect type of muscles, which has a much greater efficiency and fatigue. Most fish, according to many authors, have both types of muscle fibers, which clearly differ from each other in a number of features (color, fiber diameter, blood supply, etc.). Both types of fibers have different innervation and functional properties. In fish with different swimming speeds, the surface layer of red muscles receives unequal development, while the most high-speed species have a deep red muscle. The largest, well-detached red muscle has such active and fast-swimming fish as tuna. In sharks and swordfish, it is less pronounced. Moreover, in different species of sharks, as in bony fish, the red muscles get unequal development, depending on the swimming speed.

Superficial red musculature is present in all studied sharks (katran, blue shark, mako shark), and in different parts of the body it is expressed differently. In the head part of the body, the red muscle fibers of all studied sharks are expressed very weakly and are represented as a thin film. When approaching the tail region, that is, the most active part of the body, the number of red fibers in the myotomes increases significantly, but immediately before the tail fin it falls again. The muscle fibers here go to the tendons attached to the skeleton of the caudal fin. An isolated deep red muscle appears only in fast-swimming forms. A comparative study of the muscles of sharks of various speed groups (quatre, blue shark, mako shark) clearly shows the gradual separation of the red muscle. In the slowest swimming species studied (the Black Sea katran) the red fibers form only a thin surface film located directly under the skin; the main mass of the myotome is white fibers. In the more rapidly swimming blue shark, the muscle fibers of the superficial muscles already occupy a much larger part of the myotome. Such areas, consisting of red muscle fibers, especially in the middle part of the body, go deep into the muscle mass in the area of the horizontal septum, that is, the connective tissue septum separating the dorsal and abdominal muscles. It is believed that fish with a greater proportion of red fibers in the myotome can maintain a higher speed for longer than fish with fewer red fibers. The red musculature of the fastest of the sharks, the mako, is even more developed. Along with the superficial red muscles, a distinctly deep red muscle appears, which extends along the entire body and stands out very well among the white muscles surrounding it.

4.5 The morphology of shark skin

In Refs. [176,231], the structure of the skin of various high-speed hydrobionts is compared (Fig. 4.18). The total thickness of the skin, expressed in fractions of the body length L of the animal, is about $8.5 \cdot 10^{-3} L$ in the middle part of the dolphin's body, and for the largest high-speed fish $(0.65-95) \cdot 10^{-3} L$, that is, the dolphin skin is an order of magnitude thicker. In dolphins, the layers of continuous epidermis and dissected, but ordered papillary skin layer are sharply distinguished both in absolute value (numbers in Fig. 4.18 to the left of skin incisions) and in relation to the total local skin thickness in percent (numbers to the right of incisions). This second layer is present only in the skin of dolphins, and in high-speed fish it is absent. No less powerful and connective skin layer of dolphins, especially a powerful layer of subcutaneous fatty tissue. These differences are determined by various methods of reducing hydrodynamic resistance: in dolphins, with the help of elastic-damping skin, and in fish, due to their use of biological polymer fish mucus.

Various aspects of the structure of the skin of various types of sharks were investigated in Refs. [99,100,301−304]. When studying the histological structure of shark skin, the close contact of the skin with skeletal muscles attracts attention [303]. In all vertebrate animals, the connective tissue part of the skin is divided into three layers, with different structures

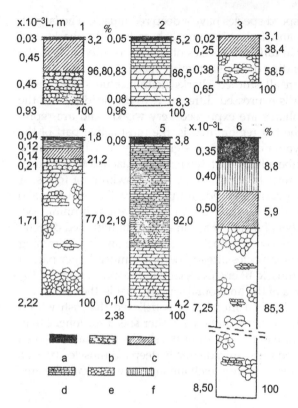

FIGURE 4.18 Layered structure of histological sections of the skin of large high-speed hydrobionts—fish and dolphins: (1) yellowfin tuna, (2) mako shark, (3) swordfish, (4) spotted tuna, (5) blue shark, (6) common dolphin. Skin layers: (a) continuous epidermis, (b) fibrous pigment, (c) connective tissue, (d) collagen bundles and fibers, (e) subcutaneous fatty tissue, (f) dissected dolphin skin papillae [176,231].

and functional significance. The main supporting function is performed by the middle, thickest part of the skin itself (corium), the so-called mechanical layer, consisting of dense connective tissue (collagen and elastic fibers). However, the structure of this layer has features characteristic of each systematic group of animals, since it is affected by differences in the tension and pressure to which the skin is exposed in its different areas. Therefore the skin of fish with rather monotonous types of movement reveals a general fundamental structure. Basically, their corium is constructed from horizontally located layers of fibrous structures, intersecting in adjacent layers at a certain angle and fastened with vertical bundles. The general structure of shark skin is the same. There are the usual layers of skin and their derivatives—secretory cells and scales. However, the histological structure of the skin and the presence of a specific placoid scale, as well as other morphological features, makes shark somewhat different from bony fish.

Some researchers believe that one of the functions of shark skin is that it performs, together with the placoid scales, the role of an external skeleton. Since the skin of sharks serves as a support for the muscles and the whole body, it is characterized by a very strong development of the connective tissue or mechanical layer, which mainly performs this support function. The strong development of this layer provides great strength. The tensile strength of shark skin is 500 kg/cm², while cowhide is about 300 kg/cm². This is also facilitated by the thin structure of shark skin collagen, which, as shown by electron microscopic study, has a special structure that can withstand very high mechanical stress.

The skin of the studied shark species (Black Sea katran, blue and gray sharks, and mako sharks) is characterized by a very thick corium and almost complete absence of the subcutaneous layer. On sections of the skin of sharks parallel to the surface of the body, it is clear that the shark's corium, as well as that of bony fish, has a network-like structure. In the adjacent horizontal layers of the front part of the body, the collagen fibers of the sharks are mutually perpendicular. In the tail region, the angle of intersection of the fibers becomes sharper (up to 70 degrees). Such a diagonal arrangement of the fibers of the connective tissue during movement allows the skin to stretch easily. When stretched, collagen fibers can, without breaking, lengthen by 10%−20%. However, this stretching is limited not only by the fibers themselves, but also by the placoid scales, which are also arranged in a spiral, following the paths of the connective tissue fibers. Thus, dense fibrous corium together with placoid scales forms a flexible, durable shell for sharks.

Shark corium consists of numerous layers of collagen fibers, located one above the other. Only in the uppermost zone of the corium are very thin collagen fibers (10−12 μm) arranged randomly, intertwining with each other, and not forming layers. Here lie numerous bases of placoid scales, which are rarely placed near the katran sharks, at a considerable distance from each other. In blue, gray, and mako sharks, the scales almost adjoin one another, separated only by thin layers of connective tissue. Directly under the bases of the scales there are a large number of thin vertical collagen fibers, as if supporting the scales. The individual fibers enter through the lower opening of the base of the scales in its cavity. When removed from the surface of the corium, collagen fibers gradually thicken and gather in bundles. The diameter of the beams is initially small (25−30 μm), then it gradually increases (84−96 μm). In the lower zone of the corium, all bundles form powerful horizontal layers (340−360 μm).

These sharks have a slightly different corium structure. A slow katran shark has a much weaker mechanical layer of corium: it does not have such powerful layers of collagen fibers as fast-swimming sharks. A feature of the blue shark corium structure is the absence of a clearly defined lamellar structure of the mechanical layer. In mako and gray sharks, the layers of collagen fibers are clearly visible.

The subcutaneous layer of the skin (the layer separating the corium from the skeletal muscles) in sharks is very weak developed compared with bony fish. This layer is most fully represented in the katran (it is found in almost all parts of the body), weaker in the blue shark (absent in the rear part of the body), and almost entirely absent in the mako and gray sharks (a thin film is present in the front part of the body). Thus, for most of the body of pelagic sharks there is no subcutaneous layer, and so there is almost no border between the skeletal muscle and corium.

In different species of sharks, the skin compound with the skeletal musculature is carried out in various ways. On most parts of the body of the slow-floating katran there is a layer of subcutaneous tissue, which prevents direct contact of muscle bundles with corium. The superficial fibers of the skeletal muscles in such areas are suitable for this layer and are attached to it with their tendons. On the back and at the rear of the body, where the subcutaneous layer is very thin, the muscle bundles approach the corium itself and are attached by tendons to the lower layers of the latter. In the tail region, the appearance in the surface layers of the muscles of powerful connective tissue strands extends from the corium.

The blue shark subcutaneal layer is thinner than that of the katran, and on the back and in the tail section it almost completely disappears. Therefore, in the front of the body, excluding the line of the back, the muscles are connected to the skin in a manner similar to that of the katran. In the caudal region and on the back, the surface fibers of the musculature come close to the corium and are attached to it by tendons. In some parts of the tail section, patterns of muscle bundles penetrating into the corium can be observed. In the mako shark, the combination of muscles with the skin has a slightly different character. The subcaneal layer is almost absent, and the muscle fibers approach directly the lower layers of the corium. The contact of the surface muscle fibers with the corium is the same as that of the blue shark, that is, simple attachment of muscles to the lower layers of the corium with tendons occurs in the anterior part of the body and on the sides. On the back and in the caudal region of the abdomen, the muscles do not simply attach to the corium, but penetrate it, as a result of which the lower layers of the latter consist of alternating layers of muscles and powerful collagen bundles. In the tail region of the body, the number of alternating layers and the thickness of the connective tissue bundles increase significantly. In the gray shark, the contact of skeletal muscles and skin is observed in the same areas as in the mako shark. However, the gray shark has fewer interleaved layers and the collagen bundles entering the musculature are much thinner. Segment musculature in sharks, as in fish in general, consists of relatively short and fairly thick muscle fibers, combined into small bundles of connective tissue shells and located between the connective tissue layers—myosepta. Among the myotome muscle fibers, there are two main types: red tonic, capable of long-lasting action fibers, and white, quickly fading nontonic fibers. The number of red fibers located on the surface of the shark's myotomy is 6% of the mass of the myotome in the front part of the body and increases to 20% in the back region, which is the most active during swimming. Thus, in the myotome of a shark, like other fish, there are two different muscular systems that can act independently of each other. In addition to the muscle fibers that make up the bulk of the myotome, fish have a thin layer of superficial, shorter, and thin fibers, located directly under the skin, between it and the superficial red fibers. The length of the surface fibers of the studied sharks remains more or less constant throughout the body. Therefore in a mako shark with a length of 130 cm, they are on average 1.6–1.8 mm. A feature of these fibers is their connection to the tendons. The contractile fibers of the muscle usually terminate in connective tissue, which passes into the tendon and takes on the stress during contraction. In the superficial muscle fibers, the transition of the muscle to the tendon occurs as follows: the tendon cords penetrate deep into the fibrous tissue with small branches that do not interrupt the longitudinal course of the fibers of the muscle itself. Thus, a strong bond is created to ensure the transfer of muscle contractions to the tendon. In the bulk of myotomy muscle fibers, tendon penetration into the muscle fiber does not occur, that is, here the connection with the tendon is carried out in a different way. The surface muscle fibers in sharks are located at a slight angle to the body surface of 8–10 degrees—almost parallel to the surface. The main mass of collagen bundles in the lower layers of the corium is located in the same plane as the surface muscle fibers. The contraction from the muscles is transmitted lower in the layers of the corium. The compressive strength of the tendon is significantly greater than that of the muscle. Superficial muscle fibers are innervated and excited in the same way as red muscles: the nerve endings are located along the entire length of the muscle fiber. White muscles have nerve endings only at the end of the muscle fiber. It is known that the more nerve endings in a muscle fiber, the stronger its contractility will be and the higher the developed tension will be.

When fish move along the sides of their body, bending waves pass. Due to the segmented structure of the muscles, the waves pass smoothly from head to tail. With the reduction of myotomes, the muscle bundles penetrating into the corium will pull the skin and automatically reduce it, causing tension—compression stresses as the wave passes. This will cause alternating changes in the geometrical values of the distance between the elements of the placoid scales (variable roughness), which can contribute to the effect of reducing resistance [303]. To identify this, the structure and functioning of the placoid scales are considered in detail [99,100].

On cross section, the placoid scale (9) consisting of dentin has the appearance of a backward-curved cone. Inside the scales there is a cavity (10), filled with cells of loose connective tissue—pulp. The cavity opens on the underside of the basal plate (2). Corium enters here along with blood vessels and nerve endings. The walls of the scales are composed of dentin, the entire thickness of which is penetrated by the tubules (11), extending from the internal cavity radially to the surface of the spike. The basal plate (2) also consists of dentin and is also permeated by the tubules. Dentin is formed by corium cells, which permeate the connective tissue of the scales. The composition of the dentin is close to that of bone, but is more dense. The dentin layer of denticules consists of a solid mineral, apatite, clad in collagen.

The spike of placoid scales (9) is distinguished by high strength, as it is externally covered with a layer of special enamel (6), vitrodentin, formed by cells of the basal layer of the epidermis (7). In the thickness of the dermis (4), each scale is retained by special strands of collagen fibers that penetrate directly into the bone tissue of the base, the basal plate (2). These bands are called Sharpey's fibers after their discoverer. The biopolymer secretion secreted by shark skin coats the scales, creating a special coating on the body of the fish that reduces resistance.

The spike (3, B, C) grows from the basal layer of the epidermis (7) and is surrounded on the outside by a shell (6) consisting of enamel. In the growing dermal tubercle consisting of dentin, the inner part of the spike is formed, and is filled with loose pulp and numerous blood vessels and nerve endings (8), penetrating to the outer surface of the scale and located in the tubules (11). A detailed study of the pattern of the distribution of placoid scales on the body surface of the Black Sea katran showed that the placoid scale of the katran has the appearance of a trident, sitting on a rhombic base, at an angle to the surface of the body with an inclination toward the tail. The rhombic bases of scales are laid in thick rows in the corium. They are located at a set distance from one another in a checkerboard pattern. Spike scales break through the epidermis and somewhat rise above it, reaching a height of 150—200 μm.

As in the study of the dolphin skin structure, the size, inclination angles, and form of the placoid scales in various parts of the body of various shark species have been studied [99,100,301—304]. For histological studies, pieces of skin were taken from 32 areas according to the scheme shown in Fig. 4.19. To study the distribution of placoid scales along the body of a shark, the Black Sea katran was used to calculate the number of spines per 1 cm^2 according to Fig. 4.19. The smallest number of spines falls along the line of the dorsal ridge—line a in Fig. 4.19; 60—100 in 1 cm^2 (Table 4.1). The research results showed that the form of the placoid scales varies considerably in different areas. Therefore in the area

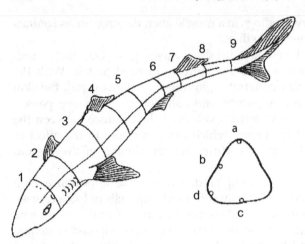

FIGURE 4.19 The scheme of distribution of areas in which material was taken for research [99].

TABLE 4.1 Number of spikes per 1 cm² of body surface area.

Measurement location	Section number in Fig. 4.19								
	1	2	3	4	5	6	7	8	9
Back crest	112	85	59	106	71	71	76	97	109
Back	132	97	103	129	91	97	118	124	115
	121	91	124	135	100	103	118	109	118
	12		109			97			
			112						
Middle line side	138	132	121	119	94	106	115	115	124
	153	138	126	118	106	112	115	2	2
	144	138	141	135	100				
		138	150	132	121				
		141							
Fold	144	138	144	126	129	106	112	115	121
	141		124	129					
Belly	135	135	135	135	129	2	118	121	121
	129	138	138	132	132		124	132	
	124	135	138						
	138	141	138						
		144							

of the head the spines are flat and wide. Further from the head and closer to the tail, spines become narrower and sharper, and they are less irregular. The shape of spines also changes along the lines of the back and abdomen. On the midline, the shape of the spines does not change as drastically. At the level of the bend of the lateral part of the body to the abdomen, along the dorsal crest and at the edges of the fins, the spines are larger, have a rounded run-in shape, are completely devoid of teeth and crests, and are very dense. It should be noted that in all parts of the body, both flat and pointed spines are found. Of considerable interest is the angle of inclination of the spike of placoid scales to the surface of the body. The data obtained from measurements of the angle of inclination of the spine tooth to the surface of the body and the angle of inclination of the rhombic base to the surface of the body are given in Table 4.2. Along the back line, the angle of inclination of the tooth remains approximately constant and varies within 30.6—36 degrees. On the midline, the angle of inclination is less (25—29.8 degrees), and only in the region of the head at point (1) is it increased (45.6 degrees).

Along the line of the abdomen, the angle of inclination experiences sharp fluctuations from 28.6 to 40.8 degrees. Thus, the angle of inclination of the spines toward the surface of the body somewhat changes in direction from the back to the belly. When moving away from the head to the tail, no patterns were observed in the change of angle.

TABLE 4.2 Angles of inclination spikes to body surface.

Measurement location	Angle of spike base inclination (degrees)	Tilt angle of the spike tooth (degrees)	Angle between the spike base and prong (degrees)
Back line			
1a	15	30.6	46
3a	17.3	36	53.3
5a	13.8	32.3	46.1
7a	13.4	24.6	48
9a	14.5	32.7	42.7
Middle line			
1b	22.6	45.6	68.2
3b	14	29.8	43.8
5b	17.3	26.3	43.6
7b	24.5	25	49.5
9b	21.8	35	46.8
Belly line			
1c	24.3	31.7	56
3c	14.4	40.8	55.2
5c	24.7	39.6	64.3
7c	19.2	28.6	47.8
9c	12.9	40.1	53

To determine the relationship between the location of the placoid scales on the bodies of sharks and their swimming speed, a study was conducted with different age groups of four types of sharks: the Black Sea katran (*S. acanthias*; three specimens, $L = 54, 97, 104$ cm), the blue shark (*P. glauca*; three specimens, $L = 75, 130, 150$ cm), gray shark (*Carcharhinus falciformis*; two specimens, $L = 87, 110$ cm), and mako shark (*Isurus oxyrinchus*; three specimens, $L = 100, 140, 160$ cm) [302]. No differences in the distribution of scales depending on age were found in the shark species studied. Studying the pattern of the location of the placoid scales along the body of the katran shark showed that the predominant direction of the tip of the spike is in a direction parallel to the lateral line. Only in the region of the back, in the front part of the body to the first dorsal fin, did the spikes enter at an angle of 5–10 degrees up from the midline. In the lateral region of the anterior part of the body, the spines enter at an angle of 30–40 degrees to the midline downwards. In the region of the eye, the direction of the spines changes sharply, forming a direction parallel to the lateral line. Only in the region of the back, in the front part of the body to the first dorsal fin, the spikes enter at an angle of 5–10 degrees up from the midline. In the lateral region of the anterior part of the body, the spines enter at an angle of 30–40 degrees to the midline downwards. In the area of the eye, the direction of the spines changes dramatically, forming turbulence. In the middle part of the body, after the first dorsal fin, all the teeth again run parallel to the midline.

In the lower part of the body, from the side of the abdomen, the spines on the head are parallel to the midline of the abdomen. In the area of the gill slits, they slightly change their direction and enter at an angle of 30–35 degrees to the midline. Behind the pectoral fins to the end of the body they are again located in parallel. Thus, on the body of the Black Sea katran, only one zone can be distinguished, covering the body of the katran in the form of a belt (in front of the first dorsal fin on the back and in the region of the pectoral fin on the ventral surface), where the tip of the spine is directed at an angle to the midline: above—in the direction of the back; below—to the stomach. On the leading edge of the fins, the placoid scales are smooth, rounded, completely devoid of carinae, and it is very tight. The width of this zone in different sharks is somewhat different (on average by 0.5 cm). Then are the scales, of typical form for this species and corresponding to the scales of that part of the body, near which the fin is located. With distance from the front edge of the fin and toward the rear edge, the scales gradually become smaller, sharper, and, finally, at a distance of 2–5 mm from the rear edge of the fin they disappear completely. The following pattern is noted in the arrangement of scales on the fins. Behind the zone of smooth scales there is a small strip of scales, directed almost parallel to the front edge of the fin, then the angle of the latter rather sharply changes and acquires the value characteristic of the fin.

On the tip of the snout of all sharks, the shape of the scales is the same: rather large, rounded, keel-less scales lie dense, just like on the leading edge of the fins. At a distance of 1.5–2 cm from the tip of the snout, scales appear with weak keels reaching up to half of the plate. Gradually, the keels become taller and longer, and the scales acquire a typical shape for this part of the body.

The pattern of the location of the placoid scales of the blue shark basically repeats the pattern at the katran. The direction of the scales parallel to the longitudinal axis of the body is also predominant. The change in the angle of the scales is observed in similar areas. In the region of the first dorsal fin, the scale orientation angle is 10–15 degrees up from the longitudinal axis of the body. On the lateral surface of the body behind the eyes, in front of the gill slits, the scales run at an angle of 25−30 degrees down from the longitudinal axis of the body. This direction is maintained above the pectoral fin and at some distance behind it. At the level of the first dorsal fin, the scale again acquires a direction parallel to the longitudinal axis of the body and retains it to the end of the caudal stem. On the ventral side of the body, the location of the scales is the same as that of the katran. On the dorsal and pectoral fins of the blue shark, the orientation of the scales is the same: at the base, the orientation angle is 10−15 degrees (down from the longitudinal axis of the body on the pectoral fins, up from the axis of the body, on the dorsal fins), then it gradually grows and at the end of the blade the scales are present along the front edge of the fin. On the upper side of the ventral fins, the angle of the scales is 35−40 degrees to the longitudinal axis of the body, on the lower side of the fins it is located almost parallel to the midline of the abdomen. In the central part of the caudal fin, the direction of the scales remains parallel; on the dorsal part of the blade, the orientation angle moves slightly down to the axis of the body, on the ventral part of the blade, down from the axis of the body.

The location of the scales of the gray shark is basically the same as that of the katran and blue shark. On the remaining parts of the body of the gray shark, the scales are parallel to the longitudinal axis of the body. The orientation of the scales on the fins of the gray shark is somewhat different from that of the katran and blue sharks. The arrangement of scales in the mako shark has some differences. All sharks clearly distinguish two main zones of change in scales relative to the longitudinal axis of the body: in the anterior part of the body and in the region of the ventral fins. In the caudal part of the body, the area of change in the orientation of scales relative to the longitudinal axis of the body is different for different species of sharks.

The change in the length and width of the plate of placoid scales in different parts of the shark body, the height of the scales, and the angle of inclination of the plate to the surface of the body were studied and the number of keels on the plate of scales in different parts of the body of sharks of different speed groups was calculated [99].

Plates of placoid scales of the katran are very large and are located at a considerable distance from one another. The results of the study of the geometrical parameters of scales are given in Table 4.3 [99]. The length and width of the scales of the katran change in the direction from the head to the tail, and the length is more significant than the width. The same pattern is observed in the direction from the back to the stomach. However, in the region of the dorsal fin, the size of scales along the length and width increases. The height of the scales increases from the back to the stomach, except for the front of the body, where it first increases toward the midline and then decreases again (280/330/260 μm). The angle of the plate of scales to the surface of the body at the katran is rather large compared with the scales of other species of sharks (25−40 degrees).

Along the head−tail line, the pattern in changing the angle is not observed. The shape of the katran scales is significantly different from the scales of other species of sharks studied by the author; in front of the body the plate of scales is quite wide, it has one high central and two poorly developed lateral keels. In the caudal region, the scales become narrower and carry one sharp and high keel. The height of the placoid scales was measured from the base of the scales to the edge of the keel.

For the blue shark, the same pattern is observed in the resizing of a plate of scales. In the direction from the head to the tail and from the back to the stomach, the length and width of

TABLE 4.3 Changing the parameters of the plate of scales on various parts of the body.

Shark species	Body length (cm)	Body parts	Front body (μm)			Middle part of the body (μm)			Tail part of the body (μm)		
			Length	Width	Height	Length	Width	Height	Length	Width	Height
Katran	85–105	Back	500	350	280	600	350	180	420	300	220
		average line	490	280	330	560	300	220	410	250	270
		Stomach	490	280	260	420	280	240	400	250	330
Blue shark	130–145	Back	420	460	220	430	450	200	350	420	200
		average line	420	490	220	420	450	220	340	420	190
		Stomach	320	320	220	350	280	200	280	280	190
Gray shark	102	Back	220	250	170	180	210	170	190	210	130
		average line	280	280	170	190	210	150	210	200	130
		Stomach	260	280	150	200	210	150	190	180	110
Mako shark	125–140	Back	130	110	90	200	120	70	120	100	80
		average line	150	130	90	200	130	80	150	100	80
		Stomach	150	130	90	180	110	80	110	90	70

the placoid scales decrease. The height of the scales almost everywhere remains the same (200–220 μm), with a slight decrease in the tail region to 190 μm. The angle of inclination of the scales to the body is very small (in some areas no more than 5–8 degrees), and the patterns in the distribution of these areas are not observed. In general, scale plates are located almost parallel to the surface of the body. The scales of the blue shark are tight; its relative width is greater than that of the katran (Fig. 4.19B). In the anterior part of the body and along the back line, some plates of scales carry five well-defined carinae each. On the abdomen plate scales are sharp, narrow, and less common than in the line of the back, and carry three keels. A feature of blue shark scales is very large plates found on the back with a large number of keels (up to six), having a fan-shaped arrangement. On all other parts of the body, the keels are strictly parallel.

Gray shark placoid scales are smaller than blue shark ones. The scales of the gray shark decrease in the direction from the head to the tail (Table 4.3) and only in the anterior part of the body along the back–abdomen line, the length and width of the scales increase [99]. In the remaining parts of the body, the size of the scales in this direction barely changes. In front of the body, the width of the scales in some places significantly exceeds the length (length 220 μm, width 280 μm). Due to this width of the scales, the gray shark often has scales with five keels in the middle part of its body. Basically, the gray shark scales have three keels. The scale plate is basically parallel to the surface of the body. And only in certain parts of the body can the inclination angle of the plate reach 5–8 degrees. The scales are very close and retain such an arrangement throughout the body. Gray shark's keels on scales are high and sharp; they are clearly visible on all parts of the body.

The mako shark placoid scales are identical to blue and gray shark scales, but much smaller in size. The change in the length and width of the plate in different parts of the mako shark body occurs smoothly (Table 4.3). The height of the scales is basically constant (90 μm) and only in the tail area does it decrease to 70 μm. The angle of inclination of the plate of mako shark scales to the surface of the body remains the same as that of the shark species discussed above. Scale plates are parallel to the surface of the body over almost the whole body; only in certain parts of the body can the angle reach 5–8 degrees. Scale plates have three keels each. The scales are located tightly in the front part of the body and less often in the tail region. Plates of placoid scales in high-speed sharks are located quite tightly, and the keels form longitudinal grooves stretching in the direction of the head-tail.

It remains to be determined whether mucous cells are present in fast-swimming sharks, that is, whether the placoid scales are covered with mucus, since the presence of the latter can to some extent alter the microrelief of the scales. Placoid scales are located parallel to the fixed surface of the body of sharks. However, since the bases of the scales lie at a certain distance from one another, then with the bends of the body there is the possibility of changing the angle of the spike to the surface of the body. The segmental structure of muscles and the location of the myotomes directly under the skin lead to the fact that their reduction causes a change in the geometric and mechanical characteristics of the skin. A more detailed study of the placoid scales of sharks showed that both scales and their structural formations, the so-called keels, undergo significant changes [99,100,303].

The geometric parameters of the keels of the placoid scales are given in Tables 4.4 and 4.5 [99,303]. The distance between the keels and the height of the keels do not remain constant in different parts of the body. They increase in the direction from the head to the region of the first dorsal fin (Fig. 4.19, 4 and 5), and then decrease again to the tail. The distance between the keels in the tail region is somewhat less than in the front part of the body. The height of the keels is the smallest in the region of the head, then increases to the middle part of the body and again gradually decreases to the tail. On the tip of the snout and on the front edges of the fins, the scales have a rounded, ovoid shape and are completely devoid of carinae.

However, already at some distance from the tip of the nose, weakly expressed keels appear in the scales at the back of the scale [303]. With further distance from the front of the body, the keels become higher and, finally, reach the value characteristic of each type of shark (Table 4.5). It can be assumed that the placoid scales and their keels serve to some extent as a means of retaining the mucous substance secreted by numerous secretory cells of the skin of sharks.

In the skin of sharks, the following layers are distinguished: the epidermis, which lies on the basement membrane, a powerfully developed corium, or cutis, and subcutaneous connective tissue, located on the skeletal muscles. The epidermis has the appearance of multilayered epithelium. The cells of the deepest layer of the epidermis, lying on the basement membrane, have a cylindrical shape and are located tightly one next to the other perpendicular to the surface of the body. During life, they retain the ability to form, by dividing new layers of cells, the germinal layer. Therefore the epidermal cells gradually move outward, being subjected to morphological and biochemical changes. In the middle

TABLE 4.4 The distance between the keels of the spikes in different parts of the body of the sharks.

Shark species	Body length (cm)	Body parts	The distance between the keels in different parts of the body								
			1	2	3	4	5	6	7	8	9
Katran	105	Back	155	160	165	170	163	152	142	135	–
		average line	138	140	148	150	147	140	137	130	–
		Stomach	130	132	138	135	133	130	127	120	–
Blue shark	150	Back	120	122	121	122	124	120	115	107	102
		average line	106	110	117	122	123	119	110	104	96
		Stomach	111	110	114	115	116	110	109	99	93
Gray shark	110	Back	62	63	65	68	70	72	67	59	55
		average line	64	66	68	69	73	70	69	61	58
		Stomach	63	64	67	69	72	68	66	59	54
Mako shark	155	Back	53	56	57	60	68	66	59	55	53
		average line	55	57	58	61	63	65	62	58	53
		Stomach	53	55	57	59	60	59	54	52	50

TABLE 4.5 The height of the keel spikes in different parts of the body.

Shark species	Body length (cm)	Body parts	The height of the fins in different parts of the body (μm)								
			1	2	3	4	5	6	7	8	9
Katran	105	Back	75	87	97	99	100	98	96	93	91
		average line	90	101	105	108	110	107	103	101	99
		Stomach	98	106	110	114	115	112	110	107	104
Blue shark	150	Back	32	35	37	45	48	49	50	46	44
		average line	34	36	45	46	47	47	49	43	41
		Stomach	39	41	44	47	49	49	47	45	43
Gray shark	110	Back	21	23	25	28	33	34	34	30	27
		average line	22	25	27	30	32	33	32	31	30
		Stomach	23	26	28	30	32	34	34	30	28
Mako shark	155	Back	15	18	22	24	25	25	26	24	23
		average line	16	20	20	23	25	26	27	25	22
		Stomach	18	22	25	27	28	27	25	24	23

layers of the epidermis, the cells acquire a polygonal shape. Then, approaching the outer edge of the epidermis, they become flat, acquire a scaly form, gradually die off and are washed away with water, giving way to newly growing cells.

In the outer layers, mucous cells exist in a large number, secreting to the surface of the body. They are formed in the deep layers of the multilayer epidermis of the epidermal cells. Together with the accumulation of secretions within their bodies, they grow rapidly and become larger than the indifferent epithelial cells. Then they begin to move from the lower layers to the surface of the body. Since, in the process of mucus formation, the protoplasm of the cell and the nucleus are regenerated, the cell produces mucus only once and then dies. Sharks have only one type of mucous cells—serous glandular cells—containing a homogeneous or granular secretion and opening out only on the surface of the skin. Cells with homogeneous secretions with varying intensity perceive color and are more or less vacuolated. Mucous cells are present in all studied areas of the body, however, clearly defined patterns of their distribution are not observed. On average, 1 mm^2 of the body has 25–32 cells. A slight increase in their number is observed in the tail area (up to 40 cells/1 mm^2). Measuring the size of mucous cells in all parts of the body showed that the mucous cells are oval. They are much larger than the cells of the epidermis. They reach 58 μm in height and 37 μm in width, and do not change significantly depending on the area of the body. Such a large number of mucous cells, evenly located throughout the body of the shark, suggests that the entire body of the animal is covered with mucus. However, it remains unclear whether the spines of the placoid scales are covered with mucus or if their tops extend over the mucous cover.

The next layer is the skin itself—the shark corium is very powerful. It consists of collagen and elastic connective tissue fibers. In the layers adjacent to the epidermis, bundles of fibers lie tight to one another and parallel to the surface of the body. Bundles in alternating layers intersect crosswise. Below are vertical beams, moving from one layer to another. The deeper layers of corium are made up of fibers arranged more loosely. Often, the thick horizontal deep layers of the dermis seem to separate into a special lower layer of the dermis, separated from its middle part by a layer of loose connective tissue rich in blood vessels. The distribution of individual layers of the skin along the body of a katran was investigated in accordance with [99].

In the tail region, the corium is separated from the skeletal muscles by a very thin subcaneal layer. This sometimes gives the impression that the corium borders directly on the skeletal muscles. Of particular interest are the detected muscle bundles of skeletal muscles, penetrating in certain areas of the body in the subcutaneous layer (points 2a, 3a). The presence of muscle bundles, penetrating into the skin of the shark, indicates the possibility of arbitrary deformation of the body surface in some parts of it. The corium gradually passes into the subcutaneous layer. Its layers are located, more rarely, between increasing layers of loose, unformed connective tissue, and finally, they form the basis of the subcutaneous tissue. In the subcutaneous layer are a large number of blood vessels.

The epidermis in all parts of the body retains an almost unchanged thickness with slight deviations from an average value of 80–100 μm. More pronounced changes, depending on the part of the body, are observed in the thickness of the corium. On the back line 1a–9a the corium has the greatest thickness (972 μm in the head region, 728 μm in the tail region). On the middle line (1b–9b) the thickness of the front corium is more uniform (426–492 μm); in the area of the tail, it increases sharply and reaches 824 μm (points 8b, 9b). Across the abdominal line, the thickness of the corium is a fairly uniform 426–492 μm, in the tail region it increases to 862 μm (points 8c and 9c).

The thickness of the corium undergoes the greatest changes along the line of transition of the lateral part of the trunk to the stomach (fold line d). This figure shows that at point 2d, the corium is the thinnest at 239 μm. The skin in this area has a peculiar structure, as there are many sensory organs here. When approaching the tail, the thickness of the corium begins to increase and reaches 650 μm. Points 7d, 8d, 9d are missing, since the fold line to the tail disappears.

The thickness of the subcutaneous tissue changes most dramatically. This is probably due to the specificity of the structure of the skeletal muscles, on which lies the subcutaneous tissue. The greatest thickness of the subcutaneous layer is observed in the region of the head, where it reaches 842 μm in places. When approaching the tail, the subcutaneous layer begins to thin and, finally, remains only in the form of a thin film (up to 40 μm) separating the corium from the skeletal muscles. The change in the thickness of the skin and its layers suggests that the density of the skin and its elastic properties also vary in different parts of the body.

4.6 The structure of swordfish skin

In Refs. [64,70,158–163], the results of Koval's studies are presented. Fig. 4.18 compares the structure of the skin of some species of fast-swimming aquatic organisms: tunas, sharks, dolphins, and swordfish. The structure of the skin in sailing plants and especially in swordfish is more complex than in other types of fish [163,215,230].

4.6.1 Longitudinal folds on the skin surface

In the process of the evolution of fish, the roughness of the outer surface of the skin decreases with increasing swimming speed. In Ref. [158], a longitudinal roughness was found on the skin surface of the fastest swordfish (Fig. 4.20).

The phylogenetic group Xiphiidae comes from the primitive sombroid fish, which are characterized by the presence of smooth skin. Therefore such ribbing of the skin and roughness of the integument arose in connection with adaptations to high swimming speeds. This suggests that the ribbing and roughness of the integument is expedient and represents a mechanism that helps reduce resistance during fast swimming. The ribbing of the body surface is formed by skin folds, which are located along the body of the animal from the head to the tail fin. The size of these folds from head to tail and from back to stomach does not remain the same. The head, immediately behind the gill cover, in the middle line, for which we conditionally take the line running from the middle of the eye to the fork of the caudal fin, has small folds. In the area of the midsection of the body, they become larger and decrease to the tail again. The same changes in the size of the folds are observed in the direction of the back to the stomach. Therefore, in the midsection, near the midline of the back, the number of folds per 1 cm of the cross-section length is seven to eight, and on the conventional midline of the body it is three to four. At the same time, the height of their crests reaches 0.7 mm (Fig. 4.20C), and by the midline of the abdomen their number increases again to eight to nine, with a simultaneous decrease in the height of folding. Reports of such a ribbed surface of the skin of swordfish in

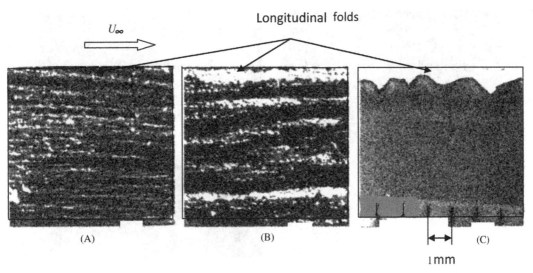

FIGURE 4.20 Longitudinal folds on the skin surface of swordfish: (A) small longitudinal roughness in the region of the midline of the back, threefold magnification; (B) roughness in the region of the conditional midline of the body, sixfold magnification; (C) transverse skin incision in the area of longitudinal folding, sixfold magnification [42,70,158].

the literature are not known. At the same time, in all eight studied animals of different sizes, the longitudinal ribbing was well pronounced and had the above regularity [158].

4.6.2 Pores and the system of subcutaneous channels on the skin surface

On the skin surface of the swordfish, numerous pores open up, and irregular channels are located in the thickness of the dermis (Fig. 4.21). Pores are observed throughout the body, head, and gill covers, and leave from the original channels that are located in the thickness of the skin at a depth of 1 mm from the surface of the body. The number of pores per 1 mm^2 is on average one to two, and their diameter is 0.07−0.2 mm. Studies have shown that pores lead to irregular channels.

They are well detected if a razor blade is used to remove the surface layer of the epidermis and dermis of the skin. On the dorsal and ventral side of the body, 2−3 cm from the midline of the back and abdomen, the channels have a mesh structure (Fig. 4.22B). On the lateral surface of the trunk, they are located in almost parallel rows in the dorsoventral direction perpendicular to the folding of the skin (Fig. 4.22C). The number and size of pores in an area of 10 mm^2 in various parts of a swordfish body with a length of 2.6 m are given in Table 4.6. The density of pores and their size on the body of fish in different parts of the body are almost the same and, only on a narrow strip, on the top line behind the protruding dorsal fin, is their number about twice as large.

The numerous pores are connected by hypodermic channels, which are located in the dense connective tissue stratum at a depth of about 1 mm from the surface of the body (Fig. 4.21C). The average diameter of channels is 0.5 mm up to a maximum of 1 mm. Channels smaller than

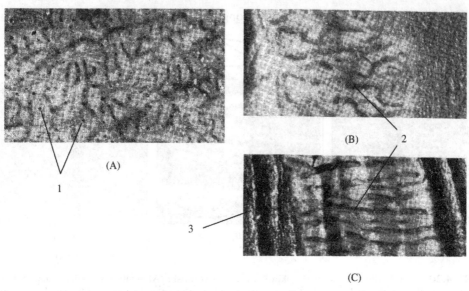

FIGURE 4.21 The location of pores (A), fivefold magnification; channels in the region of the midline of the back (B), sixfold magnification; and channels in the region of the center line of the body (C), sixfold magnification: (1) pores, (2) channels, (3) folds on the surface of the swordfish skin [42,70,158].

FIGURE 4.22 The cross section of the subcutaneous canal, 300-fold magnification: (1) the base of the scales, (2) channel, (3) mucus-forming cells [162].

TABLE 4.6 The number and size of pores in various parts of the swordfish body.

Cross sections of the body as a percentage of length, recalculated from the lower jaw of a swordfish											
30%		40%		55%		70%		80%		90%	
N	d (mm)	N	d (mm)	N	d (mm)	N	d (mm)	N	d (mm)	N	d (mm)
Dorsal line (back)											
6	0.13	13	0.08	13	0.08	12	0.08	9	0.09	6	0.09
Middle line											
6	0.1	7	0.1	7	0.1	6	0.1	6	0.09	5	0.09
Ventral line (stomach)											
6	0.1	6	0.09	6	0.08	6	0.08	5	0.07	5	0.08

0.2−0.3 mm are located on the head and gill covers. On the lateral surface of the body, the channels are directed from above downward, perpendicular to longitudinal folds of skin. Between channels there are often channels connecting them into a uniform interconnected system. On the back, top line, head, and gill covers, the channels are situated chaotically, without a definite orientation. The channels of the left and right sides of the body are connected, neither on top, nor on the bottom lines. There are two independent systems on both sides of the body. The number of channels in different parts of the body of a swordfish does not change and is on average six to seven for each 1 cm. On histological specimens, it can be seen that the walls of the channels are lined with typical multilayered epidermis of the skin type. As well as over the entire surface of the body, there are many secretory cells in it, which makes one suspect the presence of mucous substances in the intravital state in the channels. This is confirmed by the fact that when you press on the skin of a live swordfish specimen, a mucous substance is released from the pores. An approximate calculation of the internal surface of the channels shows that it roughly doubles the total secretory surface of the body.

During the histological study of the pore system and channels, it was found that the internal cavity of the channels lines the epithelial layer, consisting of two main types of cells: the epithelial and secretory (mucus-forming) cells (Fig. 4.22). The epithelium has uneven development along the perimeter of the canal. The thinnest layer is located in the upper part of the channel and at the exits from the pores. Here the epithelium is composed of two or three layers of epithelial cells, among which rarely are secretory cells. The thickness of the lining epithelium gradually increases toward the base of the canal and consists of the largest canals of five to six rows of epithelial cells. At the same site, the number of secretory cells increases significantly. The size of the secretory cells in the epithelium of the channels does not differ from the secretory cells of the epidermis.

Below the base of the scales (1) are numerous channels (2), which on the inner surface have a stratified epithelium (3) (Fig. 4.22). The thickness of the epithelium changes significantly: in one place it consists of five to six layers of epithelium, in the other from two to three layers. Like the epidermis of the skin, in the canals between epithelial cells, there are many cells secreting mucus into the canal cavity.

4.6.3 Scales in swordfish skin

In the skin of juvenile swordfish there are scales with a developed relief (Fig. 4.23A), clearly visible on histological specimens. The location of the scales on the body and their

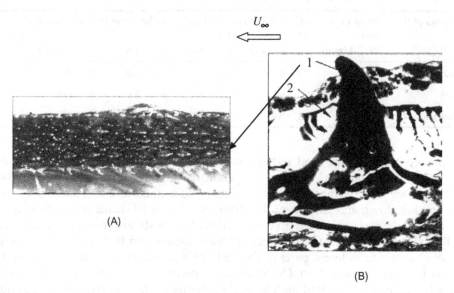

FIGURE 4.23 Swordfish scales: (A) on the body of a fish with a length of 1.2 m, threefold magnification; (B) on the body of a fish with a length of 2.6 m, 350-fold magnification [64,70].

structure are fully described in Ref. [215]. The bases of the scales (1) are lying in a dense unformed connective tissue basis (2) of the skin. The flakes of an irregular shape have an upward-directed tooth, which pierces the epidermis and rises somewhat above the surface of the skin. These teeth are located both on the tops of the crests of the folds, and in the recesses between them. The largest spines are located on the head and border the orbit. Along the body the scales are located in several longitudinal rows. The dorsal, two lateral, and one abdominal rows are most clearly visible, between which are smaller scales covering the entire body. As the fish grows, the scales change. In a swordfish with a length of 0.69 m, the scales have the form of small triangles, which are located on top of each other like tiles. On the a stomach area lines of scales in the form of bone hillocks remain. At this stage of development of the fish, thorns and spines disappear from the head, tail stem, and lateral parts of the body. It is assumed that the scales of swordfish disappear completely at a length of about 1.5 m. However, our studies have shown that scales with conical protrusions passing through all the layers of skin lying above are found on the body of an adult swordfish 2.6 m long (Fig. 4.23B).

4.6.4 Slots on the body surface

A number of slots can be traced along the midline of the back of the swordfish. Their chain begins at the base of the third ray of the dorsal fin and ends at the additional dorsal fin (Figs. 4.24 and 4.25). The length of the slots in different parts of the body is different. Small slots are located in the immediate vicinity of the dorsal fin, increasing sharply behind it.

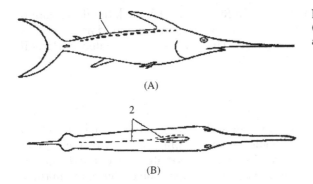

FIGURE 4.24 The location of the ampoules (A) and slots (B) on the swordfish body: (1) ampoules, (2) slots [70,160].

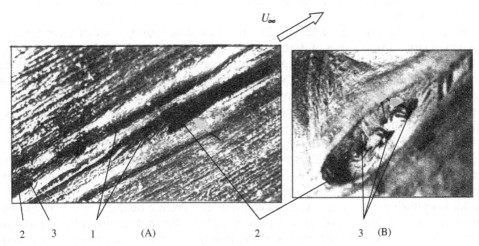

FIGURE 4.25 The location of slots (A) and cavities (B) in the swordfish skin: (1) slots, (2) cavities, (3) ducts [70,160].

Between large slots there are small slots. The minimum size of slots for swordfish 1.57 m long is 2 mm, and the maximum is 10 mm. With the increase in the size of swordfish, the size of slots also increases. Therefore at a swordfish length 2.6 m the maximum size of the slot reaches 18 mm.

Behind the dorsal fin, in the line of the back, an adult swordfish has a small longitudinal ridge, which is a remnant of a reduced large dorsal fin, which in fry extends almost to the caudal fin and decreases significantly during ontogenesis. Since the cushion is located strictly in the center of the line of the back, the slots are somewhat displaced on the lateral surfaces of the fish body and are located, as it were, in the chess order on both sides of the projecting dorsal fin and cushion. In swordfish with a length of 1.57 m, 18 slots were found on the left side and 22 slots on the right side. In the thickness of the skin, the slots are somewhat narrowed and form a flat bottom where the cavities are located, in which there are from one to four small rounded holes (ducts) with a diameter of 0.5–2 mm.

These ducts extend from large cavities—ampoules. In Refs. [45,55,65, etc.], the flow around various types of cavities and the formation of vortex structures in the boundary layer of a plate using cavities were experimentally investigated. The results are presented in Parts II and III of the monograph.

4.6.5 Ampoules in swordfish skin

Ampoules are round, oval, or bean-shaped with a slight flattening in the cranial and caudal regions (Fig. 4.26). In most cases from one ampoule the ducts come out in two slots, one of them being large and the other small, and they are located on opposite sides of the demarcating roller. Ampoules of swordfish skin are mainly distributed along the center of the midline of the back, however, there are cases when the ampoule is shifted to the right or left side due to the presence of cartilage tissue in this area—most likely, the remains of the bases of the rays of the dorsal fin.

In swordfish with a length of 1.57 m, 24 ampoules were found, some of them (14 pieces) are relatively large (up to 7 mm in diameter and up to 15 mm long), and some are small (up to 2−3 mm in diameter and 4 mm long). With increasing fish size, the size of the ampoules also increases significantly. Thus, a 2.5-m swordfish has a maximum ampoule diameter of 13 mm with a total length of up to 30 mm. The small ampoules are located at the base of the protruding dorsal fin, and the large ones are located behind the dorsal fin. The cavity of the ampoules has a cellular structure that is outwardly similar to a ball of interlaced threads (Fig. 4.26B). A histological study of the structure of various parts of the ampoules, the excretory ducts, and the walls of the slots revealed that their surface was lined with a powerful multilayered epithelium. The number of cell layers in a 2.5-m swordfish specimen examined in some areas reached 40. Most epithelial cells turn into secretory cells, with the result that the entire lining epithelium takes the form of a continuous secreting field (Fig. 4.27). The presence of numerous connective tissue outgrowths that form the cellular structure of the ampoule cavity significantly increases the secretory surface. A huge number of secretory cells at various

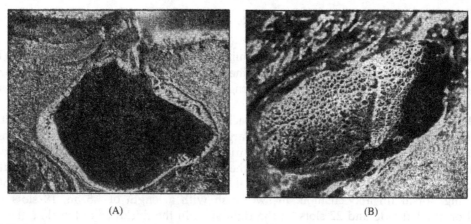

(A)   (B)

FIGURE 4.26 Cross section (A), magnification ×8, and longitudinal section (B) ampoules in the swordfish skin, ×5 magnification [64,70,160].

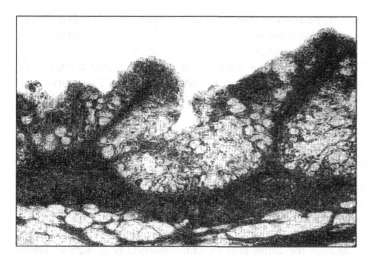

FIGURE 4.27 Secretory apparatus of swordfish skin, 150-fold magnification [64,70,160].

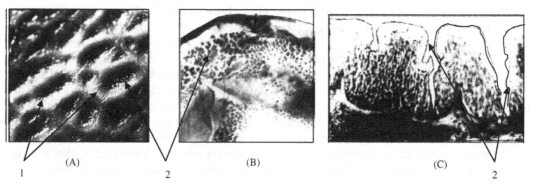

FIGURE 4.28 The structures of the skin crypts in the gill slits (A), the structure of the crypts on the inside of the gill covers (B), sixfold magnification; a cross section of the secretory apparatus of the crypt structure (C), 300-fold magnification: (1) holes, (2) cells of the crypt-like structure [64,67,70].

stages of their development suggest that in the lining epithelium of ampoules and even in the excretory ducts there is active and gradual mucus formation, as a result of which there is always a supply of mucus in the ampoule cavity. Cellularity of the ampoule cavity is created by the uneven development of dense unformed connective tissue, which forms branching ridges and individual pinpoint or fungoid outgrowths. In histological sections, this structure is somewhat similar to cryptoid processes of the intestine or stomach (Fig. 4.28).

The dense unformed connective tissue in all parts of the ampoules is vascularized vigorously. Thin blood vessels penetrate even into separate connective tissue outgrowths. Under the layer of dense unformed tissue (rigid base of the ampoules) there is mainly loose connective tissue with separate groups of fat cells. In the upper parts of the ampoules (in the area of the excretory ducts) a layer of dense unformed connective tissue merges with the connective tissue of the skin. This suggests that ampoules are skin derivatives. Special muscle bundles in contact with the walls of the ampoules were not detected. It can only be noted that in some cases at the base of the ampoules there are bundles of skeletal muscles, on the side of which, with their contraction, a partial deformation of the walls of the ampoules and a change in their volume and shape are possible. In the fry of swordfish with a length of 7 cm serial sections of the slits and ampoules was not found. Thus, we can conclude that this formation in the skin of swordfish appears in the process of ontogenesis.

4.6.6 Skin structures in the area of gill covers

In the area of the gill slits and on the inside of the gill covers there are specific structures that resemble crypts of the stomach or intestines in histological sections (Fig. 4.28A). However, this is not a solid field of crypts, but various separate forms (rounded, triangular, polygonal) islands, which are delimited by several projecting rollers. On the gill covers, the crypts are dispersed throughout the inner side by small islands (Fig. 4.28B). There are only small areas in the center of the gill cover, where there are no crypts. The cellular composition of the crypt-like skin structures of swordfish does not differ from the cellular composition of the skin epidermis and lining epithelium of the canals.

On the surface of the body, cryptographic structures are arranged in a narrow strip (3—4 cm) along the edge of the gill slit, slightly covering the space under the gill cover (Fig. 4.28C). The cellular composition of cryptographic structures does not differ from the cellular composition epidermis of the skin and lining epithelium channels. It is represented by two types of cells: epithelial and secretory. The size of epithelial cells is 5—6 μm. Secretory cells are located in the upper layers in large numbers and in some areas can reach 20 μm. The thickness of the epithelial layer on the ridges of the protruding rollers of the connective tissue is small (three to four layers of epithelial cells), and in the areas bordered by the roller, there may be 30—40 rows of epithelial and secretory cells. In places with a highly developed epithelial layer, there is also the largest number of secretory cells, where they are located in several rows.

4.6.7 Structure of the skin

The structure of the skin of swordfish is much more complicated than that of other fish, due to the presence in the skin of the above macroformations. The epidermis of a swordfish lies in a continuous layer on all parts of the body and only in certain places does it break through with conical scales or is interrupted by pores. In places where the pores reach the surface of the body, the epidermis bends inward into the pores and passes into the lining epithelium of the subcutaneous channels. The cellular composition of the epidermis consists of 8—10 rows of epithelial cells, in the thickness of which at various stages of development there are numerous secretory cells of oval form (1) and individual pigment cells that have migrated from the underlying layers of the skin (Fig. 4.29). The general structure of the epidermis of swordfish does not fundamentally differ from the structure of the epidermis of most fast-swimming bony fish, which have only one type of secretory cells. The epidermis lies on a thin basal connective tissue membrane (2), behind which there is a well-defined fibrous pigment layer (3), previously found in skombroid fish (pelamid, tuna) swimming in the range of

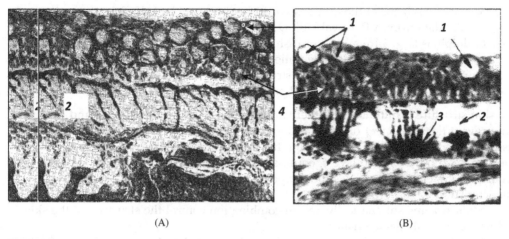

(A)   (B)

FIGURE 4.29 Cross section of swordfish skin: (A) epidermis with mucus-forming cells, 250-fold magnification; (B) secretory apparatus of the epidermis, 250-fold magnification: (1) secretory cells, (2) basal connective tissue membrane, (3) fibrous pigment layer, (4) layer of elongated cells [64,70,160].

numbers $Re = 10^7$ [89]. The structure of the fibrous pigment layer in all these fish is similar and consists of thin, loosely arranged fibers in an amorphous mass. Depending on the skin area in this layer (3), there is a greater or lesser number of pigment cells, which are often located on the bundles of connective tissue connecting the basement membrane with dense unformed tissue of the skin itself into a single system (Fig. 4.29A). The thickness of the fibrous pigment layer often exceeds the thickness of the epidermis and the underlying basement membrane. On the basement membrane (2) is a layer of elongated cells (4), where there is constant cell division. Due to this layer, the cells in the upper layers are renewed. Following this layer is a layer of normal (polygonal shape) cells, between which in certain areas of the body of adult fish separate cells begin to appear with pronounced secretory signs. Such cells subsequently, in the overlying layers, become mucus-forming. Depending on the size of the fish, the number of layers of epithelial cells varies considerably. Thus, in swordfish 2.6 m in length, there are up to 12 layers in individual parts of the body with a total thickness of the epidermis of up to 60 μm, and in fish 0.7 m in length, the epidermis consists of one or two cell layers.

Under the fibrous pigment layer there is a second, more powerful layer of connective tissue, compared to other types of fish, in which the bases of the dent-like scales lie (Figs. 4.22 and 4.23B). In the second connective tissue layer there are extensive areas filled with large cells with well-pronounced granularity. These cells accompany the small blood vessels and nerve fibers, surrounding them from all sides. Under this layer is a layer of well-developed fatty tissue, in which powerful bundles of collagen and fine fibers lie loosely. Collagen bundles (three to four layers) are strictly oriented to the longitudinal axis of the body with 45 degrees between the individual layers. On the border between the locomotor muscles and the layer of subcutaneous fatty tissue are numerous blood vessels and occasionally there separate muscle bundles. In swordfish, as in other fish, the musculoskeletal muscles consist of individual myomeres. In the absence of continuous flaking of coatings while reducing myomeres, a change in skin elasticity occurs.

Fig. 4.30 shows a scheme of the skin of swordfish, developed by V.V. Babenko on the basis of experimental data from A.P. Koval [70]. Despite significant differences in the structure of the skin of dolphins (Fig. 4.16) and swordfish, there are many similarities in the structure of their skin, developed in the process of evolution to save energy spent on swimming. In particular, a comparison of the features of the structure of the skins of dolphin and swordfish allowed us to find out a number of identical properties of their structure [37,230]:

- The skin structure is multilayered; there are specific structures of muscles and tendons that prevent the movement of the skin under the action of shear stress and pressure distribution along the body, arising from the flow around the body;
- The skin is saturated in two layers, equidistant surface of the body, with a specific circulatory system that regulates the temperature of the body and the mechanical characteristics of the skin;
- The body's systems automatically and in combination control the structure of the skin and its mechanical characteristics;
- In the skin there are two layers of membranes, equidistant surface of the body, affecting the dynamic elastic/damping properties of the skin.

In Section 2.3 the body shape of various high-speed hydrobionts is given, in particular, in Figs. 2.8 and 2.13, which show the body shape of swordfish, sailfish, and tuna. For other fast-swimming fish (tuna, sailfish), the dorsal and lateral fins retract into special pockets during active swimming and exit from it only when maneuvering, while in swordfish, the dorsal fin constantly acts above the body surface and remains in the same position for all swimming

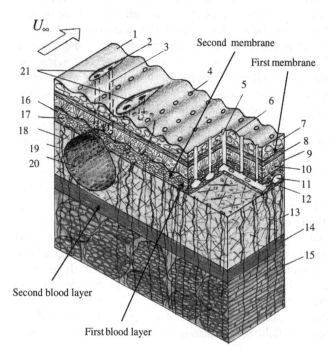

FIGURE 4.30 Scheme of swordfish skin: (1) longitudinal folds on the skin surface, (2) slots on the back, (3) pore excretory channels, (4) pores, (5) vertical pore channels; (6) secretory cells, (7) epidermis, (8) the first layer of the thin connective membrane, (9) fibrous pigment layer, (10) connecting fibers, (11) system of hypodermic channels, (12) connecting channels, (13) underskin layer of fatty cellulose, (14) layer of blood vessels, (15) locomotor musculature, (16) the second layer of thick connective membrane, (17) large cells with circulatory and innervation systems, (18) vertical vessels of the circulatory system, (19) innervation system, (20) ampoules, (21) cavities [70].

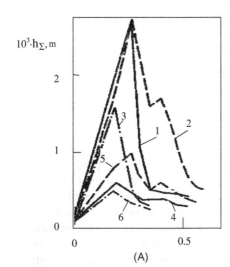 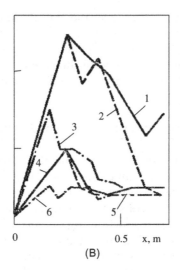

$10^3 \cdot h_\Sigma,\, m$

(A)   (B)

FIGURE 4.31 The distribution of skin thickness along the body of fast-swimming fish in the horizontal (1–3) and vertical (4–6) body planes. (A): (1, 4) striped tuna, $L = 0.63$ m; (2, 5) spotted tuna, $L = 0.76$ m; (3, 6) spotted tuna, $L = 0.49$ m; (B): (1, 4) bigeye tuna $L = 0.86$ m; (2, 5) long-necked tuna, $L = 0.83$ m; (3, 6) pelamid $L = 0.6$ m [42].

modes. The presence of any protruding formation on the body at high speeds of movement causes significant disturbances in the boundary layer [70]. In the region of the connection of the fins and the body of the swordfish, pairs of longitudinal vortices will arise. To dampen these disturbances, the structures discussed above in the skin of a swordfish are developed. Also important are the cavities found in the skin slots, which promote the interaction of mucus secreted through the ducts with the CVSs of the boundary layer. During fast swimming in the region of the dorsal fin and behind it, a vacuum is formed on the body in accordance with the pressure distribution shown in Fig. 2.5. The mucous substance that has accumulated in the ampoules, due to this, through the slots is released on the body, where it dissolves in the incoming flow. Thus, the swordfish in the process of adapting to fast swimming has developed the above set of devices, including a body shape to reduce drag.

Fig. 4.31 shows the distribution of skin thickness along the body in the ventral and dorsal planes in fast-swimming fish. The skin gradually thickens from top to bottom toward the sideline. Fig. 4.31A shows the distribution of the skin of the scaleless tuna species. If we take into account the difference in the length of the fish, we get a good comparison of the distribution of the dolphin's skin (Section 4.3, Fig. 4.16) [17]. Despite the difference in the structure of the skin in different species of aquatic animals, the distribution of skin thickness along the body correlates with the speed of their swimming. Fig. 4.31B shows the distribution of skin thickness along the body in other types of high-speed fish. The bigeye tuna is scaly, with an increase in mucus beyond the mid-section. In pelamida, the transitive structure of the skin compared with the first group of fish (scaleless) has a slimy cover in the tail of the body, and a similar structure of the skin to long-fin tuna. In these fish species (Fig. 4.31B), the skin thickness is greater than that of the scaleless species (Fig. 4.31A).

A completely different distribution along the body of skin thickness is seen in swordfish. The skin thickness of a swordfish with a length of 2.0 m (discounting the sword, 1.4 m), 0.3 m in height, and 40 kg in weight, throughout the body was approximately the same: $(2–2.5) \cdot 10^{-2}$ m. The skin thickness slightly decreased in the middle line in the tail section. Behind the gill slit at a distance of $(3-8) \cdot 10^{-2}$ m, the skin thickness reached $3 \cdot 10^{-2}$ m at the top and bottom, and $8 \cdot 10^{-2}$ m in the middle part, and about $1.5 \cdot 10^{-2}$ m in the tail part. The xiphoid should take into account the nature of the flow around the body to the gill slit and after it.

Polymer's injection optimization with the help of sword-shaped tips was experimentally investigated in [49 et al.]. The results are shown in parts II, III.

4.7 The structure of the body and skin of penguins

The first information about the exceptional hydrodynamic qualities of penguins was given in Refs. [101,131,209]. Numerous fundamental morphological and hydrobiological studies of penguins were performed by Rudolf Bannasch [7,77−83, etc.]. In particular, in the Antarctic, in a specially constructed hydrodynamic channel, he performed extensive investigations of the penguin's swimming characteristics, which confirmed the extremely low energy consumption during diving (the program of the German Antarctic included the study of the morphology, swimming kinematics, metabolic physiology, and telemetry) [79−81, etc.]. Penguins of average size of the *Pygoscelis* species with a body length of 0.65−0.7 m (measured during swimming) during the search for food swim daily up to 100 km at an average cruising speed of 2.3 m/s. Rapidly they reach maximum speeds of 4.5 m/s. Emperor penguins with a length of 1.0−1.1 m can reach a maximum speed of more than 7 m/s. Unlike fish and dolphins, the penguin corpus does not directly participate in the production of thrust and only fluctuates slightly during the stroke cycles of the lateral fins. The body of the penguin could optimally be formed in the process of evolution as an almost purely nondeformable body. This facilitates the modeling of the penguin's body and makes it interesting for technical applications. Rudolf Bannasch investigated the penguins *Pygoscelis adeliae*, *Pygoscelis papua* (gentoo penguin), *Pygoscelis antarctica*, *Eudyptula minor* (little penguin), and *Aptenodytes forsteri* (emperor penguin).

When comparing the shape of the body of penguins of various sizes, changes can also be observed in the size of the beak, head, and body. For example, with increasing size, the head becomes relatively smaller. Thus, the "wave" on the contour line of the front of the body becomes less pronounced. In addition to biological reasons, it might be of interest to discover if these changes have a hydrodynamic effect. The body shapes of the three penguin species resemble each other (Table 4.7). Table 4.7 gives the following designations: A—frontal area (m²), d—diameter of frontal area (m), l—body length (m), x_d—distance to the maximum thickness from the end of the beak (m), l/d—relative length (ratio of length

TABLE 4.7 Body geometry.

Designation	l/d	x_d/l	A	$d = \sqrt{(4A/\pi)}$
P. antarctica	4.54	0.44	0.0195	0.158
Pygoscelis adeliae	4.35	0.47	0.0208	0.163
P. papua	4.00	0.44	0.0270	0.186
Body rotation	4.23	0.44	0.0214	0.165

to thickness), and x_d/l—the relative location of the maximum body thickness. A slight dorsal–ventral asymmetry was evident when viewed from the side. At maximum thickness, the cross section was almost circular. The generalization of the penguin body geometry made it possible to represent the body shape in the form of a spindle with laminar flow with high values of maximum thickness positions and thickness ratios [126,127]. However, the structure of the beak and some roughness at the beginning of the plumage suggest that the transition from a laminar to a turbulent boundary layer is caused in the frontal part of the body itself. In addition, the "wavy" outline of the front of the body looks somewhat unusual.

The body shape of the penguin is not constant: it depends on the kinematics of movement, speed of movement, and depth of immersion. In addition, the feather cover is malleable and deformable. The outer covering during maneuvering dampens and stabilizes large vortex structures. It is not possible to fully model such a form. The body of a living hydrobiont is always significantly different from a nonliving one. In the latter case, all body systems cease to function, including no filling of the lungs and circulatory system, the turgor (elasticity) of the muscles and skeleton of the body disappears, the elasticity of feather cover disappears, etc. Therefore the body shape of the deceased penguin differs significantly from the living penguin. Similarly, the body shape of a penguin on the surface of ice or the ground differs from the shape of the body during swimming. It is advisable to analyze the body shape of a penguin during navigation when moving by inertia at the moment when the fins are in a neutral position or approaching the surface of the body—this position is presented in Fig. 2.7 (Section 2.3).

In penguins, the tail and feet form a peculiar triangular structure. In some types of aircraft, the stabilizer is made in the form of a similar triangular shape. The tail may have a hemispherical shape or form a sharp plane. As a result, the tail shape is formed in the form of three planes located at an angle of approximately 120 degrees between them. Depending on the maneuver, the angle changes, and the planes become arcuate. It is important that in penguins the structure of these tail elements has a strictly ordered moving longitudinal structure, which changes its shape under the action of hydrodynamic pressure and the corresponding loads. This form of tail is effective for eliminating the separation of flow around the body in the presence of a positive-pressure gradient in the tail part of the body.

The fins have one kink. In penguins, the fins are fixed in the head part of the body, approximately in the region of one-third of the body length, without taking into account the length of the tail during movement. The body of the penguin is maintained during movement by Archimedean force. Therefore this arrangement of the penguin drive is determined by the stability of the movement. The shape and profile of the penguin fins was studied in detail in Ref. [81]. The cross-sectional area of the fins in the area of attachment on the body is significantly reduced, which reduces the interference resistance of the fins.

In a water channel built in Antarctica, Rudolf Bannasch conducted experiments on live penguins, which swam in the channel under water. The walls of the channel were made of glass, while filming was carried out simultaneously from the top and side [78,82]. Analysis of the kinematics of movement of the penguin side fins shows that when the fin moves down, a part of the fin located near the body before section 1 moves in the process of oscillation of the fin at very small angles of attack—the figure shows a bone hinged on the penguin skeleton. In sections 1 and 2 and further to the end of the fin there are four articulated joints of the fin bones. This allows the penguin during oscillation to change the twist of the blade-fin along its span beyond section 1—the attack angles of different sections are different and increase in the direction of the end of the blade-fin according to the indicated hinges of the fin bones. At the same time, by the end of the fin, the deflection of the entire plane of the fin increases in span, like that of the dolphin's fin (Figs. 2.7B and 2.18A). The length of the cross section of the fin blade toward the end edge decreases. This leads to a decrease in the washed area of the fin along its span. Therefore, at a constant angle of attack of the fin toward the end edge in each cross section, the lift force will decrease. This twist of the penguin fin blade allows for a fin scope to have a constant value of lift. Due to this, the efficiency of penguin side fins significantly increases.

After reaching the lowest point of the transverse trajectory of the fin, the angles of attack are shifted, and the lateral fin begins to move upwards, and the angles of attack become negative. This allows you to maintain a constant thrust throughout the entire trajectory of the fin. The features of kinematics and the trajectory of movement of the dolphin's tail fin are shown in Figs. 2.30, 4.17, and 4.18. The trajectory features of the penguin side fins are very similar to the analogous features of the caudal fin of dolphins (Section 4.2).

FIGURE 4.32   Feather cover for penguins.

The structure of the feather cover of the penguin body is shown in Figs. 2.7 and 4.53. The feather cover on the body has an ordered longitudinal direction, which is similar to the direction in the structure of the skin of fast-floating hydrobionts. In penguins, the size of such a longitudinal order is substantially larger than in the skin of dolphins, sharks, swordfish, and tunas. This is due, primarily, to the difference in the Reynolds numbers corresponding to these species at characteristic swimming speeds. In addition, it was shown in a model experiment that when flowing around elastic covers, the size of the vortex structures in the boundary layer increases significantly [44,70,111,170].

The shape and structure of CVSs in the transition and turbulent boundary layers in the flow around rigid and compliant surfaces have been experimentally investigated [44,70,111,170]. Based on the results obtained, it can be assumed that the feather cover of the penguin's body is an analogue of elastic covers like dolphin skin, and the size of the longitudinal order indicates that the flow pattern in the boundary layer of penguins is the same as in the transition boundary layer of elastic surfaces.

Fig. 4.32 shows the shape of the feather cover on the fins of penguins. It is seen that when approaching the trailing edge of the fins, the dimensions of the longitudinal order with respect to the stream flowing around the fins increases, and new structures appear, resembling transverse grooves. Since the rear edge of the fins has a rounded surface, this structure suggests that in this case the flow separation at the rear edge is eliminated.

Bannasch not only performed numerous hydrobionic investigations of penguins, but produced models of the body of penguins, which he experienced in Berlin's large hydrodynamic tube [78,79,81−83]. Based on the results obtained, he introduced the results obtained in various fields of technology [77, etc.].

The results of experimental studies on the morphological features of fast-swimming hydrobionts given in this chapter allowed us to formulate technical problems of control methods for CVSs of the boundary layer [51,57,59,70] and combined methods for reducing resistance [54,60,70]. Part II contains the results of an experimental investigation of these problems, and Part III gives examples of practical implementations of the investigations performed.

Interaction of aquatic animals systems with environment

5.1 Control of the shape of the body and fins of aquatic animals

In experimental hydrobionics, the environment in which the hydrobiont moves should be considered, as in the technique, as a physical continuum. Various techniques for controlling the boundary layer of devices have been developed in the technique [85]. On the other hand, it is necessary to take into account that in organisms of hydrobionts adaptations arise to dynamic interaction with the environment [66, 67].

Hydrobionts are the most complex organisms that have analyzers of the parameters of the physical properties of the environment and driving patterns, which lock on the automatic control systems of the motor-propulsion complex and the adaptations to reduce energy costs. Hydrobionts have a number of distinctive features - this is the abundant innervation of body systems with its automatic control, the presence of a single propulsion complex, polyfunctionality and interconnectedness of various systems with a high degree of reliability of their work, etc. [43]. Therefore, in the study of hydrobionts, it is always necessary to take into account the specificity of living organisms, which consists in the non-stationary body movement, in the instantaneous reflex reaction of the whole body or its individual systems to external stimuli, and in other features causing the specificity of interaction between living hydrobionts and the environment with the oncoming flow. For example, a change in water pressure with a diving depth deforms the body of some hydrobionts, automatically changing its buoyancy and facilitating diving and ascent [12,20,207]. Adaptation of aquatic organisms to the environment is that the density of their skin is approximately equal to the density of the environment, and this is a prerequisite for the skin to stabilize the disturbances of the boundary layer. In addition to these features of the interaction of hydrobionts with the environment, it is necessary in further studies to also take into account that in hydrobionts gravity is balanced by Archimedean force, which also developed a number of adaptations in the body systems.

If, in experimental hydrodynamics, rigid objects are the objects of study, in experimental hydrobionics there are living organisms and their analogues, which should simulate some peculiarities of hydrobionts. Hydrobionts have a number of features that distinguish them from similar technical devices, including the abundant innervation of body systems with automatic

129

control, the presence of a single motor-propulsion complex, multifunctional and interconnectedness of various systems with a high degree of reliability of their work, etc. (Section 1.2).

Body energy is sufficient for swimming at low speeds (Chapter 3). When diving at cruising speeds, the energy is insufficient and adaptations to reduce frictional resistance developed during evolutionary development should be included in the work. Obviously, such devices act within the interactions of all body systems. An essential role is played by the shape of the body and fins (Sections 2.3 and 2.4), the specific structure of skin covers (Section 4.3, Fig. 4.25), as well as the irregularity of the kinematics of the swimming of aquatic organisms. The mechanical characteristics of the skin become optimal from hydrodynamic positions at high swimming speeds.

Below are described some examples of the functioning of body systems with their interconnections with the environment.

In general, the following forces and moments act on the dolphin's body [12,20]: the force of gravity G, directed vertically, is constant in magnitude and is applied at the center of gravity; buoyancy force F, directed vertically upwards, variable in magnitude and applied at the center of the magnitude; hydrodynamic force, which can be represented by three terms along the axes of coordinates—lifting force Y, drag Q, and lateral force R; thrust force of the mover of the caudal fin P, variable in magnitude and direction; moment of thrust propulsion M_p; the moment due to the eccentricity of the centers of gravity and size, M_F; moment of hydrodynamic forces arising from the flow around the body, M_Y; hydrodynamic moment of the lateral fins M'_Y; hydrodynamic moment of the caudal fin M''_Y.

In the velocity coordinate system, the origin is at the center of gravity of the body. The axis OX is directed along the velocity vector, the axis OY is perpendicular to the axis OX in the vertical plane, the axis OZ is directed to the right so that the coordinate axes form the right coordinate system. A system of the differential equations of movement of a dolphin for a flat trajectory in a vertical plane is now described. In this case, we assume that the vertical trajectory of movement consists of two sections—a curved entrance–exit section with a certain constant radius of curvature and a straight-line section of movement at a certain constant angle to the trajectory $\theta =$ constant. Due to the fact that the dolphin thruster—the tail fin—performs sinusoidal movements with different frequency and amplitude, the center of gravity of its body will also move along a certain sinusoid. The side fins of the dolphin perform not only the functions of rudders, but also are simultaneously stabilizers. In the first approximation, we consider the motion of the center of gravity of a dolphin along the axial line of the sinusoid of actual motion (Fig. 5.1).

The system of differential equations of the dolphin motion in the first section of the motion path will look like:

$$m_\Sigma \frac{dv_x}{dt} = m_\Sigma \frac{dv}{dt} = P\cos\phi - Q + (G - F)\sin\theta + Y''\sin\theta,$$

$$m_\Sigma \frac{dv_y}{dt} = m_\Sigma v\frac{d\theta}{dt} = \frac{m_\Sigma v^2}{r} = P\sin\phi + Y - Y' - (G - F)\cos\theta + Y''\cos\phi,$$

$$I_\Sigma \frac{d\omega_z}{dt} = M_P - M_F + M_Y + M'_Y + M''_Y,$$

$$\frac{d\theta}{dt} = \omega_z.$$

(5.1)

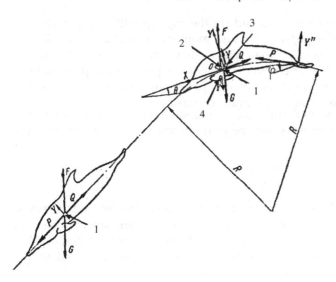

FIGURE 5.1 Forces on the dolphin trajectory: (1) center of gravity, (2) center of magnitude, (3) hydrodynamic focus of the body, (4) hydrodynamic focus of the lateral fins [12].

Here we have the following notation: $m_\Sigma = m + m$, where $m = \frac{G}{g}$ — is dolphin mass, and m' is the added mass of water;

$\frac{dv_x}{dt}$ — is acceleration along the axis OX,

$\frac{dv_y}{dt}$ — is acceleration along the axis OY;

Y and Y' are, respectively, the magnitudes of the lifting forces created by the hull and the flippers;

Y'' is total hydrodynamic force of the caudal fin arising from the sinusoidal movement of the tail, its magnitude and direction are variable in time;

φ is the angle between the force P and the speed v; Θ is the angle between the velocity vector v and the horizontal plane;

$\frac{d\Theta}{dt} = \omega_z$ — is angular velocity of rotation of the body around the axis OZ;

R is the radius of curvature of the trajectory;

$I_\Sigma = I_Z + I'_Z$, where I_Z is the moment of inertia of the dolphin body, and I'_Z is the associated moment of inertia.

For movement in the second section, it is assumed that the total hydrodynamic moment is zero and the resultant lifting force is applied at the center of gravity. The system of differential equations will be:

$$m_\Sigma \frac{dv}{dt} = P - Q + (G - F)\sin\theta,$$

$$Y = (G - F)\cos\theta.$$

(5.2)

In the system of Eq. (5.1), 10 variables are unknown (v, ω_z, Θ, φ, F, M_P, M_F, M'_Y, M''_Y); in system Eq. (5.2), two quantities (v and F), since $\Theta =$ const. Thus, if the system of Eq. (5.2) can be solved, then for the system of Eq. (5.1) it is necessary either to accept a number of assumptions, or to find kinematic relations between eccentricities, as well as patterns of change in the angle φ, or to set the optimal values of unknown values. Due to the peculiar structure of the dolphin's body and the specificity of its movements, the determination of patterns between eccentricities and for the angle φ will be a complicated and time-consuming process, and the added mass and moment of inertia will in general be variables. Consequently, even numerous assumptions that make it possible to significantly simplify the systems of differential equations do not yet provide the possibility to solve the dolphin motion problem with sufficient accuracy. To solve this problem, it is necessary to conduct numerous theoretical and experimental studies.

A submerged body is subject to normal atmospheric pressure and an overpressure of a water column. At depth h, the overpressure will be $h/10$ atmosphere. Then the absolute pressure will be

$$ATA_h = \frac{10 + h}{10} \text{ atm.} \tag{5.3}$$

The increase in pressure for a diving dolphin primarily affects the air body cavities, which are light, and air bags located under the blowhole. Large pressures during deep-sea diving will act on the dolphin, changing the volume of its body so that the internal pressure of the animal is equal to the pressure of the environment. This is evidenced by the following morphological features of the dolphin:

1. The animal's breast tag is small compared to the body. A significant number of ribs are free, and not jointed with the sternum;
2. The ribs are jointed movably, so that the skeleton can be severely deformed;
3. Light one-lobed, strongly elongated;
4. The specific cartilaginous structure of the respiratory and circulatory systems is adapted to high pressures;
5. The oblique location of the diaphragm allows you to evenly transmit the pressure of the internal organs to the lungs;
6. Consumption of food at a great depth is possible only if the pressure in the pharynx, esophagus, and stomach is equal to the external hydrostatic pressure.
7. Soft elastic tissue of the body transfers external hydrostatic pressure on the internal organs, otherwise the dolphin's body must have exceptional strength.
8. Blood pressure in the skin and oral cavity is equal to external hydrostatic pressure. Due to the incompressibility of blood, external hydrostatic pressure is transmitted through the circulatory system to all organs of the body.

Dolphins have an excellent mechanism of thermoregulation, which allows them to maintain the temperature inside the body almost constant with significant changes in the temperature of the environment [87,150,151,309–311]. Therefore the change in air temperature in the lungs and air sacs during diving can be neglected. It also does not take into account the heating of inhaled air to the internal body temperature. Dolphins can dive to a depth of 100 m, where pressure is measured in dozens of atmospheres. Therefore, the Boyle–Mariotte equation can be used in the calculations. When a dolphin is immersed to a certain depth, it will not only change the volume of air in the lungs and air sacs, but also the gases that are in its body in an undissolved state. Suppose that, according to the Boyle–Marriott law, only the volume of air in the lungs and air sacs changes. Due to the combination of a number of ribs with the sternum, external hydrostatic pressure will be transmitted to the lungs mostly through the diaphragm. It can be assumed that the internal organs of the abdominal cavity at the same time will move somewhat forward, changing the eccentricity between the center of gravity and the center of magnitude. In the calculations, the isothermal law of the change in the state of the gas in the lungs and air bags is used. Further study of dolphins, in particular their thermal conditions, will allow either to approve this assumption or to consider this process as adiabatic. According to Boyle—Mariotte's law, the volume occupied by a gas varies inversely with the pressure acting on it at a constant temperature:

$$p_1 V_1 = p_2 V_2. \tag{5.4}$$

If $p_1 = 1$ atm., and $p_2 = \frac{10+h}{10}$ atm., then at a depth h

$$V_h = \frac{10 + V_1}{10 + h}, \tag{5.5}$$

where V_1 is the volume of air filling the lungs and air bags of the dolphin on the surface of the water. Thus, when a dolphin is immersed, the volume of air cavities decreases and therefore the volume of the whole body decreases. The density of the dolphin's body and its buoyancy depend on the depth of the dive and other variables. We consider the volume of the dolphin's body (V_T) as the sum of the volume of air cavities (V_h) and the volume of the carcass (V'), which does not contain, according to the accepted assumption, gases that are in an undissolved state. The volume of the carcass practically does not change when diving and can be expressed as

$$V' = V'_T - V_1, \tag{5.6}$$

where V' is the body volume of the dolphin on the surface of the water after inhaling. We write the expression for the volume of the dolphin's body at a depth h:

$$V_T = V_h + (V'_T - V_1). \tag{5.7}$$

Substituting the value from formula (5.5) into equality (5.7), we get

$$V_T = V'_T - \frac{hV_1}{10 + h}. \tag{5.8}$$

Then the average density of the body at the depth h will be

$$\rho_T = \frac{G}{V'_T - \frac{hV_1}{10+h}}. \tag{5.9}$$

Knowing the volume of the V_T body at any depth, we determine the buoyancy force:

$$F = \rho_B V_T = \rho_B \left(V'_T - \frac{hV_1}{10 + h} \right). \tag{5.10}$$

Then excess buoyancy will be expressed as:

$$F - G = \left(\rho_B - \rho_T \right) \left(V'_T - \frac{hV_1}{10 + h} \right). \tag{5.11}$$

Thus, having measured the dolphin's weight, volume of the body and air cavities on the sea surface using formulas (5.5), (5.8), (5.9), (5.10), and (5.11), the following values can be determined for any depth by calculation: body density, buoyancy, and excess buoyancy. For a quantitative assessment of the considered phenomenon we carry out the following calculations. As shown in Fig. 5.2, the dolphin body is approximated by three simple rotating bodies.

The volume of the body is determined using:

$$\pi(0.2)^2 0.55 + \pi(0.2)^2 + \frac{1}{3}\pi(0.2)^2 0.85 = 0.134 \text{ m}^3$$

and the buoyancy force on the surface of the water is:

$$F' = \rho_B V'_T = 1.025 \cdot 0.134 = 0.1375 \text{ m} = 137.5 \text{ kg}.$$

Taking the volume of air in the lungs and air bags on the water surface $V_1 = 10$ L, using formula (5.10) we determine the buoyancy force at depths $h = 10, 20, 30, 60,$ and 90 m, and using the formula (5.5) we determine the volume of air cavities on the same depths. The results of the calculations are given in Table 5.1. Fig. 5.3 shows the dependence of the buoyancy force and the volume of air cavities on the depth of immersion in a

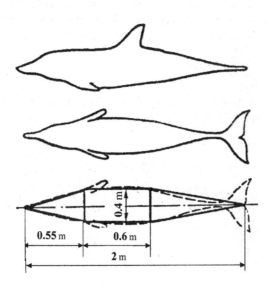

FIGURE 5.2   An approximate calculation of the volume of a dolphin body [12].

TABLE 5.1   Buoyancy depending on immersion depth.

Immersion depth (m)	Buoyancy force (kg)	Volume of air cavities (m³)
10	132.3	0.005
20	130.7	0.00333
30	129.8	0.0025
60	128.6	0.00143
90	128.1	0.001

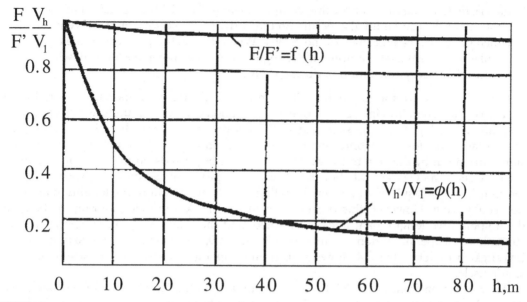

FIGURE 5.3 The change in the buoyancy force and the volume of air cavities from the depth of immersion in a dimensionless form when a dolphin dives [12].

dimensionless form. Calculations show that for large inclinations of the trajectory, that is, at $\Theta \approx 90°$, the force P practically coincides in direction with the force G, and the force F with force Q.

According to Ref. [109], the force is $P = 19.5$ kg. From the dependence $F/F' = f(h)$ we can see that, when immersed, the force F decreases at a depth of $h = 30$ m, $\Delta F = 7.7$ kg, which is approximately $\frac{1}{3}P$. When floating back up, the force F increases by the same amount and will coincide in direction with by force P. The value of the increment of the buoyancy force ΔF for dolphin energy consumption during diving is greater with the greater the angle Θ.

Based on the calculations performed, we can draw the following conclusions:

- Excessive buoyancy of a dolphin—a value depending on a number of factors. Its value is inversely proportional to the depth of immersion.
- Having data of dolphin weights, body volumes, and air cavities measured on the sea surface, it is possible to determine by calculation the volumes of the body and air cavities, body density, buoyancy force, excessive dolphin buoyancy, changed due to excessive pressure.
- The change in buoyancy due to the compressibility of air in the lungs and air bags allows dolphin to reduce the energy costs of diving, ascending, and staying at depth.
- When the volume of air cavities is 10 L, the limiting depth of the dolphin considered in the example is the depth of 60–90 m, which is fully consistent with the data from scientific observations of the depth of immersion of dolphins in natural habitats.
- As shown in Section 4.2, as the speed of movement increases, pressure in the arterial circulatory system is effectively regulated, which leads, in particular, to regulation of

the elasticity of the caudal fin using complex vessels (Fig. 4.12). Regulation of the elasticity of the caudal fin also occurs on the horizontal trajectory of its movement, which increases the efficiency of the mover. When diving, the pressure in the arterial circulatory system is also automatically regulated by the above mechanism, which also reduces the energy consumption during diving.

In dolphins, except for air cavities, the skin muscle is of great importance for reducing energy consumption. In Fig. 2.28 (Section 2.7) is a photograph of an underwater video of a dolphin swimming. It can be seen that in the middle of the body there is a longitudinal fold. A large role for the stabilization of the BL of fluid is played by a change in skin tension, carried out by the skin muscles. At high swimming speeds on the body of a dolphin, large tension and stress of tangential shear appears in accordance with Fig. 2.5 (Section 2.3), as a result of which skin buckling and displacement in the longitudinal direction could occur. However, this prevents tension of the skin muscles, and on the body of the dolphin, two longitudinal folds are formed, and the shape of the body retains a well-streamlined shape. In Section 4.1, in Figs. 4.8C and 4.9, the structure of the skin muscle is investigated in detail. Fig. 5.4 shows a diagram of options for reducing muscle bundles of skin muscle.

As can be seen from Fig. 4.8C, the muscle bundles on the lateral parts of the skin muscle are directed and arranged obliquely from bottom to top in the upper part of the body and from top to bottom in the lower part of the body from the head to the tail. During low or cruising speed swimming, dolphin skin muscle is reduced so as to maintain an optimal body shape (Figs. 5.1 and 5.4); therefore the body shape is smooth. As the speed of movement increases, the muscle bundles contract, and this increases with the swimming speed in accordance with Fig. 5.4 (2). In this case, on the side of the body, two longitudinal folds are visible, which are formed due to the specificity of the fastening of the muscle bundles in this area according to the structure of the skin muscle (Section 4.1). When maneuvering, there may be other variants of skin muscle contraction, for example, as shown in Fig. 5.4 (3).

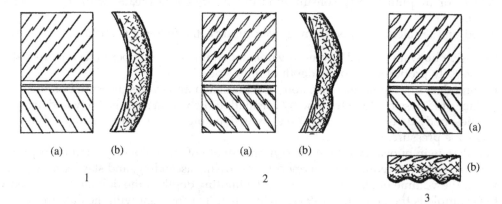

FIGURE 5.4   The folding pattern on the skin of a dolphin with the help of skin muscles: (a) muscle bundles, (b) cross section of skin and skin muscle, (1) muscle bundles are relaxed, (2) reduction of all muscle bundles, (3) one of the options for reducing individual sections of muscle bundles [3].

The same pattern of contraction of muscle bundles can also occur when the outer surface of the skin is vibrated [146].

Calculations have shown that due to a slight change in the diameter of the body and the thickness of the skin, the mechanical characteristics of the skin and its properties, which stabilize the BL, can significantly change.

In Fig. 4.53 (Section 4.7), it can be seen that, on land, the shape of the penguin's body differs significantly from the shape of the penguin's body during swimming under water (Section 2.3, Fig. 2.7). When the penguin dives, the shape of the lower body has a greater bulge compared to the shape of the upper body. When flowing around such an asymmetrical body, a negative lift force occurs, which is directed downwards and facilitates the process of diving. With the same asymmetrical body shape, the penguin floats up almost vertically at a zero angle of attack, in which there is no body lift component, and the air blanket around the body facilitates ascent. In hydrobionts, the diving and ascent processes are also provided by the relocation of the lateral fins at an angle of attack, at which a component of lifting force appears on them, directed in the corresponding direction. Since penguins have lungs, their diving has the same processes as in dolphins. Thus, hydrobionts with air cavities in the body regulate their change in body shape.

Figs. 2.17–2.19 (Section 2.4) show data on the range of the caudal fin of aquatic organisms, in particular, the dependence of the magnitude on the body length of the aquatic organisms. Fig. 2.18 shows photographs of the caudal fins of high-speed hydrobionts, from which it can be seen that the span of their caudal fins may vary depending on their speed of swimming and the nature of the movement of the fin on its trajectory.

The regulated hydroelastic effect of cetacean fins is considered in detail in Refs. [175,231]. For example, with fast swimming and the enhanced muscular work of the dolphin, the arteries of the complex vessels of the caudal fin expand and decrease the lumen of the veins that surround them. In this case, the outflow of blood from these veins is completely reduced. Thus, the entire hypodermal circulatory network is filled, and under these conditions, thermoregulation increases and at the same time the rigidity of the caudal lobes increases. This increases the blood supply of the circulatory network extending from the outer surface to the inside of the fin (Section 4.2, Fig. 4.14B). It should be noted that the elasticity of the caudal fin changes automatically as the dolphin's swimming speed changes. The elasticity of the caudal fin automatically changes on both sides alternately as it oscillates, and the fin's elasticity increases on the pressure side (Fig. 5.5). The arrows indicate the direction of the velocity head acting on the blades of the dolphin's fin on its rectilinear trajectory. The kinematics of movement of the caudal fin are given in Section 2.7. In position (I), the musculature raises the tail fin up. Under the action of the velocity head, conventionally indicated by an arrow on the tail fin profile, set at an angle of attack, a resultant thrust force developed by the fin occurs. In this position, the lower

I   *II*

FIGURE 5.5   Diagram of the trajectory of the dolphin's tail fin.

surface of the fin profile is under the action of a force determined by the velocity head. In accordance with Ref. [226], the fin's elasticity under the action of a specific circulatory system increases.

According to Figs. 5.5 and 5.6, in the extreme positions of deviation of the caudal fin on its trajectory, the magnitude of the velocity head decreases to zero. In this case, the elasticity

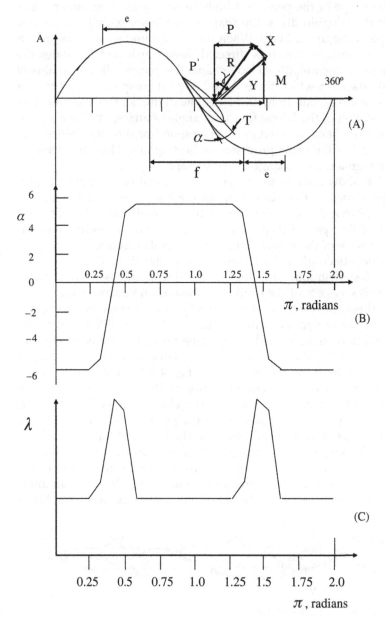

FIGURE 5.6 Scheme of movement of the caudal fin along its trajectory: (A) diagram of hydrodynamic forces, (B) changes in the angle of attack, (C) change in the span of the tail fin.

of the fin may decrease. In position *II*, the musculature lowers the tail fin. In this part of the trajectory of the fin, the velocity head acts on the opposite side of the fin profile.

Fig. 5.6A shows the magnitude of the amplitude *A* of the oscillations of the caudal fin on a sinusoidal trajectory of its movement. The sinusoid curve along the ordinate axis depends on degrees or π in and corresponds to Fig. 2.30A, on which the sinusoid is located along the *x*-axis, on which the frame numbers are plotted, respectively, from 11 to 55 frames. Plots *e* in Fig. 5.6A correspond to the areas where the blades of the fin are repositioned at opposite angles of attack α, and plots *f* to the working stroke of the caudal fin. The tail fin is conventionally shown in the form of a symmetrical profile located to the trajectory of movement at the angle of attack α. In fact, when diving at a trajectory under the action of a hydrodynamic head, the curvature of the blade profile may change. A standard wing load distribution (Section 2.3, Fig. 2.5) arises along the blade profile, in particular, a lift force *Y* is formed along with the pressure distribution along the profile, as well as viscosity resistance *X* (Section 2.2, Fig. 2.1), which determine the summing force *R*. Motion of the fin along the trajectory is provided by the force *P′* of the musculature of the caudal fin directed downwards, and the force *P* of the musculature preventing the blade profile from turning upwards under the influence of the force *M*, which determines the moment characteristics of the profile. The resulting force *R* is also determined by the sum of the vectors *P* and *P′*. The thrust force *T* is formed by the sum of the vectors *Y* and *M*.

Fig. 2.30 (Section 2.7) shows the pattern of change in the angles of attack on the trajectory of movement of the caudal fin according to the results of the analysis of dolphin swimming kinograms [144]. Fig. 5.6B shows the pattern of relays in fin angle of incidence α, obtained with a simplification of the data shown in Fig. 2.30c. In the region of the extreme positions of the fin in the region *e*, there is a sharp relay of the angles of attack, and the relay begins when approaching the region *e* and ends around two-thirds of the length of the region *e*, therefore even before the end of region *e*, the fin movement reaches the optimum angle of attack ($\alpha = 5-6$ degrees).

Fig. 5.6 shows the scheme of change swing of the fin λ on the trajectory of its movement. The maximum values of the span of the fin correspond to the measured values of λ of live hydrobionts obtained while catching aquatic organisms when measured on the deck or in a special bath. During the high-speed movement of a hydrobiont, depending on the speed of navigation on the part of the trajectory on which the thrust is formed, the magnitude of the swing decreases—in Fig. 5.6 this span corresponds to the lower horizontal straight line. When the fin is repositioned, the angle of attack decreases to zero, the loads on the fin profile decrease, and as a result the span of the fin increases to a maximum value. In this case, the inductive resistance of the caudal fin is reduced, which allows an insignificant decrease in the traction characteristics and a constant thrust on the trajectory to be maintained.

It can be assumed that the elasticity does not automatically change for the entire fin [175,231], however the elasticity changes alternately on the side of the fin that is under the action of the velocity head. The change can be smooth in accordance with the law of oscillation of the fin. It can also be assumed that under the action of the velocity head on the pressure surface the curvature of the plane of the fin profile changes.

Based on the results of experimental studies of tail fin hydrobionts, technical analogs of fin thrusters and underwater vehicles with fin thrusters [10,24,25,35,50,103,155,182,213,236-238, 299 etc.] were developed, the device of which is given in part III.

5.2 Features of control of the skin of aquatic animals

In Refs. [2–4], the nervous system was investigated in almost all parts and systems of the dolphin body. The nerves that regulate all of the most important physiological processes in the skin, and the blood vessels form, together with the skin, a single morphofunctional complex. One of the reasons for the hydrodynamic perfection of dolphins is that they actively (reflexively) control their skin, which contains specific receptors associated with the central nervous system. A free and relatively simple nervous system apparatus and complex encapsulated and nonencapsulated receptors were found in the skin of the studied dolphins. The desquamation of the outer layer of the epidermis occurs quickly, therefore only 7–15 layers of cells have time to harden. The bulk of the epidermis is made up of spinous cells. Nerve fibers, penetrating into the epidermis from the papillae or from the papillary layer, are directed toward the surface of the epidermis, usually obliquely to the level of the prickly layer. Nerve fibers do not reach the stratum corneum, since the keratinization of cells and desquamation of the epidermis is ahead of the growth of nerve fibers.

The intraorgan circulation is one of the main regulators of dolphin heat exchange. Dermal papillae in amounts of up to 5–7 billion for a dolphin of length 2.5 m and an area of skin about 3 m^2 have their own circulatory system consisting of two to five vessels of the type of arterioles with numerous anastomoses between them, very close to the skin surface (1–2 mm), actively participate in heat exchange (Section 4.1, Fig. 4.6). This also contributes to a wide network of blood vessels, passing directly into the epidermis and reporting the blood systems of the adjacent papillae.

Vascular capillaries usually originate from arterioles located in the cushion bed. A large arteriole passing along the roller gives one or several vessels to each of the papillae. This is the "main" type of blood circulation. However, in some cases, one main vessel cannot be isolated. At the same time, a powerful capillary network, the so-called vascular plexuses (Fig. 4.6), is defined in the system "dermal ridges—epidermal scallops," as well as in the adjacent part of the subpapillary layer. Often they are joined by nerve bundles and fibers, forming in the complex neurovascular plexus. Without a doubt, this kind of plexus, located in close proximity to the surface of the skin and not protected by a heat-insulating layer of adipose tissue, takes an active part in heat transfer. The nerve elements that make up the neurovascular plexus, changing the lumen of blood vessels (Section 4.3, Fig. 4.26), monitor the heat transfer in this area of the dolphin skin, depending on the total heat production of the body. In particular, it can be assumed that receptors of the type of unencapsulated glomeruli [2–4] that occur in the subpapillary layer [2–4] along the bundle can record temperature conditions in a certain part of the skin and on its surface.

In the membrane and adjacent layers of the dermis (Section 4.3, Fig. 4.25), tree-branching receptors and types of tendon spindles that carry a proprioceptive function are found. These receptors have the function of regulating the amount of displacement of the skin in any area.

It is often possible to see Refs. [2–4] how the same nerve fiber sends its branches simultaneously to the epithelium, the connective tissue of the cocotool layer, and to the walls of the blood vessels, forming polyvalent-type receptors. Obviously, in this case, more superficial epidermal branches fix disturbances in the BL in a given area of the skin and generate a nerve impulse. The resulting impulse along the nerve fiber is sent not only to the central nervous system, but also spreads through axoplasm to other branches of the fiber, ending in the papillae and under the adventitia of the vessels. Next, we can assume the following biophysical mechanism. Under the action of this pulse, the osmotic pressure is redistributed in the intercellular fluid of the papillae, the lumen changes and the blood fills the vessels of the papillae and the papillary layer (Section 4.3, Fig. 4.26). This leads to a change in the elastic-damping properties of the skin in this area. In this case, the dampening properties of the skin change almost synchronously with the change in hydrodynamic conditions in the BL. If the noted phenomena are adequate, then the impulse going to the central nervous system will be subliminal and not cause a reflex. If the compensatory mechanisms of the skin itself are not enough, the reflex arc will be activated and additional "antiturbulence" mechanisms will be mobilized: contraction of the skin muscle, redistribution of blood in the vessels of the deeper skin layers and subcutaneous fatty tissue, change in the density of the fat component of the net layer, etc. Often the terminals of the nerve fibers, intertwining with each other, form thick plexuses, localized mainly in the zone of the papillae scallops and in the subpapillary layer. Terminal branches of plexiform type nerve apparatus often form a tortuous path, spirals, and curls. The latter can simulate their appearance-sensitive nerve glomeruli, characteristic of the area of the papillae and the subpapillary layer. In the area of such a plexus, in many cases, a dense network of blood capillaries and arterioles is determined, forming in the complex neurovascular plexuses (Fig. 4.6). The last, which are located in close proximity to the surface of the skin and are practically unprotected by a heat-insulating layer of adipose tissue, can take an active part in the thermoregulation of the animal. The nerve elements that make up the neurovascular plexuses, changing the lumen of blood vessels, monitor the heat transfer in this area of the dolphin skin, depending on the overall heat production of the body. Thus, receptors of the type of nonencapsulated glomeruli that are found in the postcapillary layer, apparently, can fix temperature changes in any part of the skin and on the surface. The same can be said about some other receptor devices observed in the papillae, and in some cases even penetrated into the epidermis.

For the skin of dolphins, end-cylindrical flasks are characteristic. The structure of such a bulb is now described in detail. On the axis of the bulb passes a bundle of neurofibrils, between which there is a homogeneous axoplasm. A bundle of neurofibrils is immersed in the protoplasm of the inner bulb. Along the periphery lie the nuclei of Schwann cells. The inner flask is surrounded by one or two plates of connective tissue with fibroblasts. Trailer flasks lie in connective tissue papillae, and under epidermal scallops. With their long axis, they are located mostly obliquely to the skin surface, and less often parallel or perpendicular to it [2–4].

The detected tendon spindles are proprioceptors, that is, they determine the spatial location of an organ or part of it at some point in time. If this is so, then the detection of receptors of this kind in the membrane, which intimately fused with the connective tissue elements of the reticular layer of the dermis, can only be explained by the mobility of the latter and, therefore, the need to control the amount of displacement of the skin on any part. Tendon spindles act as biological sensors for pressure pulsations in the BL. In general, more intense innervation of the ventral and lateral surfaces of the skin of dolphins is noted.

Fig. 4.2 (Section 4.1) show the location of the motor muscles along the body. With the tension of the motor muscles and dermal muscle, the volume, cross-section and shape of the body can change due to the high mobility and specific structure of the skeleton and lungs filled with air.

Fig. 5.7 presents the scheme of sectional distribution of body systems and their interconnection [46,50,62,66,67]. There is a functional connection between the skeleton of a dolphin and its moving ribs with the musculoskeletal muscles, dermal muscle, and all skin, including the circulatory and lymphatic systems and innervation. In the area of each vertebra, blood vessels, lymphatic vessels, and nerve trunks are directed vertically upwards. This allows the shortest way to ensure blood flow and automatically section by section (along each section of the vertebra) to regulate blood flow in the outer layers of the skin. Fig. 5.7 also shows two layers saturated with horizontal layers of circulatory systems: the papilla derma 2 and skin muscles (6). The skin muscle layer can regulate the mechanical properties of the skin layers located above this muscle layer, and the tension of the motor muscles cyclically, in tune with the work of the propeller, change the stress of all skin.

During the oscillatory work of the propulsion unit, the mobility of the body is provided by a specific skeleton structure. At the same time, the skin muscle and membrane strands in the skin are activated so that a propulsive wave runs over the body surface. When the mode of motion is changed the frequency of oscillation of the tail propulsion changes. The

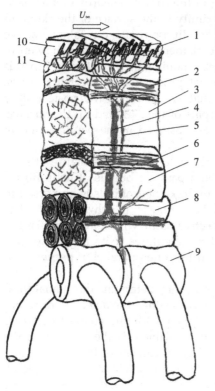

FIGURE 5.7 The sectional arrangement of the circulatory and innervation systems of the cetacean body along the body: (1) papillary dermis, (2) subpapillary dermis, (3) reticular dermis, (4) circulatory system, (5) innervation system, (6) skin muscles, (7) subcutaneous fatty tissue, (8) propulsion muscles, (9) skeleton, (10) epidermis with microfolds on the outer surface, (11) dermal ridges.

propulsion muscles are associated with tendon cords with the skin muscle, which is connected by membranes with all layers of the skin. When the oscillation frequency changes, the tension of the skin and the diameter of the whole body change, affecting the body resistance. Thus, in addition to the direct functions of creating a thrust, the motive muscles affect drag reduction and maneuverability.

Fig. 5.8 shows the dependence of the length of the self-regulation section of cetacean skin damping on the relative maximum swimming speed and the Reynolds number [229,231] [36,229,231]. The length of the self-regulation section is expressed by the ratio L/n, where L is the length of the body, and n is the number of vertebrae of the animal. The partitioning value of the vascular system of autoregulation of cetacean skin damping is confirmed by the relationship between the average section length L/n and the relative maximum swimming speed U_m/L, as well as the Reynolds number. In the largest and most high-speed killer dolphin with a length of 6 m, the average length along the body of a separate autoregulation section of skin damping is $L/n = 0.12$ m.

Fig. 5.9 shows a diagram of the interaction of the BL with the skin of a dolphin [46,50,62,66]. In the upper part of the figure, a picture of the development of coherent vortex structures (CVSs) in the BL of a flat plate is given. The features of the development of the CVS at various stages of the transition are discussed in detail in Section 2.3 and illustrated in Fig. 2.4. Fig. 2.5 shows the distribution of load parameters on an axisymmetric rigid body of rotation, having a shape similar to a dolphin. The lower part of Fig. 5.9 shows the structure of the upper layers of the dolphin skin, on the surface of which is shown the flow pattern of Tollmien-Schlichting (T-S) plane waves (2) in a laminar BL and the longitudinal vortex pattern (5) in the transition BL. Considering the above features of the circulatory system and innervation of the upper layers of the skin, it can be assumed that at moderate swimming speed the T-S waves will be damped by the system of transverse rows of dermal papillae (8). The pulsation field of the pressure of the T-S waves will

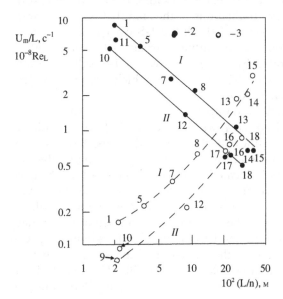

FIGURE 5.8 Dependence of the longitudinal section of the autoregulation section of cetacean skin dampening on the relative maximum swimming speed and Reynolds number: (I) fast, (II) slow cetaceans, (2) U_m/L, (3) Re_L; (1) short-beaked common dolphin, (5)-bottlenose dolphin, (7) pilot whale, (8) orca, (10) porpoise, (11) Dall's porpoise, (12) beluga, (13) sei whale, (14) fin whale, (15) blue whale, (16) humpback whale, (17) gray whale, (18) sperm whale [231].

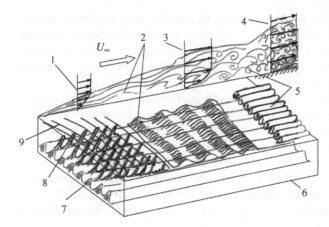

FIGURE 5.9 Interaction of the boundary layer with the dolphin skin: (1) velocity profile of the laminar boundary layer (BL), (2) Tollmin-Schlichting waves, (3) velocity profile of the transient BL, (4) velocity profile of the turbulent BL, (5) longitudinal vortices in the transient BL, (6) dolphin skin, (7) dermal ridges, (8) dermal nipples, (9) epidermis with microroughness on the skin surface.

flow into the streamlined surface. Based on the principle of receptor regulation (Section 1.2) and the above data on the system of innervation of dolphins, it becomes clear that the pressure field of the BL captures receptor endings located in the dermal papillae.

In order to eliminate pain as a result of evolution, automatic regulation of dolphin body systems has been developed. To eliminate pain, it is necessary to act on the flow patterns in the BL in order to reduce the pressure field caused by the CVS of the BL.

The distance between the transverse rows of the dermal papillae in the longitudinal direction can be represented as flat sinusoidal waves. Moreover, the wavelength of such sinusoidal waves can vary through a wide range, as the dermal papillae are inclined at different angles depending on the location along the body (Section 4.3, Fig. 4.22) and are automatically regulated by heating the papillae due to their blood supply and innervation. With the corresponding pressure in the dermal papillae, the pulsating field of the pressure of the T-S waves in the BL will cause flat elastic oscillations in the skin that will flow from below into the BL. As a result, the pulsating pressure field of the T-S waves interacts with the flat pressure field generated in the BL from the streamlined surface. At resonant interaction of T-S waves with plane waves generated by the skin, waves of T-S will be dampened or quickly transformed into longitudinal CVS of the BL, which will reduce the painful effect on the innervation of the dolphin's skin.

Formed in the BL, longitudinal CVS (5) will be dampened by similar structures in the skin, which are formed by continuous longitudinal rows of epidermal elastic partitions. Inside the skin, the epidermis consists of longitudinal elastic partitions located between the rows of the dermal papillae and the longitudinal rows of the corresponding epidermal papillae, between which the dermal papillae are located (Section 4.3, Figs. 4.18, 4.20, and 4.24). Just as in the case of T-S waves, the structure of the upper layers of the skin cover allows various combinations of longitudinal CVS sizes to be formed in the skin, which will be introduced into the BL from the bottom of the skin and will dampen a wide range of sizes of the longitudinal vortices in the BL. The mechanism will be the same: the pressure from the BL will cause elastic deformation in the structure of the skin, which will form the same type of disturbance generated from below into the BL. The longitudinal vortices (5) of the BL will interact with the longitudinal perturbations generated from the skin. Under resonant interaction, the perturbations of the BL will either be completely dampened, or the speed of their development will decrease significantly. This will lead to the elimination or significant reduction of pain in the skin of dolphins.

At high and maximum swimming speeds, the specified passive method of interaction of external disturbances with the skin can be active. With increasing velocity, the magnitude of the dynamic pressure disturbances of the BL increases. In this case, the deformation of the skin will be stabilized by dermal rollers (7) (Section 4.3, Fig. 4.23). Fig. 4.25 (Section 4.3) shows a photograph of the dermal papillae, which have changed their shape under the action of increased pressure in the circulatory system. Fig. 5.10 shows a diagram of the formation of longitudinal vortices in the BL under the influence of various variants of the shape of the dermal papillae. The top of the figure shows the location of the vortices in the BL, and the bottom shows the profile of the longitudinal velocity U formed inside it in the transverse direction (along the OZ axis). The OX axis is directed along the dolphin body, the OY axis is perpendicular to the skin surface, the OZ axis is directed along the body cross-section line. According to Ref. [148], "hollows" and "peaks" mean alternating along Z and along the OX axis areas where local fluid flows are directed to and away from the body surface, respectively. The wavelength λ of the vortex system determines the periodicity of the variation in Z of not only the average velocity, but also the pressure and thickness of the BL δ. All this causes various stresses in the skin of dolphins. Therefore, for example, in the places of "peaks" and "hollows" in the skin, there are stresses on the

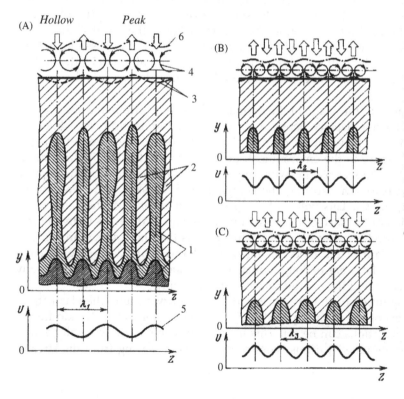

FIGURE 5.10 Diagrams of longitudinal vortices in the boundary layer on the skin of dolphins: (1) longitudinal rows of dermal ridges with dermal papillae, (2) longitudinal rows of epidermal septa, (3) skin surface, (4) longitudinal vortices, (5) distribution of the longitudinal component of velocity U (z), (6) change in the thickness of the boundary layer Blood supply to the longitudinal rows of the dermal papillae: (A) - alternately, (B) - absent, C - of all rows [14,70].

separation and compression. The velocity distribution $U(z)$ indicates the presence in the skin of shear—periodic shear stresses on Z along the X axis. In addition, longitudinal vortices cause shear—shear stresses in the skin that are directed along the Z axis and alternately in this direction tightening and stretching the skin. With a high density of nerve endings in the skin of dolphins, such a complex stress state of the skin may have led to the development in phylogenesis of the longitudinal structures of the skin, reducing irritation in the skin in accordance with the principle of receptor regulation [56].

All variants of Fig. 5.10 suggest a stabilizing effect of the structure of the skin on the BL vortex system. The mechanism of interaction of the longitudinal vortex structures of the BL and the skin can be represented as follows. The rows of dermal ridges due to the inclination of the papillae and greater compliance as compared to the rows of the epidermis better dampen the normal stresses.

Therefore, the rows of "peaks" and "hollows" should coincide with the rows of the dermal papillae. This will partially eliminate the irritation of the skin receptors by BL disturbances. The vortex systems corresponding to such an interaction condition differ from each other by a wavelength multiple of 2, as can be seen from Figs. 5.10A, B, C, where $\lambda_1 = 2\lambda_2 = 2\lambda_3$. Then the normal stresses acting on the surface will cause deformation and a change in the inclination of the dermal papillae, as shown schematically in Fig. 5.10. However, the deflections from the surface, shown by dashed lines, are conditional; therefore, the mechanism of formation of vortices cannot be identified with the artificial creation of a system of longitudinal vortices [148]. In addition, it is necessary to take into account the peculiarities of the interaction of the skin of live dolphins with the BL. Thus, due to the nonstationarity of the flow around the dolphin's body, the "peaks" and "hollows" can change places, which helps to relieve irritation in the skin. It is possible that the shown vortex systems may exist in the BL both simultaneously and alternately, depending on the dolphin's swimming regime, and small-scale eddies should correspond to higher swimming speeds.

In the absence of a change in the shape of the dermal papillae (Fig. 5.10B), the shear wavelength λ_2 is determined by the distance between a pair of longitudinal vortices. Almost λ_2 will correspond to the distance between the dermal papillae. Since the pressure in the papillae corresponds to the pressure in the environment, the deflection in the longitudinal direction will be the lowest. With an increase in swimming speed, the blood filling of the papillae and their temperature increase. Due to this, the compliance of the skin in the area of the papillae is increased (Fig. 5.10C). This will lead to a steady maintenance of the value of λ_3. The depth of the deflection of the surface of the skin in the region of the rows of the dermal papillae depends on the speed of swimming. Fig. 5.10A reflects the photo shown in Fig. 4.25 (Section 4.3). With this variant of the blood filling of the dermal papillae, λ_1 significantly increases and, accordingly, the scale of longitudinal vortices increases, which significantly reduces the friction resistance and the painful sensation in the skin.

Depending on the magnitude of the pressure in the circulatory system, the elasticity of the systems of the cavities of the dermal papillae (8) (Fig. 5.9) is regulated. In addition, a certain law of pressure pulsations can be set. This makes it possible to directionally generate in the BL the above-mentioned various types of disturbances arising from the streamlined surface.

Experiments have shown that the friction resistance coefficient of a plate is minimal when CVS with a vortex axis prevails in transversal (along the OZ axis) or longitudinal (along the OX axis) directions (Part II). Thus, it can be expected that if you maintain a stable existence of these two forms of disturbing movement, it is possible to reduce pain in the innervation system of the skin of dolphins, as well as reduce energy consumption per control BL. Longitudinal vortex systems are most favorable for reducing friction resistance in a wide range of Reynolds numbers.

In Ref. [69], experimental investigations into the interaction of specified three-dimensional longitudinal perturbations introduced into the BL with coherent structures of the BL were performed. The results are set out in Part II. The spectrum of disturbances in the BL of a liquid is quite wide. In this connection, the dolphin's skin should be "tuned" to a wide range of disturbances in the BL and to the energy-carrying frequencies of the disturbance spectrum, and when the speed of the animal changes, the setting should also change.

On the basis of the obtained results of experimental studies of the regulation of the structure and mechanical characteristics of the skin of dolphins, technical analogues were developed [35, etc.], the device of which is given in Part III.

In Section 4.5, the peculiarity of the morphology of shark skin is considered. It is shown that all the main features of shark skin are similar to the features of the skin of dolphins. Fig. 4.35 shows the direction of the scallops of the placoid scales along the body of a shark, and Fig. 4.36 shows a photo of the placoid scales of various shark species. Fig. 4.33 shows a transverse section of the placoid scale of a shark. Each placoid scale inside contains capillaries of the circulatory system and nerve endings. The angle of inclination of the scales varies along the body and may vary depending on the speed of swimming. The placoid scales outside are covered with a layer of mucus and form a specific elastic-dampening cover, like the skin of dolphins. Since the range of swimming speeds in sharks is narrow compared to dolphins, the settings of the stabilizing properties of the structures of the skin of sharks are also inferior to those of dolphins. It is important that the regulation mechanism in sharks differs from that in dolphins, however the structures of shark skin are also supported by longitudinal vortex systems in the BL.

In swordfish, the specific structure of the gills and the production in them of aqueous solutions of low concentration of mucous substances generated from cracks in the BL contribute to the formation of longitudinal vortex structures in the BL. The stability of these vortices is enhanced by longitudinal folds on the surface of the skin, filled with mucus entering the surface of the skin through the pores. Fig. 5.11 shows a diagram of the formation of longitudinal CVSs in the BL of the skin of swordfish [70]. The interaction of the CVSs of the BL and the CVSs generated by the skin of high-speed hydrobionts leads to the formation of new stable forms of longitudinal CVS. Thus, using the example of a swordfish, a combined method of reducing friction resistance developed as a result of evolution was discovered. It consists of the simultaneous use of the interaction method of longitudinal vortex structures, a multilayer elastic cover, the use of aqueous solutions of low concentration of mucous substances, and the multislit injection of these solutions.

Fig. 5.12A shows a photograph of a group of penguins in the initial phase of diving. It is seen that the bodies of the penguins, as it were, are in the air envelope, and behind them is a trace containing air bubbles. R. Bannash investigated the kinematics of penguin

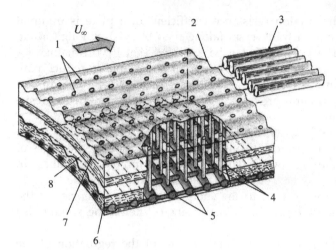

FIGURE 5.11 Stability of longitudinal vortices on the skin of hydrobionts at maximum swimming speeds: (1, 2) Neutral stability curves according to Görtler and Smith calculations [116,276], (3) curve 5=39 for concave rigid surfaces, (4) calculation of Görtler stability for dolphins, (5) for the katran shark, (6) for the blue shark, (7) for the mako and gray sharks, (8,9,10) for the xyphoid fish (swordfish), (11, 12) for the sailboat [32,39]. The parameters of hydrobionts and their speed of movement for calculating the dependences G*(**) shown in graphs (A) and (B) are given in the text

swimming in a water channel built in Antarctica. Based on the obtained kinograms, he claimed that during immersion in water, penguins seem to be capturing air, while the air bubble can stably remain behind the head and, if necessary, be replenished with exhaled air [79]. On the basis of his experiments, as well as images from other authors, it can be concluded that the feather cover of penguins, like that of sharks, is not passive. While staying on the surface inside the feather cover, air accumulates, which increases the insulating properties of the outer layer of the penguins. Penguins have muscles that control the position of feathers and, accordingly, the elastic-dampening properties of the skin [80]. When moving in water, the feather cover does not allow air bubbles to float, as they are held under the feather cover. In addition, when penguins dive, the increasing hydrostatic pressure of the aquatic environment "presses" the feather cover to the body and facilitates the retention of air bubbles under the feather cover.

In this case, an effective elastic-dampening cover is formed, which reduces resistance. When diving, the air located under the feather cover is not used. Outside air is captured, which is clearly seen in Fig. 5.4a, b. Movies have recorded the process of penguins jumping from water onto the surface of ice, where the thickness of the ice can be several meters. In Fig. 5.12, it is clearly seen that when rising to the surface, the penguins with the help of muscles "lift" the feather cover, and the air from under the feather cover escapes and envelops the body of the penguins with a thin cover of microbubbles. This also contributes to the reduction of hydrostatic pressure of water. The films show that in this case the movement of penguins sharply accelerates, and they fly out of the water to a considerable height with great speed. In Fig. 5.12 it can be seen how a concentrated trail containing air bubbles remains in the unseparated wake of the penguins in the water. Currently, a large number of experimental and theoretical investigations into the effect of injected microbubbles on drag reduction have been performed. It was shown that in this case it is possible to obtain a significant decrease in friction resistance of up to 80% [84,102, 110, 123, 124, 135, 185, 186, 196, 197, 202, 203, 284].

In these photos it can clearly be seen how the shape of the body is deformed, acquiring the optimal shape, with drag reduction. At the same time, the elongated shape of the head

(A)

(B)

(C)

FIGURE 5.12   Diving (A and B) and rising to the surface (C) of penguins.

with its beak corresponds to a short xiphoid tip, which, like in the swordfish, also contributes to drag reduction. Thus, penguins use a combined method of the drag reduction.

5.3 Investigation into the stability of coherent vortex structures of the boundary layer of aquatic animal skin

The roughness height on the surface of the hydrobiont skin (Sections 4.3, 4.5–4.7) is such that the body surface remains hydraulically smooth up to the speed of movement $U_\infty = 10\,\text{m/s}$ [31]. With a uniform motion along the hydrobiont, a variable-amplitude wave moves with speed $U/c = 0.96$, where c is the speed of the wave traveling along the body. On the convex side of the body, the transverse microcaps are smoothed, and on the concave, a strip of microspaces with a wavelength λ between adjacent microspaces is formed (Section 4.3, Fig. 4.19). The frequency caused by these microfolds of longitudinal near-wall velocity pulsations in the BL can be determined by the formula $f = c/\lambda$. The values of f and λ depend on the curvature of the body, and are determined by the swimming speed.

Head oscillations at maximum characteristic swimming speeds occur with an amplitude of $(30-40)\delta$ and cause a periodic change in the thickness of the BL δ on it, and, therefore, the distance of the critical layer δ_{cr} of the BL from the surface of the body. For a rigid plate, $\delta_{cr} = 0.3$, and for an elastic plate, this distance is shorter than for a rigid plate and is 0.15–0.2 [170]. In the region of the critical layer, the pulsating velocities are maximal and the most intense interaction of disturbing motions occurs in the BL [70,111]. With nonstationary flow around the microfolds on the skin surface, a flat wave perturbation motion is generated, which is directed directly into the critical layer.

To estimate the hydrodynamic functions of transverse microfolds for various species of hydrobionts, the following calculations were performed. According to Ref. [9], the shape of rigid bodies of rotations with contours corresponding to the body shapes of the various hydrobionts considered was chosen (Section 2.3, Figs. 2.15 and 2.16). For the selected contours of an axisymmetric body of revolution, the parameters of the BL are determined by calculation, in particular, the distribution along the body of the extrusion thickness δ^* [9]. For several characteristic swimming speeds in the area of the midsection of the body, the value δ^* and the corresponding Reynolds numbers Re^* were determined. For these velocities, taking into account the above ratio, the values of c and frequency f were calculated, as well as the dimensionless frequency of disturbing motion $\beta_r \nu / U^2$, where $\beta_r = 2\pi f$, and ν is the kinematic viscosity of water (assuming $\lambda \approx$ constant along the body) [170]. The results are shown in Fig. 5.13, which, for comparison, plotted neutral curves for a rigid plate and two analogues of the skin of hydrobionts [170]. Classic neutral curves when wrapped around a rigid plate are experimentally determined in Refs. [258,259]. In Refs. [21,31,32,39,170, etc.] limit neutral curves were experimentally constructed that limit the region of unstable oscillations, when finite disturbances are introduced into the BL or when nonlinear or nonstationary fluid flow affects fluid flow.

Fig. 5.13A shows the usual (1, 3, 5) and limit (2, 4, 6) neutral curves for rigid (1, 2), polyurethane foam with outer elastic film (3, 4) and polyurethane foam without film (5, 6) plates [170]; oscillations generated by the outer covers of dolphins (7–10) and katran (11) and mako sharks (12). For curve (7) $l = 2$ m, $\bar{x} = 0.312$, $\lambda = 2.4 \cdot 10^{-4}$ m; (8) 2 m, 0.312, $10 \cdot 10^{-4}$ m; (9) 2 m, 0.412, $2.4 \cdot 10^{-4}$ m; (10) 2.6 m, 0.312, $2.4 \cdot 10^{-4}$ m; (11) 1.05 m, 0.312, $5 \cdot 10^{-4}$ m; (12) 1.55 m, 0.312, $2 \cdot 10^{-4}$. Fig. 5.13B shows the neutral curves in the coordinates of the dimensionless wave number $\alpha\delta^*(Re^*)$ for the rigid plate (1), polyurethane foam (2) [170] and dolphins (3–7): (3) $l = 1$ m; $\bar{x} = 0.312$; $\lambda = 2.4 \cdot 10^{-4}$ m; (4) 2 m; 0.312; $2.4 \cdot 10^{-4}$ m; (5) 2 m; 0.412; $2.4 \cdot 10^{-4}$ m; (6) 2.6 m; 0.312; $2.4 \cdot 10^{-4}$ m; (7) 2 m; 0.312; $10 \cdot 10^{-4}$ m.

The BL of the hydrobionts is nonstationary; therefore it is more reasonable to compare the obtained data for the skin with the limiting curves of their analogues (curves 2, 4, 6). From Fig. 5.13 it follows that the flat disturbances generated by the microfolds in the area

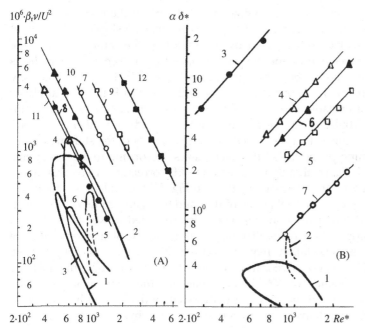

$10^6 \cdot \beta_r \nu / U^2$

$\alpha \, \delta *$

(A)

(B)

$2 \cdot 10^2$ 4   6   8 10^3 2   4   6 $2 \cdot 10^2$ 4   6   8 10^3 2 $Re*$

FIGURE 5.13 Neutral curves for dimensionless frequencies (A) and wave number (B) for flow around plates and hydrobionts: conventional (1, 3, 5) and limit (2, 4, 6) neutral curves for rigid (1, 2) polyurethane foam with an outer film (3, 4) and polyurethane foam without film (5, 6) plates [170]; oscillations generated by the outer covers of dolphins (7–10) and katran (11) and mako sharks (12) [39,170].

of the midsection are stable and stabilize the natural disturbances of the BL. If we perform similar calculations for the same velocities U at smaller x, since $\lambda \approx$ constant and the value $\delta *$ is proportional to x, the points shown in Fig. 5.13 will move on the graph to the left parallel to the x-axis. In this case, at maximum swimming speeds, microsheets generate unstable oscillations, and the mechanism of interaction between natural and disturbances introduced into the BL can be represented as follows. In the case of oscillatory motion of the body, a negative pressure gradient acts on the convex side, which stabilizes the BL. On the concave side of the body, at low speeds, there is no need to stabilize the BL, since the energy of the aquatic organisms is sufficient to overcome the resistance.

At maximum speeds, the BL becomes essentially nonstationary, and δ reduces. On the concave side of the body, microsheets generate unstable vibrations, which in the BL will cause quick alternation of the transition stages even to the midsection. From the results of experiments on skin analogues [170], it follows that the T-S wavelength varies less with the flow around the analogues depending on the Reynolds number (in length) than that of the hard reference. The same property is observed in hydrobionts ($\lambda \approx$ constant). In the BL, at the nonlinear stages of transition, dualism was discovered: at nonlinear stages, when longitudinal vortex disturbances were formed, flat disturbance continues to develop. According to this property (Fig. 5.13), we see that, on the one hand, flat disturbance generated by microfolds that are unstable at the beginning of the body and contribute to the formation of longitudinal vortices, and on the other hand, when advancing the microfolds to the midsection they generate stable flat oscillations that stabilize longitudinal vortex perturbations [170].

It can be expected that microscopes have a number of other hydrodynamic functions. For example, the spectrum of unstable oscillations in the BL is quite wide. It is known [170] that the most "dangerous" for the transition are the oscillations of the BL in the region of the second branch of the neutral curve. From Fig. 5.13A it can be seen that the slope of the straight lines, which characterize the oscillations generated by microfolds, is the same as that of the second branch of the neutral curve. Consequently, at any speed of movement, microsheets generate oscillations close to the second branch of the neutral curve, that is, most intensely affecting the development of disturbances in the BL. It also follows from Fig. 5.13 that elastic surfaces reduce the spectrum of unstable vibrations, shift along the flow, and stretch the transition region from laminar to turbulent flow. This reduces adverse gradient loads on the body of a hydrobiont [26]. It can be assumed that the pulsations of the speeds of the shift are dampened due to the elastic-dampening properties of the microglazes. According to measurements on hydrobionts, in Ref. [173] it is stated that oscillations in the BL correlate with oscillations of the caudal fin. This is confirmed by the calculations given in Fig. 5.13. Oscillations of the second type, equal to twice the frequency of oscillations of the caudal fin, and significantly affecting the flow pattern in the BL beyond the mid-axis, were also recorded in Ref. [173]. In Ref. [146], measurements of the propagation of oscillations on the surface of a hydrobiont are given. During slow swimming, the oscillation frequency on the skin surface was 130–140 Hz, and during jumps it was 115–230 Hz. According to our data, when applying a single impulse, fluctuations with a frequency of 600 Hz propagate on the surface of plexiglass, 130 Hz on the surface of the analogue of the skin of a hydrobiont, and 157 Hz when temperature controlled. By changing the mechanical parameters of the skin, hydrobionts are able to influence the field of pressure pulsations of the BL.

Thus, during the movement of hydrobionts in their BL, three types of disturbing motion are generated: from the action of microfolds on the skin surface, from its vibration as an elastic-viscous material, and due to the work of the tail propulsion device. The spectrum of generated disturbances is significantly different, for example, at $U = 10$ m/s, in the first case by tens of kilohertz, in the second by hundreds of hertz and in the third by tens of hertz. Therefore, according to the mechanism of complex interactions of disturbing motions [41,154], in the first case, the skin-generated disturbances will interact with the disturbances of the BL, and in the second and third cases the impact will be on the energy spectrum of the BL, especially on its low energy-carrying frequencies.

The effect of microfolds on the BL, at its maximum at the beginning of the body, decreases as the microfolds move to the midsection, as the movement of the body of a hydrobiont along sinusoidal law can be viewed as occurring in the same sinusoidal wake fluid caused by the oscillation of the head. Therefore the body is flown around a plane-parallel nonstationary flow. With an increase in the thickness of the BL, the role of microfolds weakens, while the second and third disturbance factors of the BL act constantly along the whole body. At $\bar{x} = x/l > 0.4$, the subsequent stages to the transition in the BL develop, where x is the current coordinate along the body, and l is the body length. At the beginning of the body, the formation of longitudinal vortex disturbances occurs rapidly in the BL under the influence of the indicated susceptibility phenomena [70,111,154]. Subsequently, these vortex systems develop and stabilize under the influence of centrifugal forces caused by the curvature of the body during its oscillatory motion.

Currently, quantitative data on the degree of deformation of the surface of the skin or the degree of irritation by its local pressure gradients (especially when changing the swimming regime) is insufficient. There are also no experimental data on the structure of the flow on elastic surfaces in the considered range of Reynolds numbers. All this does not allow us to unambiguously determine the choice of the scheme for the calculation of the stability of longitudinal vortices. Therefore the assessment of the stability of the longitudinal vortex disturbances of the BL during the flow around the skin of a live dolphin was carried out according to the "less stable" scheme of Fig. 5.10A (Section 5.2), as having a longer perturbation wavelength.

The stability of the BL during oscillatory motion of the body of a hydrobiont in concave areas was estimated by the dependence of the Görtler parameter $G^* = \frac{U_\infty \delta^{**}}{\nu} \sqrt{\delta^{**}/R}$ on the dimensionless wavenumber $\alpha\delta^{**}$, where δ^{**} is the thickness of impulse loss, R is the radius of curvature of the surface, $\alpha = 2\pi/\lambda_z$ is the wave number along the transverse coordinate z, and ν is the kinematic viscosity coefficient [116,313–315]. In the calculations, the values of $\lambda = 2z_o$ were used, to determine which the results of measurements of the longitudinal microroughness of the outer covers of aquatic animals were involved. For a dolphin, the curvature of the body was determined by Cousteau kinograms (Section 2.7, Fig. 2.28 and Fig. 5.14). R_i values were determined by the contour of the outer surface of the concave part of the body of a hydrobiont, which was considered as a set of circular arcs (Fig. 5.15). In calculations, the smallest values of the radii were taken, which shifted the calculated points to the instability region on the Görtler diagram. The value δ^{**}, as well as δ^*, was determined on the basis of a comparison of the forms of the body of revolution [9] and the corresponding forms of the bodies of hydrobionts (Section 2.3, Fig. 2.16).

For example, for the dolphin, the δ^{**} and δ^* values were taken according to the calculations made in Ref. [9] for the model of the body of rotation No. 32; for the katran shark, No. 22; for the blue shark, No. 25; and for the mako shark, No. 32 with correction No. 25. Necessary values of swimming speed were chosen according to Fig. 2.23, Section 2.5.

The results of the $G^*(\alpha\delta^{**})$ calculations in the form of straight lines connected by the calculation points are depicted on the stability diagram (Fig. 5.16). It can be seen that the vortex systems, determined by the parameters of the skin structure, are stable for the characteristic speeds of movement, since all the dependences $G^*(\alpha\delta^{**})$, except curves (8–12).

Fig. 5.16A characterizes the flow of hydrobionts with the following parameters: curve (4) dolphin, $l = 2$ m, $U_\infty = 10$ m/s; (5) katran shark, $l = 1.05$ m, $U_\infty = 5$ m/s; (6) blue shark, $l = 1.5$ m, $U_\infty = 10$ m/s; (7) mako shark (circles), $l = 1.55$ m, $U_\infty = 20$ m/s and gray shark

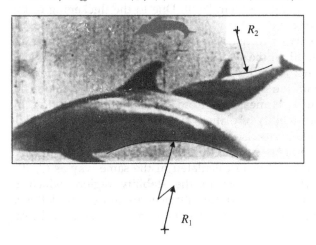

FIGURE 5.14 Photo of swimming dolphins at different body oscillation positions.

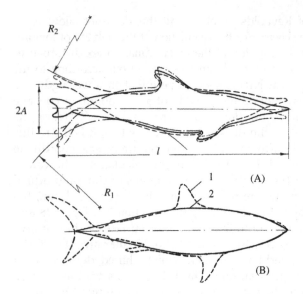

FIGURE 5.15 The scheme of oscillatory motion of hydrobiont (A) and approximation of hydrobiont (1) by a body of rotation (2); (B): l is the body length of the hydrobiont, A is the oscillation amplitude of the caudal fin, R_1 and R_2 are the radii of curvature of the concave body parts [32,39].

(triangles), $l = 1.1$ m, $U_\infty = 10$ m/s; (8) swordfish $l = 1.8$ m (with a nasal needle), $U_\infty = 20$ m/s, (9) the same as (8, 10), but δ^{**} was calculated from the gill slit, (11 and 12) a sailboat, l and U_∞ are the same as for (9 and 10).

For Fig. 5.16B, the following calculations were made: (1) Görtler [116]; for shark skin, quails: (2) $l = 0.54$ m, $U_\infty = 5$ m/s, (3) 0.54 m, 2 m/s, (4) 1.05 m, 5 m/s; mako sharks: (5) 1.55 m, 20 m/s, (6) 1.55 m, 15 m/s, (7) 1.0 m, 15 m/s, (8) 1.0 m, 10 m/s; blue shark: (9) 0.75 m, 7 m/s, (10) 1.5 m, 10 m/s; gray shark: (11) 1.1 m, 10 m/s, (12) 1.1 m, 6 m/s, (13) 0.87 m, 4 m/s. The calculations were made for swordfish at the origin of coordinates in the area of the midsection (8), between the midsection and the gill slit (9), and in the area of the gill slit (10).

The peculiarities of the body shape of a swordfish [215] determine the method for calculating δ^{**}. For curve (8), the value of δ^{**} was determined from the equivalent body of rotation with a nasal needle at $U_\infty = 20$ m/s and $l = 1.8$ m [308]. Due to the thickening of the body of the swordfish from the sides for the (9) δ^{**} curve, it was calculated on a plate with $U_\infty = 25$ m/s and $l = 1.8$ m. The instability of vortex systems can be explained by the existence of a different flow structure in the BL of the swordfish: either an overestimated value of U_∞, or the fact that the specific structure of its branchial branch leads to an intensive injection of solutions of biopolymers into the BL. Then δ^{**} should be calculated under the assumption that the current coordinate x_1 is measured from the gill slit. In addition, by analogy with the results concerning the stability of plane waves on elastic surfaces [170], it can be expected that the instability region will also decrease and shift for the subsequent stages of the transition. Since no experiments similar to those in Ref. [308] were carried out for the sailboat, the dependencies (11, 12) were calculated in the same way as (9, 10). The curves characterizing faster sharks lie deeper in the stability region, which is explained by the orderliness of their skin scales (α) and surface curvature. The latter is inherent in all considered species of aquatic animals and follows from the empirical relation $U_\infty \sim 1/A \sim R \sim 1/G^*$.

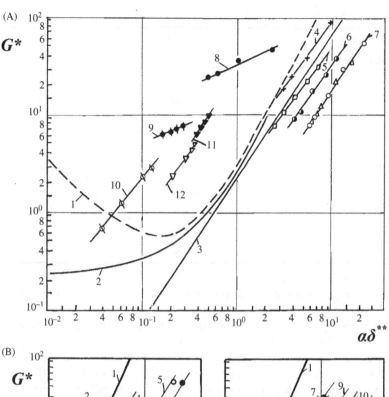

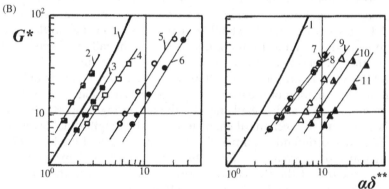

FIGURE 5.16 Stability of longitudinal vortices on the skin of hydrobionts at maximum swimming speeds: (1, 2) Neutral stability curves according to Görtler and Smith calculations [116,276], (3) curve 5 = 39 for concave rigid surfaces, (4) calculation of Görtler stability for dolphins, (5) for the katran shark, (6) for the blue shark, (7) for the mako and gray sharks, (8,9,10) for the xyphoid fish (swordfish), (11, 12) for the sailboat [32,39]. The parameters of hydrobionts and their speed of movement for calculating the dependences $G^*(\alpha\delta^{**})$ shown in graphs (A) and (B) are given in the text.

The same limits of change in G^*, despite the different swimming speeds of hydrobionts, are explained by the different growth patterns δ^{**} along the body, which is related to its shape: the front part of the body of slowly swimming aquatic animals increases the values of δ^{**} and, therefore, G^*, and for fast-swimming animals G^* increases due to greater speed, since $G^* \sim U_\infty (\delta^{**})^{3/2} \sim U_\infty^{1/4}$.

The inaccuracy of determining the δ^{**} values does not affect the conclusion about the stability of the vortices, since the change in δ^{**} moves the points on the diagram of the Görtler stability along the lines $P = \text{constant}$ without intersecting the neutral curve. The errors in the calculation of the curvature $1/R$ also have little effect on the results of the calculation. In addition, the same calculations were performed for

TABLE 5.2 Shark speed and shark body length.

No.	Kind of animal	l (m)	U_∞ (m/s)
1	Katran shark	0.54	3.5
2	–	0.54	2
3	–	1.05	5
4	Mako shark	1.55	20
5	–	1	10
6	Blue shark	0.75	7
7	–	1.5	10
8	Gray shark	1.1	10
9	–	1.1	6
10	–	0.87	4

various species of aquatic animals with variable parameters l, U_∞, the values of which are given in Table 5.2.

A comparison of the characteristics obtained as a result of the calculations shows that an increase in U_∞ can lead to instability of the vortex system. The same tendencies appear with a decrease in body length; therefore, for a blue shark, an increase in l by a factor of 2 and U_∞ by a factor of 1.4 does not change the stability characteristic on the Görtler diagram. In addition, a decrease in the λ values from the midsection to the tail in all species of hydrobionts contributes to the preservation of a stable system of vortices in the BL along the body length. The mechanism of interaction of the flow with the external covers only to a small extent consists in the deformation of the skin surface, as shown in Fig. 5.10 (Section 5.2). This mechanism is more complex and is explained by the absorbing properties of the covers of aquatic animals. Different in mechanical characteristics, the longitudinal layers of the skin (1) and (2) (Fig. 5.10) will differently dampen the pulsating energy of the BL, which will lead to the directional absorbing characteristic of the outer layers and may help to maintain a regular vortex structure of the BL such as Görtler vortices. Since the rows of dermal rollers shown in Fig. 5.10 have a higher temperature than rows (2), this should also stimulate the formation of a system of longitudinal vortices, as shown in Refs. [90,116].

Thus, an ordered structure of the flow in the BL can take place in a fairly wide range of Reynolds numbers due to the structure and optimal mechanical characteristics of the streamlined skin of hydrobionts.

5.4 Measurement of the temperature of the skin of dolphins

Temperature measurement on the surface of the skin of dolphins was carried out V.V. Babenko using semiconductor microthermal resistances of type MT-54 [15]. The main requirements for semiconductor microthermal resistance when measuring the temperature on the surface of the skin of dolphins are low inertia, high sensitivity, small size, sufficient mechanical strength, and heat insulation from the sensor body. The high sensitivity of the MT-54 made it possible to measure the temperature on the skin surface of dolphins without using an amplifier. The electrical signal from the sensor was transmitted by wire to the measuring bridge and then to the H-700 oscilloscope. In parallel, the signal was fed from the measuring bridge to the indicator device, which allows monitoring of the correctness of recording signals on the oscilloscope.

The technique of working with dolphins during experiments was carried out in accordance with Sections 1.2 and 5.1. Experiments to measure the temperature of the surface of the skin were carried out on three types of dolphins: the short-beaked common dolphin, the bottlenose dolphin, and the harbor porpoise. They were placed in a special box filled with water. Fig. 5.17 shows the waveforms of temperature records on the body surface of the bottlenose dolphins and porpoises.

Temperature curves on the oscillogram were plotted from right to left. While the sensor touches the surface of the skin should be a gradual increase in temperature; after the sensor was taken away from the body and placed in water, a sharp drop in temperature was observed. The curved shape shows that the measurement data are somewhat underestimated due to insufficient contact time of the sensor with the body of the animal. This is evidenced by the asymptotic increase in temperature on the oscillograms. If the true temperature of the body surface is measured, then the temperature-recording curve should be a horizontal line or asymptotically approach the horizontal line. In most of the measured points, the error of the results obtained in this way is within the error of the experiment. Part of the measurements, mainly on the fins and ridge, shows that the measurement data are understated from tenths of a degree to a degree. Waveform analysis made it possible to determine that, at the confidence level $\alpha = 0.95$, the accuracy of the measurements was 5%.

The measurements were carried out according to a scheme that allows determining the temperature on the surface of the entire dolphin body (Fig. 5.18A). The air temperature during the experiments was in the range of 14°C–15°C. The measurements were carried out across the cross sections of the dolphin body from top to bottom. In sections 3, 5–9, the temperature of the skin surface was measured at five points on one side of the body, and in sections 10 and 11 at three points.

In addition, the temperature was measured at specific points: in section 1, in the middle of the frontal protrusion in front of the breath; in section 2 in the middle of the conditional line connecting the blowhole and the eye; and in sections 4, 8, and 12, respectively, in the middle of the lateral fin, vertical fin, and caudal fin.

During measurements, the sensor was tightly applied to the body of the dolphin for a few seconds. Recording was made on an oscilloscope. When the sensor was removed from

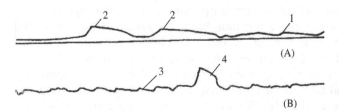

FIGURE 5.17 Oscillograms of temperature measurement on the body surface of the bottlenose dolphins (A): (1) caudal stem, (2) caudal fin; (B) porpoise: (3) middle part of the body, (4) vertical fin [15].

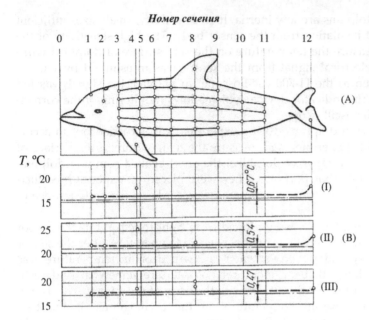

Номер сечения

FIGURE 5.18 Temperature distribution on the skin surface of dolphins: (A) scheme of measurement points on the surface of the body, (B) measurement results, (*I*) short-beaked common dolphin, *l* = 1.3 m, (*II*) two bottlenose dolphins, *l* = 1.7 and 1.6 m, (*III*) two porpoises, *l* = 0.7 and 0.8 m [15].

the body of the dolphin, the sensor fell into the water: at this time, recording was made on an oscilloscope of water temperature. Thus, the records on the tape of the oscilloscope can be judged on the temperature drop of the body surface and water. At the beginning, five measurements were made at each point. Primary processing showed that this is enough to carry out three measurements, and at some points, two or one measurement. Due to the low inertia of the MT-54, measurements throughout the body were carried out within a few minutes. Oscillogram records showed that the temperature at the surface of the body did not rise at the end of the experiment as compared with the beginning of the experiment. This indicates that the dolphins did not overheat while in a fixed position during the measurements. It can be assumed that the conditions of the experiments correspond, at least, to the conditions of slow swimming of dolphins. Control measurements carried out according to this scheme showed that there was no asymmetry of temperature on the surface of the body; therefore further measurements were carried out on one side of the body. The results of temperature measurements on the fins in one cross section of the fin showed a slight decrease in temperature in the section by the end of the fin (by the end of the chord).

Fig. 5.18B shows the waveforms of the measurements: solid lines—the results of measurements in sections 3, 5–11; points—measurements at specific points of sections 1, 2, 4, 8, 12; interpolation, dotted lines; dash-dotted—water temperature during the experiment.

The data obtained show that in a stationary position, the temperature difference on the surface of the skin of dolphins and water did not exceed 1°C. The temperature difference between the surface of the skin and the water on the ridge, side, and tail fins is much higher than in other parts of the body. For example, in the bottlenose dolphin, the temperature difference reaches 4°C on the lateral fins; in the short-beaked common and bottlenose dolphins on the ridge, this difference is slightly different from the temperature difference between the body surface and water (in the short-beaked common dolphin, there is no temperature difference). In dolphins in the fins and in the ridge are located complex vessels, which are an effective mechanism for thermoregulation. Measurements have shown that the lateral and caudal fins as moving parts of the body, being constantly in the water, are the main mechanisms of heat exchange in the animal. The ridge, when swimming dolphins near the water surface and while jumping in the air, is in the air and much less in the water compared to other fins, thereby its role as a thermostat has decreased dramatically due to the fact that the thermal conductivity of air is significantly lower than water. The research results show that the bottlenose dolphin and the short-beaked common dolphin can control the thermoregulatory properties of their fins. According to the results of these measurements, the temperature difference between the surface of the skin and water is proportional to the power supply of the animal: the high-speed hydrobiont has a larger temperature difference.

5.5 Thermoregulation of aquatic animal skin

Investigations into thermal regulation of dolphins were performed in Ref. [87,291], and the biophysical principles of blood circulation in Ref. [254]. In Ref. [221], some results of studies of the circulatory system of the external layers of the skin of hydrobionts are given, in particular, the heat conductivity coefficient k of the fin whale skin specimen (*Balaenoptera physalus*), the numerical value equal to $5 \cdot 10^{-4}$ cal/s $\cdot$ cm $\cdot$ grad, was calculated. In the same units, k is up $1.5 \cdot 10^{-3}$ for water, $6 \cdot 10^{-4}$ for wool, $5 \cdot 10^{-4}$ for wood, and $9 \cdot 10^{-5}$ for felt. The specific heat flow q through the unit of skin surface with a thickness of $d = 1.8$ cm was estimated using the formula:

$$q = \frac{k}{d}(T_w - T_f),\qquad(5.12)$$

where T_w is the temperature inside the body, and T_f is the temperature of the medium.

For the dolphin at $T_w = 36°C$ and $T_f = 10°C$, the value $q = 72 \cdot 10^{-4}$ cal/cm$^2 \cdot$ s $= 260$ kcal/m^2•h was determined.

The value of k in Ref. [221] was methodologically determined incorrectly, as the sample was frozen for 5 months, and then thawed for 3.5 hours during transportation in an airplane. With such sample preparation, cells could be partially destroyed, which could distort the value of k. The thermal conductivity of living tissue is different from the thermal conductivity of dead tissue. In addition, the value of k should be considered complex: consisting of the thermal conductivity of the layers of the skin and circulatory system.

The value of k cannot be constant and completely identical for all species of aquatic animals. In addition, due to specific distributions of skin thickness (Section 4.3, Fig. 4.27) and the circulatory system [4], the value of k along the body of dolphins may be different.

Measurements performed in Refs. [15,311] showed that dolphins can control the heat flow passing through the integuments in different parts of the body. It was found that the mechanical properties of the integuments [29,33] depend not on the ambient temperature, but on the state and behavior of the hydrobionts.

In Ref. [221], the outer integuments are represented by three layers: the epidermis, the dermis, and the hypodermis. Thermoregulation in the skin is due to the specific structure of the circulatory system only in the epidermis, when the venules are very tightly intertwined with arterioles. Below is a different mechanism of thermoregulation. In accordance with this, special significance is attached both to the specific structure of the circulatory and nervous systems in the outer integument, which regulates the value of k, and to the specific structure of cutaneous muscles, which affects the values of k and d.

Fig. 5.19 shows a diagram of the circulatory system of the skin of dolphins in accordance with Figs. 4.24 (Section 4.3) and 5.7 (Section 5.2). The layer of subcutaneous fatty tissue (8) and the layer of hypodermis (6) are saturated with blood vessels and capillaries. In the layer of the dermis (5), branched vessels and capillaries are very small. The hypodermis and dermis, along with others, penetrate the vessels that have a predominantly vertical direction. In the skin there are two specific layers in which the vessels are located, mainly parallel to the surface of the body. The first, the main one, is located in the region of the conditional border of the epidermis (3) and the subpapillary layer (4), and the second is in the region of the dermis and hypodermis, with the first layer saturated with most venules.

In this diagram, the epidermis, venules, and arterioles are conditionally spaced apart. In the dermal papillae, the venous and arterial capillaries are jointly and tightly intertwined. The presence of two horizontal layers of blood vessels in the skin is similar to the thermal curtains used in the technique.

In accordance with Fig. 5.19, we consider both extreme cases of the work of system thermoregulation of cetacean skin.

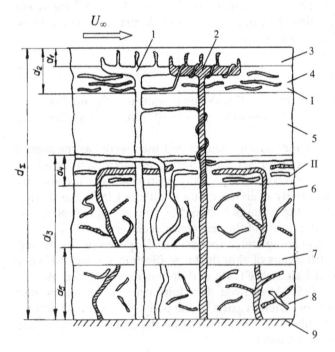

FIGURE 5.19 A scheme of the blood circulation in dolphin skin: (1) venules, (2) arterioles, (3) epidermis, (4) nasal layer, (5) dermis, (6) hypodermis, (7) skin muscles, (8) subcutaneous fatty tissue, (9) skeletal muscles; (I) the first and (II) the second layers of the horizontal arrangement of vessels (arterial and venous networks are conditionally spaced) [38].

5.5.1 Increased heat flow through the skin

If we assume that the blood circulatory system in the heat transfer is not involved, then the specific heat flow through the skin can be written as:

$$q_\Sigma = \frac{k_\Sigma}{d_\Sigma}(T_w - T_{w0}), \tag{5.13}$$

where k_Σ is the integral heat-conductivity coefficient of the entire skin, depending on the structure and thermal insulation properties of the skin material; T_{w0} is the temperature on the surface of the skin. In the case when blood enters only the d_3 layer, in which intensive blood circulation takes place, it can be considered that the temperature of layer II of the horizontal arrangement of vessels is numerically equal to the temperature inside the body. The heat exchange between arterioles and venules occurs in layer II, above which the temperature of the venous blood is equal to the ambient temperature, that is, T_{w0}. Thus, it can be assumed that there is no temperature difference on the outer surface of the layer $(d_3 - d_4)$ and, therefore, there is no heat flow. Heat is transported through the vertical dilated arterial vessels directly into the d_4 layer. Therefore, we can write the specific heat flux through the skin using the expression:

$$q_1 = \frac{k_1(T_{w3} - T_{w0})}{d_\Sigma - d_3}, \tag{5.14}$$

where T_{w3} is the temperature on the upper surface of the d_3 layer, and it differs slightly from temperature T_{w4} on the lower side of the layer d_4, which is equal to the temperature T_w.

In reality, there is no heat flow only through the collagen and elastic fibers of the layer $(d_3 - d_4)$.

If we consider these fibers and the circulatory system of this layer as a whole, then the heat flow through this layer will pass. But since in this layer there is no temperature difference between the circulatory system and its skin, the assumption adopted above can be considered fair. In the general case, depending on the amount of blood filling of the d_3 layer, the value of the thermal conductivity coefficient k of this layer will vary. For example, the value of k for water is an order of magnitude greater than for skin.

Since heat exchange between venous and arterial blood occurs in layer II, the temperature of the d_4 layer will differ somewhat from the layer temperature $(d_3 - d_4)$. Given the above, we can write $T_{w0} << T_{w3} < T_{w4}$. Denoting the thermal conductivity coefficient of the d_4 layer due to the increased blood filling through k_1, we write the specific heat flow through the d_4 layer:

$$q_2 = \frac{k_1(T_{w4} - T_{w3})}{d_4}, \text{ or since } T_{w4} \approx T_w, \ q_2 \approx \frac{k_1(T_w - T_{w0})}{d_4}, \tag{5.15}$$

and the refined total heat flow through the entire skin will be $q_1 + q_2$. Since $(d_\Sigma - d_3) << d_3 << d_\Sigma$, then $q_1 >> q_\Sigma$, and all the more $(q_1 + q_2) >> q_\Sigma$. In this case, the specific heat flow through the skin increases even more. The fact is that layer I of the horizontal arrangement of the vessels will be filled with "fixed" blood. And although the diameter of the vessels will be greatly reduced, the thermal conductivity of this layer is higher than the thermal conductivity of the dermis layer 5 $[d_\Sigma - (d_2 + d_3)]$. Thus, we can assume that $q_1 = q'_1 + q''_1$, where

$$q'_1 = \frac{k(T_{w3} - T_{w2})}{d_\Sigma - (d_2 - d_3)} \text{ И } q''_1 = \frac{k_2(T_{w2} - T_{w1})}{d_2}. \tag{5.16}$$

Here T_{w2} is the temperature on the lower surface of the d_2 layer; $T_{w1} \approx T_{w0}$ is the temperature on the bottom surface of the d_1 layer. In the general case, the total specific heat flow across the entire skin layer is written:

$$q'_\Sigma = q'_1 + q''_1 + q_2. \tag{5.17}$$

If we accept the assumption that $T_{w2} \approx T_{w0}$, then $q'_1 \approx 0$ and the specific heat flow will increase due to a decrease in wall thickness.

Consider the second case when arterial blood by increasing the diameter of the arteries and blood pressure circulates vigorously not only in the d_3 layer, but also in the d_2 layer. Then the heat exchange between arterial and venous blood is carried out mainly in layer I. Therefore, we can assume that $T_{w3} = T_w$, and $T_{w2} \approx T_{w3}$. The temperature ratio is written as follows: $T_{w0} \ll T_{w2} < T_w$, and the total specific heat flow through the entire skin layer will be expressed by the formula

$$q''_\Sigma = q_3 + q_4, \tag{5.18}$$

where $q'_3 = \frac{k(T_w - T_{w2})}{d_\Sigma - (d_2 + d_3)}$ is the heat flow through the dermis; $q'_4 = \frac{k_1(T_{w0} - T_{w2})}{d_2}$ is the heat flow through the layer of the epidermis (3) and the subpapillary layer (4). If we take $T_{w2} \approx T_w$, then $q''_\Sigma = q_4$, or

$$q''_\Sigma = \frac{k_1(T_{w0} - T_w)}{d_2}. \tag{5.19}$$

Since $T_{w2} \approx T_w$, in this case the specific heat flow will be particularly intense. If we take, for example, the average layer thickness d_2 for a short-beaked common dolphin as $2.5 \cdot 10^{-3}$ m, and the layer thickness $d_\Sigma = 2.5 \cdot 10^{-2}$ m, then only by changing the thickness of the thermal resistance can the dolphins change the specific heat flow from the skin surface by an order of magnitude. If we take into account that the value of k due to blood filling may also change by an order of value, in general, the value of d_Σ may change by two orders of value.

5.5.2 Reduced heat flow through the skin

At first glance, it can be expected that the minimum specific heat flow will be when, due to the skin muscles being clamped, the blood circulatory system of the skin does not participate in heat transfer and then the equality is fulfilled:

$$q_\Sigma = \frac{k(T_w - T_{w0})}{d_\Sigma} = q_{min}. \tag{5.20}$$

However, the system of the heat-regulating system of dolphin skin allows reducing heat flow through the skin in another way. If the venous pressure is sharply increased and the diameter of the venules has also increased, and the diameter of the arteries and the arterial pressure have decreased, then there is a circulation of venous blood between layers *I* and *II*, by analogy with the heat curtain. Since the intensity of the current of venous blood increases in comparison with arterial blood, due to the capillaries located in the epidermis, and also due to the fact that the papillary layer is heavily saturated with venous capillaries, super cooling of the d_2 and $[d_\Sigma - (d_2 + d_3)]$ layers will occur. We can assume that $T_{w2} = T_{w0}$, and T_{w3} differs little from T_{w0}. Then, in the old terminology, $q_4 = 0$ and $q_3 \approx 0$. The value of q_Σ in this case is written

$$q'''_\Sigma = q_3 + q_4 + q_5, \tag{5.21}$$

where $q_5 = k_3(T_w - T_{w3})/d_3$. Considering that $q_4 \approx q_3 = 0$, we obtain $q'''_\Sigma \approx q_5$.

It can be assumed that the value of q_5 will significantly increase due to the small value of d_3. However, due to the fact that k_3 is a complex value, the heat coming from the inside to the d_4 layer returns back due to heat exchange in this layer between the venules and arterioles. Therefore, the relation $k_3 << k$ is fulfilled and, thus, we get:

$$q_\Sigma < q'''_\Sigma. \tag{5.22}$$

Dolphins can further reduce heat transfer through the skin. In this case, heat transfer in the d_4 layer increases, arterial blood cools in the d_4 and d_3 layer by increasing the cold transfer by the venous system into the skin, and the $[d_3 - (d_4 + d_5)]$ layer is supercooled. At the same time, the role of heat insulator is performed by subcutaneous fatty tissue, which, due to its specific structure, has $k_4 << k$. The relation $T_{w4} \approx T_{w0}$ is fulfilled. We can assume that $T_{w5} \approx T_{w0}$, where

T_{w5} is the temperature on the upper surface of the d_5 layer. Then the total specific heat flow is written:

$$q_\Sigma^{IV} = q_6 + q_7, \tag{5.23}$$

where $q_6 = \frac{k_3(T_{w5} - T_{w0})}{d_3 - (d_4 + d_5)}$ is the specific heat flow through the hypodermis layer; $q_7 = \frac{k_4(T_w - T_{w5})}{d_5}$.

Since k_3 and k_4 are substantially less than k, it becomes clear that the relation $q_\Sigma^{IV} < q'''_\Sigma < q_\Sigma$.

To calculate the total specific heat flow q_Σ^i at different swimming speeds, it is necessary to know the corresponding temperatures T_{wi} and thermal conductivity coefficients k_i. Using these data, it is possible to determine the total heat flow through the surface of the dolphin's body and thereby determine its role in energy exchange and in ensuring the optimum mechanical characteristics of the external covers.

A new explanation for the possible mechanism of thermoregulation through the outer covers of aquatic animals is proposed, which allows increasing the heat flux by approximately two orders of magnitude from its average value given in Ref. [221] or reducing the heat flux by creating two layers of a thermal curtain. Heat flow can be regulated by changing the coefficient of thermal conductivity of the skin, due to changes in the blood supply of individual layers of the skin, due to the effect on the vascular system of the skin muscle,

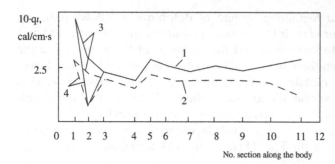

FIGURE 5.20 An example of calculating the specific heat flow through the unit of length of the dolphin skin of the short-beaked common dolphins (1) and the bottlenose dolphins (2); (3) is from below and (4) above the head.

as well as due to changes in skin thickness during various swimming regimes [33]. The mechanical characteristics of the outer covers of aquatic animals can be regulated over a wide range under different modes of movement by changing the temperature of individual layers of the skin or of the skin as a whole.

Fig. 5.20 shows an example of calculating one of the variants of the specific heat flow through the unit length of the skin of the short-beaked common (1) and bottlenose (2) dolphins.

Curves (3) and (4) indicate the calculations, respectively, from below and above the head. The calculations were performed using the above formulas [38].

The calculations are based on measurements of the temperature of the outer surface of the skin, performed on various types of live dolphins (Section 5.4, Figs. 5.17 and 5.18) [15]. In Fig. 5.20, the numbers of sections along the body are defined in accordance with Fig. 4.26 (Section 4.3). The calculations were carried out on the backs of the dolphins. In the head area, calculations were also performed along the abdomen. The patterns shown in Fig. 5.20 were determined by the selected distribution of skin thickness along the body.

This chapter 5 discussed the features of the interaction of aquatic organisms with their habitats.

As a result of this interaction in hydrobionts the following occur:

- Control by form bodies and fins;
- Control of the geometric parameters of the skin;
- Adaptation of the mechanical characteristics of the skin by tightening the skin with muscles and skin fluctuations;
- Regulation of the mechanical characteristics of the skin with the help of the circulatory system and innervation;
- Specific regulation of body temperature and skin;
- Optimization of the mechanical parameters of the skin;
- Regulation of vibrations on the surface of the skin, etc.

Experimental investigations into the boundary layer and mechanical characteristics of dolphin skin

6.1 Measurement of the longitudinal averaged and pulsation velocity of the boundary layer of dolphins

Based on the studies performed in earlier chapters, unique features of the organisms of aquatic organisms to interact with the environment have been identified, which allow them to save energy and significantly reduce the resistance from the environment when swimming at different speeds. For a better understanding of the detected features of hydrobionts, experimental studies on live dolphins were performed in accordance with the methodology of experimental investigations of hydrobionts developed in Chapter 1.

At the Institute of Hydromechanics of the National Academy of Sciences of Ukraine, Kiev V.M. Shakalo, under the direction of L.F. Kozlov, carried out an experimental investigation of the longitudinal velocity of the boundary layer on dolphins under various conditions of their movement [164–166,173,175,246,247,265,266]. The results of theoretical studies of the corresponding electronic equipment for experimental studies of the characteristics of the boundary layer of dolphins are given in Refs. [164,265,266]. In the equipment, a wire hot-wire anemometer was used as a flow mode sensor in the boundary layer. Based on theory of electricity [290], methods and results of measurements of the flow regime in the boundary layer of a model moving in an aqueous medium are described in Ref. [170]. Measurements were carried out in the Crimea at the Karadag biostation. The experiments were conducted in a closed coastal pool, aviary, and network corridor. The pool dimensions were: $24 \times 5.5 \times 3.5$ m; aviary: $40 \times 8 \times 3$ m; network corridor: $30 \times 2 \times 2.5$ m.

When measuring the flow regime in water, the initial turbulence was measured, which was determined by the critical Reynolds number for the point of the beginning of the transition of the laminar boundary layer to the turbulent boundary layer on the pop-up plate. The flow regime on the plate of the pop-up device was determined using telemetry equipment [268] intended for measurements on dolphins [173].

Experimental Hydrodynamics of Fast-Floating Aquatic Animals
DOI: https://doi.org/10.1016/B978-0-12-821025-3.00006-6

The hot-wire anemometer sensor was installed in four different sections of the bottlenose dolphin body: in the eye zone (1), in the sections in front of the pectoral fins (2), and in front of the dorsal fin (3−5), and also halfway between the dorsal and tail fins (6, 7). The different sensor locations on the dolphin are shown in Fig. 6.1. The measurements were carried out on trained and wild bottlenose dolphins. The trained bottlenose dolphins were trained to swim with the trainer and to contact with him up to the free dressing of a belt with sensors and a platform with suckers on them. Installation of sensors on wild cetaceans was carried out in two ways. In some cases, the bottlenose dolphin was caught and removed from the water, after which a device with sensors was installed on the animal. In other cases, the sensors were installed on an animal caught in the water. After attaching the sensors, the bottlenose dolphin was released into the water area for free swimming.

Floating bottlenose dolphins with the equipment installed happened naturally. Investigations have shown that animals that are not accustomed to transporting equipment at high speed quickly get used to the harness, but swim slowly, at a speed not exceeding 2 m/s. To obtain high-speed modes of swimming of the bottlenose dolphins, various methods of stimulating the animal were used. In some cases, the aspiration of the animal, on which the harness was put on for the first time, was used, to flee from it. In this case, the flow regime was measured at the beginning of the free swimming of the animal. In other cases, the animal developed a steady skill to drop a loosely fixed harness with sensors during a jerk at high speed, after which the bottlenose dolphin was released into the water area with a well-secured harness, but during a jerk it tried to dump it in the usual way. At this time, measurements of the flow regime were made. In the third method of stimulating the animals, ecological relationships were used in a herd of bottlenose dolphins. To the experimental animal-male, an aggressive-minded other large male was released into the water area—a bottlenose dolphin, which tended to conquer primacy in the herd. Fleeing from persecution, the experimental bottlenose dolphin developed great speed. At these moments, the flow regime in its boundary layer was measured. After the experiments with stimulation according to the third method, the animals were separated. In the fourth method, the habit of jumping of the bottlenose dolphin was used to coax it to jump over a net with which it was driven into a corner of the water area. The jump occurred if the network fell below the surface. The animal developed great speed before and after the jump. The described methods made it possible to measure the flow patterns in the boundary layer of the bottlenose dolphins at swimming speeds of up to 6 m/s. The equipment installed had no noticeable effect on the hydrodynamics of the bottlenose

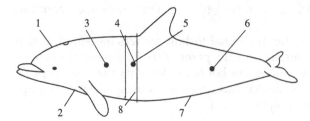

FIGURE 6.1 The location map of the hot-wire anemometer sensor when measuring the flow mode in the boundary layer of the bottlenose dolphin: (1−7) is the location of the sensors, (8) is the location of the belt with equipment [165].

dolphin. The behavior of the tamed animals on which the equipment was worn did not differ much from the usual dolphin behavior.

Calculations showed that the swimming speed of a dolphin with a turbulent boundary layer will decrease slightly if it tows a cable 5 mm thick and 100 m long. These data were confirmed during experiments with telemetry equipment that was used in the experiments. Investigations have also shown that with proper training, a dolphin can become so accustomed to the cable that there are no noticeable differences when swimming with or without the cable.

In the experiments, telemetry equipment was used to measure the longitudinal component of the velocity pulsations in the boundary layer of marine animals and the instantaneous speed of their navigation. As a measuring instrument of the velocity pulsations in the apparatus, a hot-wire anemometer sensor with a platinum filament was used. The instantaneous speed sensor is the microheader Kherkheulidze X-6 [104].

The considered telemetry system consists of two devices: operational equipment and equipment for decryption and rewriting. Operational equipment is designed to convert the longitudinal component of the velocity pulsations in the boundary layer, the average velocity of the flow of the body into electrical signals and record them on a magnetic carrier. The decoding equipment serves to reproduce the signal recorded on a magnetic medium, extract from it and register on the oscilloscope a frequency-pulse-modulated signal of the average flow velocity of the animal, a signal of the longitudinal component of the velocity pulsations in the boundary layer, and obtain the low-frequency component of the square of the signal of the longitudinal component of the velocity pulsations.

A set of operational equipment consists of two devices for installing sensors on the body of an animal, a communication cable 100 m long, an electronic device, a tape recorder, and two sources of heating the filament. The device for installation of sensors has been developed in two versions: in the form of a platform with suction cups and in the form of a belt with a buckle (Fig. 6.1, 8). The heat source of the thread can be a rechargeable battery or a special power supply.

The wire line sensor of the hot-wire anemometer is connected with a communication cable through high-frequency filter chokes to the hot-wire anemometer unit. The power supply of the block is carried out from the thread heating source. In the block of the hot-wire anemometer, the variable component of the signal of the hot-wire anemometer sensor is allocated. Calibration of the channel of the hot-wire anemometer is performed by applying to the input of the amplifier with a voltage conversion frequency of 80 Hz, controlled by the measuring device and produced by the calibration signal generator.

When the animal dives, the hot-wire anemometer sensor may be above the water surface. In this case, the heated thread of the hot-wire anemometer burns out. The equipment provides protection to the filament against burnout when the sensors exit to the air. For this purpose, normally closed contacts of the relay controlled by the relay protection unit are included in the supply circuit of the hot-wire anemometer sensor. The input of the unit is connected to an electric microworm cell. When the speed sensor reaches the surface, the resistance of the electrolytic cell increases dramatically. The supply voltage of the cell increases accordingly, leading to the operation of the relay controlled by the unit. When the speed sensor is immersed in water, the resistance of the cell and, accordingly, the supply voltage, decreases to the same value and the relay releases the contacts. The speed sensor is always installed at the same level vertically or above the thread of the hot-wire anemometer sensor. Thus, when the hot-wire anemometer sensor reached the surface, the current through the thread is always turned off and preventing the thread from blowing.

Measurement of the average value of the voltage at the output of the calibration signal generator and control of the supply voltage of the blocks was carried out with a voltmeter of average values. The development of two types of devices for installing sensors on the body of marine animals is due to the difficulty of combining the basic requirements for devices into one design: the ability to quickly install sensors in any part of the animal's body and the reliability of sensor attachment at any swimming speed.

The best option to meet the first requirement is to install sensors on suction cups (Fig. 6.1, 1−7). Reliable sensor mounting is best provided with a belt. When developing devices for measuring the pulsation characteristics of the boundary layer of an animal, one has to solve a specific design problem, due, on the one hand, to the need to maintain exact constancy of the x and y coordinates at the measurement site at high swimming speeds, any flexural-vibrational movements of the animal, and variable pliability of the skin, and, on the other hand, the requirement of minimal surface deformation by an adjacent sensor.

The design of the developed device for installation of sensors on the body of the animal using a belt is given in Ref. [173]. The sensor sleeve of the hot-wire anemometer is streamlined and U-shaped. Inside the side parts of the clip there are grooves into which the U-shaped plug is inserted and pressed down with flat springs. A vertical U-shaped holder is attached to the front ends of the fork on two rubber stretch marks. A replaceable nozzle of the hot-wire anemometer sensor with inserted wedge-shaped electrodes located in parallel at a distance of 10 mm from one another is installed in the holder. Inside the insert, flexible insulated wires are terminated to the electrodes, terminating in single plugs. Platinum wire with a diameter of 30−50 μm was soldered to the beveled surfaces of wedge-shaped electrodes. A microrotating X-6 was placed in removable fairings. The position of the sensor around the circumference of the body was varied by moving the sensor along the belt strap. The buckle on the belt for attaching the apparatus (Fig. 6.1, 8) serves to fix the animal belt worn on the body and automatically drop it from the body when a certain force is applied to the communication cable. A special feature of the device for mounting sensors on the body of the animal in the version with cable connection is a mandatory device that clears the buckle when a strictly defined force is applied to the wedge, which is determined by the cut voltage of the calibrated wire. The system of automatic belt shedding works in two cases: (1) when the animal leaves the water area, the linear size of which is limited by the length of the communication cable; (2) if necessary, to terminate the experiment.

For fixing the belt on the animal in a fixed place during the deformation of his body while swimming in the middle of the long strap a shock absorber is installed. When dressing the belt, the ribbon and loop are stretched. During the deformation of the animal's body, leading to a decrease in its diameter, the tape assembles the strap into a loop, keeping the belt taut. The size of the belt can be adjusted by choosing the extra length of the straps through the slots in the buckle and the holder of the hot-wire anemometer sensor. The hot-wire anemometer sensor is attached to the belt or platform, which is mounted on the body of the dolphin with the help of special suction cups. The platform design allows it to be installed at any place on the body surface under study having a radius of curvature of over 20 cm. The combined signal recorded on magnetic media was played on a tape recorder and entered into a filter unit, in which the signal of the speed sensor and the hot-wire anemometer sensor are selected from low noise spurious modulation.

In 1966−69 experimental studies were carried out on dolphins 1.8−2.5 m long. It is known that the errors in vario-weight measurements of the degree of turbulence in the marine environment with the help of a wire line anemometer are very large. To improve the reliability of the data, a method for processing measurement results was developed. Instrumental recording of measurements is subdivided into measurements of one "weight," for each of which a graph of the dependence of the averaged degree of turbulence $\bar{\varepsilon}_i$ on the speed of swimming is plotted. In this case, the measurements of one "weight" are short continuous records of the degree of turbulence ε_i. The scatter of points on the dependency graph for measuring each "weight" is very small. As a result, the record quite accurately reflects the dependence of the degree of turbulence ε on speed. In the range of measured speeds, the value is related to the actual value of the degree of turbulence $\bar{\varepsilon}_i$ by the following relation:

$$\bar{\varepsilon}_i = k_i \cdot \varepsilon = \frac{\bar{\varepsilon}_{io}}{\varepsilon_o} \cdot \varepsilon, \tag{6.1}$$

where k_i is some constant coefficient for each "weight" of measurements; and the "o" index indicates the degree of turbulence defined for the dolphin's swimming speed, which is most characteristic of these measurements. The actual value of the degree of turbulence ε_o is approximately equal to the average value of the quantity ε_i. Due to the large number of measurements at a characteristic speed, ε_o is determined with a sufficient degree of accuracy. The absolute value of the degree of turbulence for each weight of measurements was determined by the formula:

$$\bar{\varepsilon}_i = k_i \bar{\varepsilon}_i \tag{6.2}$$

Normalizing the absolute values with regard to formula (6.2), we obtain the resulting averaged graph that displays, with the required degree of accuracy, the dependence of the actual degree of turbulence ε on the navigation speed U_∞ (Fig. 6.2). The dependence of the degree of turbulence ε on the average dolphin swimming speed was constructed for cases of accelerated, uniform, and slow movements with speed fluctuations whose frequency was 0.14− 0.3 Hz (Fig. 6.2) [165]. The hot-wire anemometer sensor was installed on the side of the dolphin in front of the dorsal fin. The x and y coordinates of the measurement point were respectively 900 mm from the tip of the nose and 1 mm from the body surface ($x/L = 0.5$; $L = 1.8$ m). The degree of free turbulence in the water area was 0.5%.

The degree of turbulence in the boundary layer with uniform, accelerated, and slow motion of the dolphin are in the same range of variation of experimental points. This indicates that the flow regime in the boundary layer of a dolphin depends little on the periodic nonstationarity with a frequency of 0.14−0.3 Hz. The transition of a laminar flow to a turbulent one in the boundary layer of a dolphin and a rigid model begins at approximately the same Reynolds numbers Re_x.

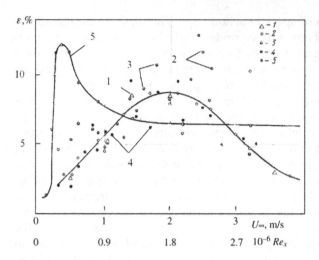

FIGURE 6.2 Dependence of the degree of turbulence in the boundary layer of a dolphin on the averaged swimming speed and Reynolds number: (1) measurements in 1967, (2) acceleration, (3) uniform swimming, (4) braking (measurement in 1969), (5) rigid axisymmetric body of rotation [170].

However, the length of the transition region on a dolphin is much wider than on a rigid model. In addition, with $Re_x > 2.7 \cdot 10^6$, the dolphin shows a significant decrease in the level of turbulent pulsations almost to the level of the laminar flow regime, while on a solid model the corresponding value decreases slightly. However, it can be noted that in some cases, a high level of turbulent pulsations is maintained at high Reynolds numbers. This circumstance, apparently, is connected with the fact that a dolphin can include a mechanism for controlling turbulent pulsations at vital moments of swimming. We also note a good agreement on the graph of the results of the corresponding measurements performed in 1967 and 1969. Thus, dolphins have mechanisms that allow them to tighten the transitional flow regime in the boundary layer, and in a developed turbulent boundary layer to significantly reduce the level of turbulence. With increasing swimming speed, regardless of the mode of movement, the degree of turbulence in the boundary layer of the dolphin decreases. This indicates the mechanisms of the dolphin's body systems that contribute to the reduction of friction resistance.

Below are the results of experimental studies of instantaneous swimming speed and flow pattern in the bottlenose dolphin. The studies were carried out in the sea on a living dolphin 2.45 m long. The coordinates of the boundary layer x and y at the measurement site were 1250 and 1 mm, respectively, according to Fig. 6.1, point 5. The error in the velocity measurements did not exceed 4%, and the degree of turbulence was 10%. The measurements were carried out on the straight part of the trajectory of the animal. The motion length of the bottlenose dolphin was 75 m.

Fig. 6.3A presents a typical plot of the dependence of the instant swimming speed on the time of a bottlenose dolphin. Dependence built by interpolation speed values were determined by the waveform every 0.1 seconds. A more frequent measurement of speed, as shown by the test processing, negligibly refines the constructed curve. Fig. 6.3B shows a graph of similar dependence, built on a compressed time scale by averaging the instantaneous velocity over time intervals of 0.2 seconds. The graph allowed us to analyze the swimming speed of a dolphin for a long period of time. From the above graphs, it follows

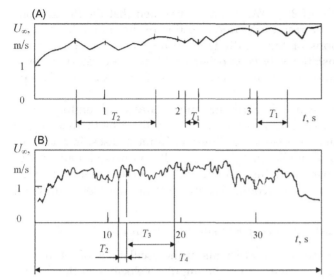

FIGURE 6.3 Dependence of swimming speed of a bottlenose dolphin on time: (A) T_1 and T_2 are periods of speed fluctuation of the first and second types; (B) T_2, T_3, T_4 are periods of oscillation of the speed of the second, third, and fourth types [166].

that the swimming of a dolphin is distinguished by a substantial nonstationarity of a very complex nature with a number of regular patterns of the periodic type. In Fig. 6.3A, the oscillations of the velocity of the first type with a period T_1 are highlighted. The frequency of these oscillations varies within 1.7–5.6 Hz. Speed modulation factor, defined by the formula $m = \Delta U / U_{av}$, where ΔU is the amplitude of the velocity oscillations, U_{av} is the average velocity over a series of oscillation periods, was 5–10%.

The fluctuations of the velocity of the second type with a period T_2 are shown in Fig. 6.3. Their frequency is 1-2.5 Hz, and the modulation coefficient of oscillations is 20-30%. The regions of inhibition and acceleration of these oscillations are one or several oscillations of the first type. Speed fluctuations of the first type can be explained by various reasons. The frequencies and amplitudes of oscillations correspond to the low-frequency components of the velocity pulsations in the boundary layer of the dolphin, which can be recorded by a low-inertia micro-rotator. Indeed, at low swimming speeds of the bottlenose dolphins with a low intensity of turbulence in the boundary layer, oscillations of the velocity of the first type are not observed. The oscillation frequencies of the second type correspond to the double oscillation frequency of the caudal stem, which is 0.8-3 Hz for a dolphin. From the synchronous graphs of the dependences of swimming speed and movement of the tail of the bottlenose dolphin on time given in Ref. [183], it also follows that the oscillation frequency of the tail stem, equal to 3.5 Hz, takes place at a low swimming speed (0.5 m/s). However, according to Ref. [243] in the case shown in Fig. 6.3A, the oscillation frequency of the tail stem should not exceed 0.6 Hz. The frequency of oscillations of the velocity of the second type, in turn, as it is seen from the graphs, does not exceed 1.2 Hz. Thus, it can be argued that the fluctuations of the velocity of the second type are more likely due to fluctuations in the tail stem of the bottlenose dolphin.

Fig. 6.3B shows the periods of oscillations of the velocity of the third and fourth types T_3 and T_4. The frequency range of the oscillations of the speed of the third type is

0.14–0.3 Hz with a modulation factor of 20–30%. It can be assumed that the fluctuations of the speed of the third type arise as a result of periodic resting of the animal in the process of swimming. Speed fluctuations of the fourth type correspond to frequencies of 0.02–0.07 Hz. Apparently, such nonstationarity is associated with the respiratory cycle of the bottlenose dolphin. The modulation factor in the oscillations of the speed of this type is 50–60%.

Periodic changes in speed are approximately harmonic only in areas with uniform averaged speed. When driving with the acceleration of the duration of the growth areas (t_{grow}) and decay (t_{decay}), the speed and their steepness are different. In most cases, in accelerating areas $t_{grow} > t_{decay}$, and in areas of inhibition $t_{grow} < t_{decay}$. If you make an attempt to approximate the speed curve for a part of the movement with a uniform averaged speed by analytical dependence, then, taking into account the principle of superposition, it can be written as follows:

$$U_\infty = U_{cp}(1 + m_1\sin \omega_1 t + m_2\sin \omega_2 t + m_3\sin \omega_3 t + m_4\sin \omega_4 t), \qquad (6.3)$$

where m_1, m_2, m_3, and m_4 are the modulation coefficients in the oscillations of the first to fourth types, respectively, and ω_1, ω_2, ω_3, and ω_4 are the angular velocities in the oscillations of the corresponding types. Consequently, the nonstationary movement of the dolphin can be divided into two types: nonstationarity, due to the work of the tail stem, and nonstationarity, associated with the axial movement of the dolphin.

According to the measurement results, the dependence of the degree of turbulence on the sign of acceleration in various types of oscillations of the dolphin swimming speed was analyzed. As a result of the research, it has been established that the main influence on the flow regime is the nonstationarity in the form of oscillations of the second type, having a frequency of 1–2.5 Hz and corresponding, apparently, oscillations of the tail stem of the dolphin. It was also established that the degree of turbulence in the boundary layer of the bottlenose dolphin reaches a maximum in the acceleration regions and decreases to a minimum in the deceleration regions in these oscillations. The probability of this process is not less than 0.9. At the same time, a certain phase shift was detected between fluctuations of the level of the degree of turbulence and oscillations of the velocity of the second type. Fig. 6.4 shows the dependences of the maximum (1) and minimum (2) levels of the degree of turbulence in accelerated (3), uniform (4), and slowed (5) dolphin swimming areas with variations of the velocity of the third type. As follows from the graphs, nonstationarity in the form of oscillations of the velocity of the third type does not significantly affect the degree of turbulence in the studied range of Reynolds numbers. The flow regime corresponding to the extreme values of the degree of turbulence is transitional from laminar to turbulent in the range $Re = 10^5 - 2.3 \cdot 10^6$ for minima and $Re = 10^5 - 2 \cdot 10^6$ for maxima of the degree of turbulence. At high Reynolds numbers, the flow regime is turbulent.

The hydrodynamic resistance of the body of revolution in the unsteady flow regime [267] showed that the friction resistance decreases with deceleration and increases with acceleration compared to the similar resistance with a stationary flow around. These results were confirmed experimentally. As is known, friction resistance obeys the law:

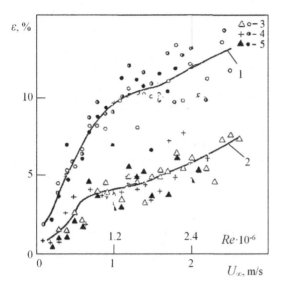

FIGURE 6.4 The dependence of the maximum and minimum levels of the degree of turbulence of the longitudinal pulsation velocity in the boundary layer of the bottlenose dolphin on swimming speed with different accelerations in the oscillations of the velocity of the second type and with different nonstationarity of the third type (explained in the text) [166].

$$R \sim \overline{u'v}' \, ,$$

where $\overline{u}'$ and $\overline{v}'$ are, respectively, the longitudinal and transverse components of the velocity pulsations in the boundary layer.

It is also known that the energy source of the transverse component is the longitudinal component. In Ref. [267], it was shown that the drag coefficient during unsteady motion is proportional to the number $N = \left[\left(\frac{dv(t)}{dt}\right)L\right]/v^2(t)$, where v is the instantaneous velocity of flow, L is the body length. Acceleration in the oscillations of the velocity of the second type, having a frequency of 1–2.5 Hz, is three or more times faster than the acceleration in oscillations of the velocity of the third and fourth types. In accordance with the foregoing, their influence on friction resistance and, apparently, on velocity pulsations is much less. This conclusion agrees well with these experiments. Consequently, the fluctuation of the mean-square value of the velocity pulsations in the boundary layer of the bottlenose dolphin is due to the nonstationary nature of the dolphin's swimming.

Fig. 6.5 shows the dependences of the averaged degree of turbulence of the longitudinal component of the velocity pulsations $\varepsilon_{\tilde{n}\partial} = \left(\sqrt{\overline{u^2_{\tilde{n}\partial}}}\right)/U_\infty$ in the boundary layer of the bottlenose dolphin on the mean swimming speed and the sign of acceleration in the third-type velocity oscillations at different x/L values of the boundary layer.

The results shown in Fig. 6.5 illustrate a good agreement with the results of experiments obtained on similar-sized animals in 1966, 1970, and 1971. This indicates sufficient reliability of the experimental data obtained. Fig. 6.5 clearly shows the statistical nature of the dependence due to fluctuations in the degree of turbulence due to the biological activity of the animal. The main factors influencing the last magnitude on the velocity pulsations are as follows. The biological activity of the bottlenose dolphin leads to different configurations of its body during movement, to different trajectories of movement, to approaching

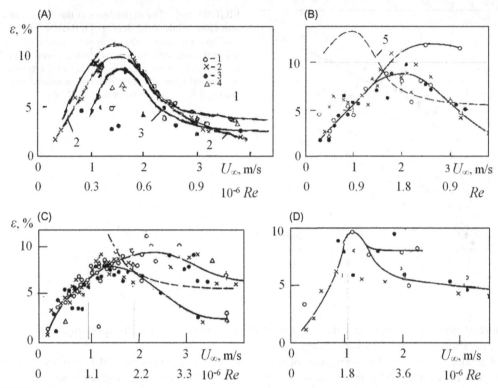

FIGURE 6.5   The dependence $\varepsilon = f(Re_x)$ at $x/L = 0.125$ (A), $x/L = 0.46$ (B), $x/L = 0.52$ (C), $x/L = 0.75$ (D): (1) acceleration, (2) uniform motion, (3) deceleration, (4) uniform motion (measurements 1966–68), (5) the bottlenose dolphin model [173].

and moving away from the free surface of water and the walls of a closed water area, to swimming in a semisubmerged position, to an arbitrary intersection of areas with different intensities of free turbulence (including own trail in various stages of degeneration). All this has a different effect on the velocity pulsations in the boundary layer. The biological activity of the bottlenose dolphin manifests itself in the form of an arbitrary change in the amplitude-frequency characteristic of the flexural-vibrational movements of the tail stem of the animal, which can lead to a significant change in the pulsation characteristics of the boundary layer [265,266]. In addition, with the biological activity of an animal, one of the hypothetical drag reduction mechanisms can be put into operation at an arbitrary time.

All values of ε_{av} in Fig. 6.5 are grouped around two dependency curves. The lower curves describe the phenomenon of a simultaneous decrease in the maxima and minima of oscillations of the degree of turbulence $\varepsilon_{av} = \left(\sqrt{\overline{u_{n\partial}^2}} \right) / U_\infty$. This phenomenon was observed in the bottlenose dolphin in the following cases.

1. On the areas of oscillations of the third type with a uniform speed of movement, containing oscillations of the second type with a small modulation coefficient. Such

sites were very rare. The probability of this phenomenon in these areas does not exceed 0.5.

2. On the areas of oscillations of the third type with a relatively small uniform swimming speed (up to 2.5 m/s), filled with oscillations of the velocity of the second type of high frequency. Such sites were also very rare. The probability of reducing the degree of turbulence in these areas is less than 0.5.
3. At high swimming speeds achieved in oscillations of the third type, including artificially stimulated fast-swimming regimes described above. The probability of decreasing extremes in this case is highest at maximum in the duration of the third type of velocity oscillations. In general, the probability of a phenomenon increases from 0.2 to 0.6 when the average navigation speed changes from 1.5 to 4 m/s.

The analysis of the oscillograms and velocity graphs shows that the phenomenon of a simultaneous decrease in the maxima and minima of the degree of turbulence ε takes place in parts of the third-type velocity oscillations, characterized by an increase in the motor activity of the animal. In the boundary layer of the bottlenose dolphins, oscillations of the degree of turbulence are also observed, in which areas with a minimum level of ε last for a long time, capturing several periods of oscillations of the second type. The level of the degree of turbulence in these minima, as a rule, approaches the level of ε with a laminar flow regime in the boundary layer. The points corresponding to these values of ε are also grouped around the lower curve. These oscillations of the degree of turbulence were observed in the following cases.

1. With a sharp inhibition in the oscillations of the speed of the second and third types. In this case, a minimum of the degree of turbulence included a section of abrupt deceleration in speed oscillations and captured subsequent acceleration and deceleration areas. Such cases were observed at average speeds of a bottlenose dolphin of 0.6–2.0 m/s. A long minimum of ε occurred in all investigated cases of sudden inhibition.
2. With prolonged inhibition in oscillations of the third type. At the same time, the minimum of the degree of turbulence fell on the stagnation section and dragged on to the acceleration section in velocity oscillations of the third type. The probability of occurrence of a long minimum degree of turbulence in the second case is less than one.
3. On the deceleration regions in oscillations of the third type with a preceding sharp acceleration. Such patterns of change in the speed of the bottlenose dolphin in oscillations of the third type are quite common. The probability of occurrence of a long minimum in these cases is less than 0.5.

6.2 The flow regime in the boundary layer of the dolphin model

To check the reliability of the results obtained in Section 6.1, similar measurements were made on a dolphin model with a hard surface. Pyatetsky created a plaster model of the body shape of a recently deceased 2-year-old dolphin. Two plaster half-molds of the dolphin's body were made. Then, in this form, the dolphin's body was cast, which was carefully measured and corrected so that it resulted in a symmetrical body. A dolphin

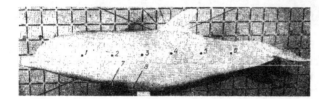

FIGURE 6.6 General view of the dolphin model with an indication of the measurement points of the degree of turbulence in its boundary layer at different x/L values: (1) 0.23; (2) 0.34; (3) 0.46; (4) 0.57; (5) 0.69; (6) 0.8; (7) 0.33; (8) 0.43 [36,37,46,246].

model with a smooth surface, reflecting the shape of the body of a live dolphin, was made on this body. The material of the model is epoxy resin reinforced with several layers of fiberglass fabric in such a way that the thinnest wall thickness of the model is 8–10 mm. Inside the model, there are several metal longitudinal stiffeners to which a steel plate extending outward is attached, which is intended to fasten the model with a profiled bracket to the towing carriage. The model is hollow, and made without pectoral fins. The nasal extremity and tail fin are made in one piece. To eliminate underwater positive buoyancy in the body of the model there are several drainage holes with a diameter of 5 mm to fill it with water. To ensure a technically smooth surface, the model was plastered from the outside, cleaned, covered with primer, and painted with white nitrokraiding with a spray gun in several layers. Fig. 6.6 shows a photograph of the model under study on the background of the coordinate grid. The main dimensions of the model are $L \times B \times H = 1750 \times 270 \times 380$ mm (H does not take into account the height of the dorsal fin).

Investigation into the flow regime on the model was performed by measuring the degree of turbulence (the longitudinal averaged component of the pulsation velocity) using a hot-wire anemometer sensor. The sensor was placed in the boundary layer of the model at points 1–8 (compare with Fig. 6.1) at a distance of 1 mm from its surface. The x coordinate of the measurement points given in Fig. 6.6 was for the point (1) $x = 0.4$ m, (2) 0.6 m, (3) 0.8 m, (4) 1 m, (5) 1.2 m, (6) 1.4 m, (7) 0.58 m, (8) 0.75 m. The model length was $L = 1.75$ m.

To measure the longitudinal averaged component of the pulsation velocity, an apparatus was used that operates according to the wire hot-wire anemometer with a constant amperage. The main fundamental features of the equipment are described in Ref. [164]. The magnitude of the longitudinal averaged component of the pulsation velocity in the turbulent flow regime was determined from the output voltage of the thermal anemometer circuit, using the following formula:

$$\varepsilon = k' \frac{\sqrt{U_\infty}}{I^3}(1 + \gamma)\left(\frac{y}{\delta}\right)^{3/4}\frac{c}{\beta}\sqrt{\overline{u^2}}, \tag{6.4}$$

where k' is the conversion coefficient of the hot-wire anemometer sensor, equal to 13.95 ± 0.1 mV^{-1} a^3 (m/s)$^{-1/2}$; U_∞ is the flow speed of the dolphin model at the measurement point, m/s; I is the heating current of the hot-wire anemometer sensor wire; γ is the coefficient of approximation obtained in Ref. [164]; y is the coordinate of the boundary layer; δ is the thickness of the boundary layer (m); c is the Scramsted correction factor, taking into account the commensurability of the working wire length and the integral scale of turbulence; β is the transfer coefficient of the hot-wire anemometer; and u is the output voltage of the hot-wire anemometer circuit (mV).

The measurements of the longitudinal averaged component of the pulsation velocity in the boundary layer of the dolphin model were carried out in a high-speed towing tank. The water depth was 3 m; the model was buried at 1.5 m. The model was fixed with a profiled bracket to a dynamometer trolley, which was set in motion using a towing system from a DC motor installed in the head part of the pool. The average speed of the trolley with a model attached to it in the working area of its movement was determined automatically by an electric stopwatch, and the instantaneous speed of movement was calculated according to the speed indicator of the drive wheel of the starting carriage, the signals of which were recorded on the oscilloscope film simultaneously with the time stamp and the readings of other sensors. The readings of the hot-wire anemometer mounted on the model being tested were transmitted via a flexible coaxial cable, one end of which was connected to the sensor and the other end to the recording equipment. The degree of

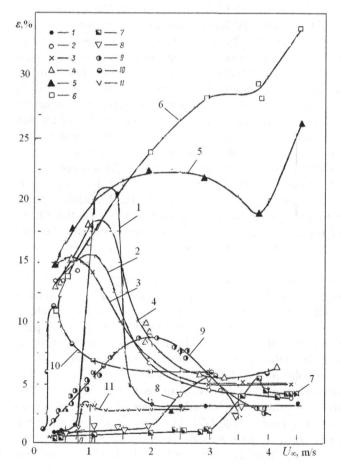

FIGURE 6.7 The dependence of the degree of turbulence ε in the boundary layer of various bodies on the averaged flow velocity. Measurement points: (1–8) dolphin model [246], (9) live dolphin [247] $x/L = 0.5$, $L = 1.8$ m, (10) model of a body of revolution [111], (11) ship model (qualitative measurements) [170], and L, T are the points of the beginning and end of the transition for (11 [36,37,46]).

initial turbulence when towing the model in the basin was 1.6% and was mainly due to the vibration of the model bracket during the trolley movement on the rails.

The results of measurements of the degree of turbulence in the boundary layer of the dolphin model are presented in Fig. 6.7. It also presents measurement data on the degree of turbulence in the boundary layer of a live dolphin (Section 6.1), as well as the value of the degree of turbulence in the boundary layer of a model of a body of revolution [164] and a model of a ship with areas of transition and transition zones at different flow rates [111]. Curve (1) corresponds to $x/L = 0.23$, (2) 0.34, (3) 0.46, (4) 057, (5) 0.69, (6) 0.8, (7) 0.33, (8) 043.

Comparison of these dependences shows that the flow regime in the boundary layer on a live dolphin and its model changes from laminar to transitional and turbulent with increasing speed. The degree of turbulence (the longitudinal averaged component of the pulsation velocity) in a developed turbulent boundary layer of a live dolphin is sometimes lower than in the boundary layer of a rigid dolphin model. Curve (10) is obtained by testing a longitudinal cylinder with an lively nose section. Curve (1)was obtained with a similar measurement on the dolphin model, which corresponds to the computational model No. 32 [9] (Section 2.3, Fig. 2.15). The dolphin model has a streamlined shape; therefore, the shape of the transition curve ε in the boundary layer shifts toward higher velocities. Curves (2−4) are obtained for large values of x/L. In accordance with the distribution of load parameters and, in particular, the distribution of pressure along the body, the maxima of these curves decrease, and the curves themselves become more stretched. Curves (5, 6) indicate the location of the sensor in the region of the separation of the boundary layer of the dolphin model. In the live dolphin in this area there is no separation in accordance with the oscillations of the tail part of the body and the sucking power of the tail propulsion unit. Curves (7, 8) are obtained when the sensor is located at the bottom of the model, where there is a negative pressure distribution. Therefore, curves (7, 8) have a gentle form, low values of ε, and are shifted toward high flow velocities. When comparing curve (9) with a similar value of x/L curves (3, 4), it becomes obvious that the distribution and the value of ε for a live dolphin are significantly less than for its rigid model. A peculiarity of the flow regime in the boundary layer of the dolphin and its model is the almost invariant location of the maxima of the ε (U_∞) curves in the case when the flow regime measurements were made in areas located on the body of the test object in the zone to the dorsal fin inclusive.

The high level of initial turbulence in testing the dolphin model suggests that the relative coordinates of the neutral point x_n/L for the plate, elliptical profiles, and the laminarized profile close to the dolphin profile [126,127], coincide with the relative coordinate of the transition point x_{tr}/L [256]. Then from Fig. 6.8 it follows that the profile of the dolphin model is significantly more laminarized than the plate and elliptical profile, with a ratio of axes equal to the relative thickness of the dolphin, and less laminarized than the laminarized profile close to the profile of the dolphin [126,127]. It also follows from Fig. 6.8 that the live dolphin has longer laminar and transitional areas than its model. However, the transition of a laminar flow to a turbulent one in the boundary layer of a dolphin occurs much earlier than on a theoretical laminar profile [126,127]. The flow around the dolphin model in the investigated speed range from 0.3 m/s and above is detachable beyond the point with relative coordinate $x/L = 0.67$. This is evidenced by both the continuous

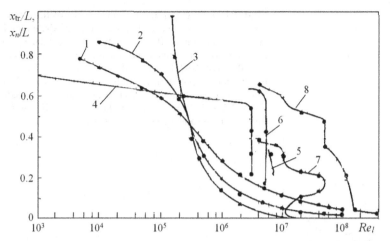

FIGURE 6.8 Relative coordinates of the neutral point x_n/L and transition points x_{tr}/L when the laminar mode transitions to turbulent on profiles of different bodies: (1) elliptical cylinder with $a/b = 4$ (x_n/L), (2) elliptical cylinder with $a/b = 8$ (x_n/L), (3) plate (x_n/L), (4) dolphin model: lateral surface (x_n/L); (5) the same but for the ventral surface (x_n/L); (6) live dolphin (bottlenose dolphin): lateral surface (x_n/L); (7) position of the neutral point on the profile close to the profile of the dolphin [126], (8) position of the transition point on the same profile [36,37,46,127].

increase in the degree of turbulence ε at measurement points (5) and (6), reaching high speeds of considerable magnitude (Fig. 6.7), and the type of oscillograms of velocity pulsations characteristic of a separated flow. In the live dolphin, as shown by our measurements, at all points studied along the body length (up to $x/L = 0.75$ inclusive), with the observed maximum average speeds (3 m/s), an unseparated flow takes place.

The flow around the dolphin model in the investigated speed range from 0.3 m/s and above is detachable beyond the point with relative coordinate $x/L = 0.67$. This is evidenced by both the continuous increase in the degree of turbulence ε at measurement points (5) and (6), reaching high speeds of considerable magnitude (Fig. 6.7), and the type of oscillograms of velocity pulsations characteristic of separated flow. On our live dolphin, as our measurements showed, at all points studied along the body length (up to $x/L = 0.75$ inclusive), with the observed maximum average speeds (3 m/s) there is unseparated flow around the body. Thus, the shape of the body of the dolphin has a certain value in reducing its hydrodynamic resistance. At the same time, the live dolphin has additional capabilities that allow it to prevent the separation of the boundary layer.

In Ref. [247], the results of studies performed on a 2.8 m long Black Sea bottlenose dolphin (Fig. 6.9) are presented. As in Fig. 6.7, a comparison was made of the results of measurements of ε along the length of the body of a live dolphin and its rigid model. The error of the $\varepsilon_{aver.}$ values found in the experiments did not exceed 18% at a confidence level of 0.95. Measurements were made of the averaged velocity and the degree of turbulence in the boundary layer. In all cases (except point 1), they were performed on the lateral (lateral) side surface of the animal's body.

Based on the results of the experimental investigations performed in the flow regime in the near-wall area of the boundary layer of the bottlenose dolphin and its rigid model, we can draw the following conclusions. The flow around a dolphin's body and its rigid model

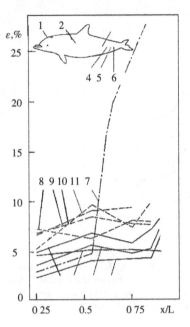

FIGURE 6.9   Dependence of the degree of turbulence in the boundary layer of the bottlenose dolphin and its model on the longitudinal relative x/L coordinate, flow velocity, and character of movement: (1–6) are points on the animal's body with relative abscissas $x/L = 0.22$, 0.56, 0.75, 0.8, 0.84, 0.89, in which the measurements were made, (7) experimental data obtained on the dolphin model at $U = 4.7$ m/s, (8–11) data from the passive dolphin swimming mode at speeds of movement $U = 1$; 2.7; 2. 5; 3.3 m/s; (12–15) data of the active swimming mode of a dolphin at the same speeds [36,37,46,247].

to the coordinate $x/L = 0.5$–0.55 is nonseparable and is characterized by a degree of turbulence varying within $\varepsilon \le 10\%$. On the dolphin's body model, a separation flow is observed at $x/L > 0.55$ (behind the mid-section), and nonseparation flow on the animal body, a continuous flow around ($\varepsilon = 5$–10%) persists along the lateral surface of the stem up to the blade of the caudal fin (for $x/L > 0.89$ experiments were not performed).

The degree of turbulence in the boundary of the bottlenose dolphin during active swimming is significantly reduced compared to its inertial motion along the whole body; at the same time as the swimming speed increases, this difference becomes more significant.

Fig. 2.15 (Section 2.3) shows calculations of the extrusion thickness δ^* in the case of the laminar boundary layer model of model No. 32 [9], which coincides well with the shape of the dolphin during inertia. When wrapped around a rigid plate, δ^* is about one-third of the thickness of the boundary layer δ. According to the data in Ref. [9], it is possible to calculate the distribution of δ along the body having the contours of model No. 32. In Ref. [170], the calculation of the coordinate y is given, on which the hot-wire anemometer sensor should be placed when testing the ship models. The calculation is made for the thickness of the boundary layer on a rigid plate. The coordinate y is calculated by the formula:

$$y = (0.05 - 0.8)\delta = (0.02 - 0.3)x\, Re^{-0.2}, \tag{6.5}$$

where x is the distance from the hot-wire anemometer sensor to the nose edge of the ship model. Calculations showed that for $x = 0.1$–2 m, $U_\infty = 0.5$–3 m/s, $\nu = 1.14 \cdot 10^{-6}\, \mathrm{m^2/s}$, the value is $y = 0.4$–1.2 mm.

In Ref. [111], the shapes of profiles of averaged velocities and longitudinal pulsation velocity are presented depending on the value of U_∞. The results show that, depending on the transition stage in the boundary layer, the value of the location of the coordinate y

at which the maximum of the longitudinal pulsation velocity is found varies. For example, in the laminar flow regime in the boundary layer, at the beginning of the body the maximum pulsation velocities are in the region of the displacement thickness δ^*, and as the velocity of the main flow increases, the thickness δ increases, and the y coordinate of the pulsation velocity maxima decreases, with a turbulent boundary layer this value does not exceed 0.2 δ^*. In dolphins, the characteristics of the boundary layer differ from the corresponding characteristics in the flow around a rigid surface. In dolphins, the boundary layer is quasilaminar.

Therefore the constant value y of fixing the hot-wire anemometer sensor led to the fact that the value ε_i was measured at different thickness of the boundary layer. Therefore, in fact, in experiments, we measured not the maximum values of the longitudinal pulsation velocity, but arbitrary values of ε_i, which leads to a significant scatter of the values of ε_i at the measurement points. It is important to compare the measured values on a live dolphin and its solid model (Figs. 6.7 and 6.9).

6.3 Models of elastic surfaces

The specific properties of dolphin skin are one of the most important factors influencing the stabilization of the boundary layer during their swimming. The structure and properties of dolphin skin are similar to polymeric materials. Polymers are characterized by large strains at low stresses, a significant dependence of stress on the time of the force and on the strain rate, a sharp dependence of the whole complex of mechanical properties on temperature, and anisotropy of the properties of polymers by volume. The physical and mechanical properties of polymers are divided into three main groups: mechanical, physical, and technological. Some physical properties of the dolphin skin are discussed in Section 5.4. Comparing the structure of polymers and dolphin skin, from the whole variety of mechanical properties, we choose the basic, necessary at the first stage of studying the skin: distribution functions $E(\tau)$ and $J(\tau)$, elastic modulus $E = \sigma/\varepsilon$, compliance $J = 1/E$, phase angle θ, the viscosity of the polymer material $\eta = E/\omega$, and the speed of propagation of surface waves c_m. In the expressions for the distribution functions, the quantity τ is a constant, called the time or relaxation period, $\tau = \eta/E$, s; in terms of viscosity, ω is the limiting frequency of oscillation.

The most illustrative properties of polymers are illustrated by mechanical models, the schemes of which are shown in Fig. 6.10. Neglecting, in the first approximation, the relaxation nature of high elasticity and assuming that the viscous flow obeys Newton's law, the dependence of stress and strain is obtained from the differential equation:

$$\frac{d\varepsilon}{dt} = \frac{1}{E} \cdot \frac{d\sigma}{dt} + \frac{\sigma}{\eta} \tag{6.6}$$

where ε is the relative deformation; t is time (seconds); E is modulus of elasticity (N/m^2); σ is tension (N/m^2); and η is viscosity (N $\cdot$ s/m^2). The mechanical model corresponding to expression (6.6) was proposed by Maxwell. The circuit of this model is circled in Fig. 6.10A by a dash-dotted line. The model taking into account the relaxation properties

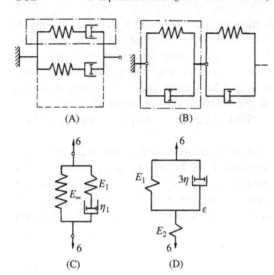

FIGURE 6.10 Mechanical models of polymeric materials: (A) parallel connection of Maxwell elements, (B) serial connection of Voigt elements (Kelvin), (C) the Dogadkin–Bartenev model, (D) the standard three-element model.

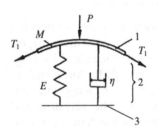

FIGURE 6.11 The scheme of a simplified mechanical model of the skin of dolphins [44].

was developed by Voigt-Kelvin (in Fig. 6.10B, it is outlined with a dash-dotted line). The differential equation of motion of this model is:

$$\eta \cdot \frac{d\varepsilon}{dt} + E \cdot \varepsilon = \sigma \tag{6.7}$$

All mechanical schemes can be reduced to two forms [44,70]. The first of them (Fig. 6.10A) consists of Maxwell elements connected in parallel, the second (Fig. 6.10B) consists of Voigt-Kelvin elements connected in series. These models characterize the work of homogeneous polymers. More information on various models of elastic-damping materials is given in Refs. [44,70,170,187,212,239,249,260].

The mechanical model of the skin of dolphins is more complex. Fig. 4.24 (Section 4.3) shows a diagram of the structure of the dolphin's skin. A simplified mechanical model of the skin element, taking into account the work of only the first and second layers of the skin, is presented in Fig. 6.11, where (1) is a membrane with an oscillating mass M $(kg \cdot s^2/m^3)$, corresponding to the first layer of the skin; (2) is the spring stiffness (modulus of elasticity) and dampening value (viscosity), corresponding to the second layer of skin; (3) is a rigid substrate corresponding to the third layer of skin; T is the tension in the membrane (N/m^2); and P is the surface pressure (N/m^2). In this case, we neglect the physic mechanical properties of the substrate, as well as the bending rigidity of the first and second layers; the oscillating mass of the second layer is not taken into account; the dampening of the second layer is assumed to be linear with respect to the ejection speed; we consider only the first three layers with the material of the skin as being isotropic. The differential equation of motion of a single element of such a model is written in the following form:

$$T\frac{d^2\varepsilon_y}{dx^2} - M\frac{d^2\varepsilon_y}{dt^2} - \eta\frac{d\varepsilon_y}{dt} - E'\varepsilon_y = P, \tag{6.8}$$

where ε_y is the displacement of the membrane in the direction of the pressure (minutes); x is the membrane displacement along the skin surface (m); η is the dampening or viscosity of the skin (H $\cdot$ s/m^3); E' is the cover stiffness, $E' = E/t$ (N/m^3); and t is the thickness of the skin.

Eq. (6.8) takes into account only the displacement of the skin element that is normal to the skin surface, since the tangential displacement has little effect on the stability of the laminar boundary layer [85]. Despite the simplifications noted above, a flexible coating that has the mechanical model considered allows us to significantly increase the Reynolds number of buckling [170]. Measurements of elasticity and other parameters of dolphin skin [44] showed that not only the first two layers of the skin, but also the underlying layers are important for stabilizing the boundary layer. According to the dolphin skin structure, the work of individual layers can be represented schematically as follows (Fig. 6.12): layers (1) and (2) are a membrane with a mass M_1 and a tension T_1 resting on pillars with elastic modulus E_1 inclined to its base at an angle [50]. Therefore the deformation of the skin should take into account the bending stress columns. Dampening is due to viscous flow of blood and lymph and the capillaries of layer (2) into the vessels of layer (3), therefore layer (2) is also inherently dampening or viscosityη. Layer (3) with horizontal elastic fibers, collagen fibers, and blood vessels can be represented as a membrane with a mass of M_2, tension T_2, and elasticity E_2. Layer (3) also has some dampening. Given the

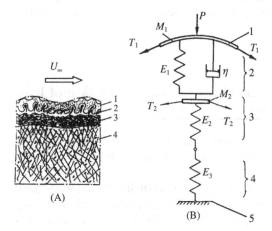

FIGURE 6.12 The structure of the upper layers of the dolphin skin and the mechanical model of the skin [50].

structure of the circulatory system of the skin, the amount of dampening can be neglected. The two membranes rest on an elastic base (layer 4) with an elastic modulus of E_3. It should be noted that the tension T_2 must be greater than the tension T_1, according to the morphological structure of these layers. The mass M_2 is a variable quantity depending on the amplitude of the disturbing motion and the magnitude of the current pressure on the surface of the skin. The hinges shown in Fig. 6.12 indicate that the bending stress of layers (3) and (4) can be neglected due to the smallness of P.

In a simplified form, the differential equation of motion of a mechanical model of a dolphin's skin can be written as follows:

$$(T_1 + T_2)\frac{d^2\varepsilon_y}{dx^2} - (M_1 + M-_2)\frac{d^2\varepsilon_y}{dt^2} - \eta\frac{d\varepsilon}{dt} - \left(E_1' + E_2' + E_3'\right)\varepsilon_y = P - \sigma_{\text{bend}} \qquad (6.9)$$

The coefficients in Eq. (6.8) are dimensional and constant values. Applying the characteristic values for the boundary layer (Re and Ca numbers, boundary layer thickness δ; speed U_∞ m/s, time δ/U_∞ seconds; pressure ρU_∞^2, N/m^2), the coefficients of Eq. (6.8) will be obtained in a dimensionless form. Thus, we write down the dimensionless parameters of mass, tension, elasticity, limiting frequency, and dampening necessary for modeling the phenomenon under consideration:

$$K_M = \frac{Re_1}{\rho}M_\Sigma, \; k_T = \frac{T_\Sigma}{\mu U_\infty}, \; k_E = \frac{E_\Sigma'}{\mu U_\infty Re_1^2}, \; k_\omega = \left(\frac{E_\Sigma'}{M_\Sigma}\right)^{1/2}\frac{1}{U_\infty Re_1}, \; k_\eta = 2\zeta = \frac{\Delta}{\pi},$$

where Re_1 is the unit Reynolds number, $Re_1 = U_\infty/\nu$ (m); ρ is water density (kg/m^3); μ is the dynamic viscosity of water (N $\cdot$ s/m^2); ζ is the dampening factor; and Δ is the logarithmic dampening factor.

For modeling, it is important to know each coefficient of Eq. (6.8). Since at present it is very difficult to measure these values, the parameters reflect the total characteristics of the skin layers. In addition to these parameters in modeling, the dimensionless velocity of propagation of surface waves in the membrane $c_{0m} = \frac{c_m}{U_\infty} = \left(\frac{T}{M}\right)^{1/2}\frac{1}{U_\infty}$ is of great importance.

The resulting Eq. (6.8) of the movement of the skin of dolphins can be used in theoretical calculations of the stability of the laminar boundary layer over a flexible surface and in the creation of artificial flexible coatings.

6.4 Apparatus for measuring the mechanical characteristics of the skin of live dolphins

Important parameters that determine the effectiveness of elastic-dampening coatings are the oscillating mass per unit area, the speed of the surface wave, elasticity, and dampening [120].

Analysis of the existing equipment designed to measure the elasticity of materials from polymers and rubber showed that it is not suitable for the task of experimentally studying the mechanical characteristics of the skin of living hydrobionts. Therefore, to measure the

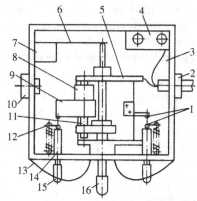

FIGURE 6.13   Photo (A; cover removed) and scheme (B) of the device for measuring the elasticity of elastic surfaces: (1) electro contacts, (2) hermetic outlet, (3) case, (4) light bulbs, (5) main bracket, (6) flat spring, (7) mounting bracket springs, (8) induction coil, (9) coil holder, (10) plug, (11) permalloy rod, (12) spring, (13) limiter, (14) limiting rod bracket, (15) restrictive rod, (16) measuring rod [44,70].

skin elasticity of the dolphin, a special device has been developed (Fig. 6.13) with dimensions $100 \times 100 \times 40$ mm.

The device is made of plexiglass, consists of a hermetic case (3), inside which is placed the main bracket (5), limiting rod brackets (14), the coil holder (9), the angle for fastening the spring (7), and the lamp holder (4). In the main bracket (5) there is a measuring rod (16), on which is attached a plate with a permalloy rod (11). The rod (11) enters the hole located in the inductance coil (8) installed in the coil holder (9). The measuring rod abuts against a flat spring (6) fixed to the angle (7). On the case there are limiters (13), made of plexiglass, which protrude from apertures, and restrictive rods (15) set in the brackets (14). To the restrictive rods attached springs (12). The housing is closed on a waterproof putty plexiglass lid. On one side of the case there is a rubber hermetic lead-in (2) for wires, and on the opposite side a plug (10).

The device functions as follows. The measuring rod (16) protrudes a certain distance beyond the plane, which is at a tangent to the limiters (13). The rod (16) is pressed into the skin of the dolphin until the limiters (13) touch the skin. With different elasticity of the skin under the action of the spring measuring rod (16) it will be deeper into the skin by different values. Since the measuring rod (16) is rigidly connected to the permalloy rod (11) moving in the coil, the depth value will be recorded by measuring the inductance of the coil. Thus, at each measurement point, the core penetration into the skin and the spring force can be known. To ensure that the measuring rod is pressed by a balanced system of external forces (pressing is performed by the hand of the experimenter), a special cut-off scheme has been developed. Structurally, it consists of restrictive rods, springs, contacts, and light bulbs. When in contact with the body of the dolphin, the limiting rods are pressed by the skin, closing the contacts. The lamps light up, fixing the nominal position of the measuring rod.

Circuit diagram of a device with a transformer output for measuring the elasticity of elastic surfaces and the cut-off scheme of restrictive rods are presented in Refs. [13,44,70,170]. In addition to the specified device, an H-700 oscilloscope, a power supply unit, a sound generator, and an indicator device were used to measure the surface elasticity.

The device was also upgraded due to the requirements to carry out measurements on curved surfaces in laboratory and expeditionary conditions, which are as follows.

1. A measuring rod was designed and manufactured, with removable spherical heads of various diameters (2.5; 4; $8 \cdot 10^{-3}$ m). This eliminates the need to disassemble the device when changing the specific load on the material without changing the spring;
2. Produced a complete sealing device;
3. Reduced the size of the support base to perform measurements on curved surfaces;
4. Designed and manufactured an electronic unit for autonomous work.

A functional diagram of the electronic unit is shown in Fig. 6.14. An electronic block is executed in the form of a device containing three functionally independent parts: the generator of sinusoidal oscillations; the electrical bridge scheme for measuring inductance with a low-quality factor (sensor); and the electronic scheme to identify the imbalance of the electrical bridge. All three nodes are assembled on a printed circuit board. The unit provides autonomous power and the ability to connect any appropriate power.

Babenko also developed a universal compact sealed device for measuring elasticity (Fig. 6.15), the principle of operation of which is basically the same as the device shown in Fig. 6.13.

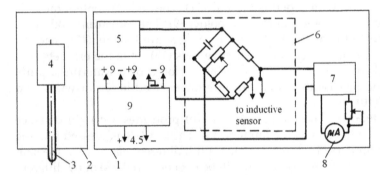

FIGURE 6.14 Functional diagram of the instrument for measuring the elasticity of elastomers: electronic (1) and mechanical (2) parts of the device, (3) measuring rod, (4) inductor, (5) RC generator, (6) measuring bridge, (7) error signal amplifier and detector, (8) milliammeter, (9) power supply [315].

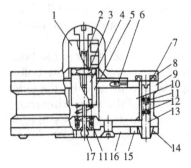

FIGURE 6.15 Scheme of the sensor - elastic meter: (1) choke, (2) permalloy rod, (3) check - net, (4) sleeve of dielectric, (5) steel cylinder, (6) light indicator, (7) seal, (8) cover, (9) frame, (10) inductance coil, (11) springs, (12) contacts, (13) limiting rods, (14) rubber film, (15) pressure ring, (16) support base, (17) measuring rod. [44,70].

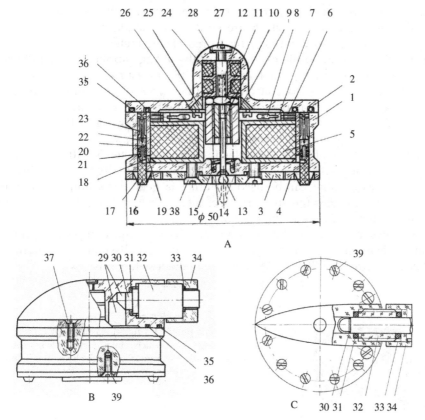

FIGURE 6.16   Round elastic meter: (A) cross section, (B) side view, (C) top view [44,70].

Fig. 6.16 shows three projections of the design of a circular elastic meter (Fig. 6.15), the case, and a number of parts made from plexiglass.

In Fig. 6.16 the following designations are given: (1) device case, (2) cover, (3) the bottom inner clamping ring, (4) lower outer clamping ring, (5) main coil of inductance, (6) plexiglass screw with a through hole, (7) the case of the light indicator, (8) light indicator, (9) outer metal cylinder, (10) internal dielectric cylinder, (11) lock nut, (12) permalloy rod, (13) measuring rod, (14 and 17) rubber thin films, (15 and 21) springs, (16) restrictive rods (four rods), (18 and 19) holes for electrical wires, (20 and 22) limiting rod contactors, (23) top contactor switch, (24) choke, (25) pressure ring for choke, (26) fixing screws, (27) screw/bolt for adjusting the permalloy rod, (28, 31, 33, 35, 36) rubber sealing rings, (29) channel for electric wires, (30) electric cable, (32) plug for hermetic connection of the electric cable, (34) captive nut, and (37–39) fixing screws.

The principle of operation of a round elastic meter (Fig. 6.15) is that, unlike a flat elastic meter (Fig. 6.13), there are *four* restrictive rods (16) in a round elastic meter. When conducting measurements, the elastic meter is pressed against the surface of the measured material so that *four* photodiodes light up and they must be lit continuously in the measurement process. Previously, the elastic meter was calibrated on standard elastic materials, the characteristics of which are studied by standard instruments [44,70]. The measuring rod (13) is pressed into the measured elastic meter to a predetermined depth with a force corresponding to the magnitude of the voltage applied to the inductor (10). A calibration curve is constructed for the force on the measuring rod as a function of the electrical voltage applied to the coil (10). On this curve the amount of stress in elastic meters with the same depth is determined. This method of measuring elasticity is more accurate than that given in Ref. [260]. The amount of movement of the measuring rod is controlled by the change in inductance of the choke (24).

The small dimensions of the round elastic meter (diameter 5×10^{-2} m) allow it to be installed in the fairing and placed on the elastomers or hydrobionts during flow tests (Fig. 6.17). The elastic meter can be used for other targets. For example, if the measuring rod slightly protrudes from the support base, then, during testing, it is possible to determine the vibration parameters of the measured surface.

Fig. 6.18 shows a device for determining the elasticity and dampening properties of a ball rebound. There are many methods for determining the dissipative properties of elastomers. A known expression [239,260] to determine the loss factor is:

FIGURE 6.17   The photo of a round elastic meter (2) (Fig. 6.15) placed in a plexiglass fairing (1), on which a micro twirl (3) is installed to measure the flow velocity. Elastic meter installed on the body of a dolphin (4).

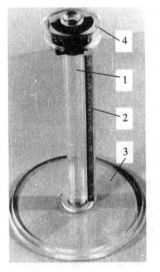

FIGURE 6.18   Device for measuring the dissipative properties of elastomers on recoil balls: (1) removable plexiglass tube, (2) measured ruler, (3) flat base, (4) removable plexiglass case with an electromagnetic coil [29,44,70].

$$\psi = \frac{\Delta W}{W} = \frac{2\pi}{Q} 2\pi tg\delta,$$

where ΔW is the energy absorbed by a unit volume of material; W is the energy saved by the unit volume of the material; and Q is a parameter characterizing the increase in amplitude with resonance. Shock load leads to the generation of oscillations of a wide frequency range, as in real complex interaction.

The device (Fig. 6.18) was developed by Babenko [29,44,70] by analogy with that given in Ref. [260]. The device consists of a hollow cylinder (1), made of transparent polished plexiglass, a rigidly fixed measuring ruler (2), and a viewfinder (not shown in the figure). The cylinder is mounted on a flat circular base (3; for measurement on flat surfaces) and may not be fixed on the base (for measurement on curved surfaces). In the upper part of the cylinder there is an electromagnet (4; to hold the steel ball) and a plumb (to check the vertical). When the electromagnet is deenergized, the ball falls on the test surface. The recoil height can be registered with the cursor ring or with a photodiode ridge mounted parallel to the cylinder, or by registering elements of shock moments with a piezoelectric transducer mounted under the surface under study. The ball is removed from the cylinder using a rod with a magnetic tip. The device does not move, so several measurements can be performed at the same point. Using balls of different mass m, the energy value ΔW can be measured. The parameter of the dampening properties is determined by the expression:

$$K = 1 - h/h_0,$$

where h_0 is the height of the fall; and h is the height of the ball rebound. At low-impact energy, the value of K can be associated with the loss tangent:

$$tg\delta = \frac{1}{\pi}\ln\frac{h_0}{h} = -\frac{1}{\pi}\ln(1 - K).$$

6.5 Preliminary results of the investigations into the skin elasticity of live dolphins

A dolphin was placed in a special box. About 10 minutes after the dolphin had become accustomed to the new setting, the experiment began. The equipment was turned on and the elasticity of the skin was measured successively along the section planes from top to bottom by five measurements in a row at each point. Babenko proposed a scheme for the location of points on the surface of the body for measuring (Section 4.3, Fig. 4.26A). After performing the measurements, the dolphin was released from the box. To process the obtained data of measurements of the elasticity of the skin of live dolphins, two types of calibration were carried out. In the first type of calibration, a measuring rod with a force P was loaded with a special device and the calibration curve of the oscilloscope load dependencies P, N was built. In the second calibration method, the measuring rod was moved with a micrometric device and the second calibration curve of the oscilloscope bunny movements was constructed from displacement of the measuring rod. Measurement data were processed using these calibration graphs, as a result of which the elastic moduli of the skin of live dolphins were determined. Analysis of the waveforms made it possible to discover that, at a confidence level of $\alpha = 0.95$, the accuracy of the measurements was within 5%. Special devices for calibrating the meter introduced a certain error. In addition, the selected meter design could be a systematic error. To take into account these errors and compare the elastic properties of the skin of live dolphins with the standard mechanical characteristics of various materials from polymers and rubbers, a third calibration of the measuring device was carried out. On standard specimens with different mechanical properties, elasticity measurements were made using a standard Shore elastic meter and a developed gauge. The resulting calibration curve is shown in Fig. 6.19. The measuring device determined the complex plastic-elastic deformation.

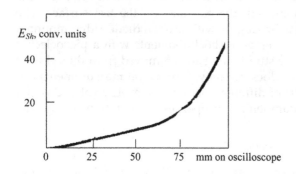

FIGURE 6.19 Calibration graphic of the measuring device.

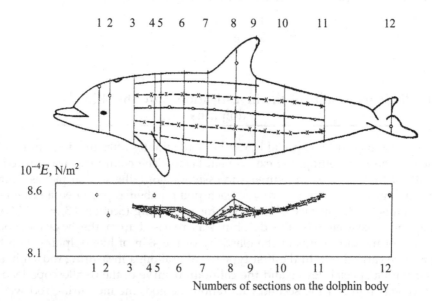

FIGURE 6.20 The scheme of distribution of the elastic modulus of the skin on the body of a short-beaked common dolphin in relation to the calculations given in Table 6.1.

The elasticity of dolphin skin is understood as a complex of properties that characterize their ability to deform mechanically and reversibly under the influence of a system of external forces.

Fig. 6.20 shows the results of numerous measurements of the distribution of skin elasticity over the body of dolphins. In September 1966, the elasticity of the depleted short-beaked common dolphin was measured, and in January of the following year, the same dolphin in the normal state. The elasticity of the skin of an exhausted animal was somewhat reduced. In two bottlenose dolphins (an adult female 2.1 m long and a male 1.6 m long), the skin elasticity was about the same. Measurements of skin elasticity in porpoises (sick and healthy) showed that the skin elasticity of a sick dolphin (more emaciated) was noticeably lower than that of a healthy one.

In the transverse and longitudinal directions of the body, the elasticity of the skin of dolphins varies. In the transverse direction in the middle of the body of the dolphin, the elasticity decreases, and in the lower part it increases. In the longitudinal direction in the middle part of the body the elasticity decreases and increases at the tail part. The elasticity of the skin varies throughout the body gradually. In the porpoise, the elasticity of the skin throughout the body is about the same. The difference in elasticity in the body is 20–30% of conventional units Shore E_{Sh}.

According to the measurements made and the values of the calculated averaged velocities under the condition of a laminar flow around the body of dolphins, the following calculations were made. Calculated values of elastic modules of the skin of live dolphins according to the Hertz formula [187]:

$$E = 0.795 \frac{P}{\Delta^{3/2} d^{1/2}}. \tag{6.10}$$

The elastic modules were taken according to measurements in the section plane (7) (Fig. 6.20). Calculated criteria Froude and Cauchy:

$$Fr = \frac{U_\infty}{\sqrt{gL}} \tag{6.11}$$

$$Ca = \frac{\rho U_\infty^2}{2E}. \tag{6.12}$$

Using the well-known relation $\nu = \mu/\rho$, the formula for calculating the elastic parameter k_s, proposed in Ref. [120], was converted to the form:

$$k_s = \frac{S}{\mu U_\infty R_1^2} = \frac{S\nu}{\rho U_\infty^2 U_\infty} = \frac{E\nu}{\rho U_\infty^2 U_\infty t} = \frac{\nu}{2Ca U_\infty t}. \tag{6.13}$$

For three species of dolphins, the elastic parameter was found by formula (6.13). Taking into account the fact that the swimming speed of dolphins is proportional to the length of their body, as well as the fact that with straight-line movement dolphins move near the surface of the water to breathe oxygen, the Froude number should be included in the elastic parameter k_s. Therefore, the formula (6.13) is converted into a formula for calculating the relative elastic parameter:

$$k'_s = \frac{k_s}{Fr} = \frac{\nu}{2CaFrU_\infty t}, \qquad (6.14)$$

according to which the corresponding calculations were made. In formulas (6.10)−(6.14), the following notation is used: P is the load acting on the measuring rod (N); Δ is the depth of immersion of the measuring rod into the skin of a dolphin (m); d is the diameter of the measuring rod (m); U_∞ is the average speed (m/s); g is the acceleration of a freely falling body (m/s^2); L is the average body length (m); ρ is the water density (kg/m^3); μ is the dynamic viscosity coefficient of water (N · s/m^2); ν is the kinematic viscosity coefficient of water (m^2/s); S is the compression rigidity (N/m^3), $S = E/t$; t is the skin thickness (m); and Re_1 is the unit number Reynolds (m^{-1}), $Re_1 = U_\infty/\nu$. In the calculations, the values of ρ and ν were taken at a temperature of 20°C. Due to the fact that measurements of elasticity were made on stationary dolphins, when their muscles were relaxed, the data of measurements of elasticity most likely correspond to slow swimming speeds. Therefore, in the calculations, the speed U_∞ was assumed to be equal to the average speed with the duration of navigation for 1 day. The results of the calculations are presented in Table 6.1.

To simulate an increase in hydrodynamic pressure with increasing speed, the spring stiffness in the measuring device was increased. At the same time, in the calculations, the speed U_∞ was taken to be equal to the average speed with a duration of swimming of 15−20 minutes. Carried out under these conditions, the calculations are summarized in Table 6.1 in the second column. The results of the calculations given in Table 6.1 are relative, because the number of dolphins studied was not sufficient to obtain reliable results and, moreover, at high speeds, the elasticity of the dolphin skin was not measured, but only modeled. However, the results obtained allow for certain qualitative characteristics the skin of different dolphin species.

M. Kramer conducted a series of towing tests of a cylindrical model with a large elongation [177−180]. The model was covered with a special coating that simulated the skin of a dolphin. The best tested coating had a stiffness of 800 psi (21.75 · 10^7 N/m^3). The coating thickness was 0.14 inches (3.55 · 10^{-3}m). Since in formulas (6.8) and (6.9) the compression stiffness is expressed through the ratio $S = E/t$, the elastic modulus of the Kramer coating will be $E = 77.3 \cdot 10^4$ N/m^2, and according to our data (Table 6.1 for the short-beaked common dolphin) $E = 10.1 \cdot 10^4$ N/m^2. The discrepancy is due to the fact that they measured

TABLE 6.1   Parameters of elasticity of the dolphins skin.

Dolphin species	L (m)	U_∞ (m/s)	$E \cdot 10^{-4}$ (N/m^2)	Fr	Ca	$t \cdot 10^2$ (m)	$k_s \cdot 10^5$	$k'_s \cdot 10^5$
Short-beaked common	1.4	6.0	8.38	1.63	0.214	2.5	1.57	0.963
	1.4	7.8	10.1	2.12	0.302	2.5	0.856	0.404
Bottlenose	1.9	6.9	7.93	1.6	0.299	3	0.815	0.51
	1.9	9.0	9.62	2.08	0.42	3	0.444	0.213
Harbor porpoise	0.8	4.5	4.55	1.61	0.222	1.8	2.8	1.74
	0.8	5.0	7.06	1.79	0.177	1.8	0.316	0.176

the elasticity of the skin of dolphins at rest, in a relaxed state. The increased stiffness of the measuring instrument spring, imitating the elasticity of a dolphin swimming at high speeds, could not fully reflect the simulated phenomenon. Therefore, for example, the elastic modulus of skin on the fins of the short-beaked common dolphin, calculated by formula (6.10) was $E = 37.3 \cdot 10^4 \, \text{N/m}^2$. In addition, it is unknown how the stiffness of the Kramer coatings was measured. If without taking into account the influence of the substrate, then its data are significantly overestimated, since according to the existing standards it is allowed to measure the elastic properties of rubber samples with a thickness of at least 6 mm. It should be noted that dolphin skin thickness is significantly greater than the thickness of the coatings investigated by Kramer, and he actually imitated only the upper layer of the skin—the epidermis. It is also unknown whether Kramer took into account the rigidity of the lacquer coating of the model, as all his models were covered with three layers of varnish. All these reasons could lead to inconsistency between the results of Kramer and the measurements taken. However, the fact that dolphins in a relaxed state have more pliable skin than Kramer's cover testifies to dolphin reserves in the sense of a change in the elastic-dampening properties of the skin and the possibilities to change the elastic properties of the skin with the help of skin muscles, which influence the stabilization of the boundary layer.

In Ref. [180], the modulus of elasticity of the skin layer (epidermis and dermal outgrowths) with a thickness of $5.08 \cdot 10^{-4}$ m, equal to $150 \, \text{psi/m}^2$ ($1.034 \cdot 10^6 \, \text{N/m}^2$) is presented. Measurements were performed on dead skin, which can introduce significant errors. Despite the significantly overestimated result of these data in comparison with the data of discounts [177], Kramer argues that the modulus of elasticity of the dolphin's skin is equivalent to the modulus of soft rubber. An analysis of the experimental data allowed us to draw the following conclusions.

The results of experimental investigations have shown that the most elastic skin in dolphins is the short-beaked common dolphin, closely followed by the bottlenose dolphin, and much less elastic is the harbor porpoise. At low swimming speeds for bottlenose dolphins and short-beaked common dolphins, the elastic properties of the skin are $E_{Sh} = 30-40$ conventional Shore units, at high swimming speeds, above 40 units E_{Sh}. By state standards, these data correspond to soft rubber, at high speeds approaching medium-hard rubber. At average speeds corresponding to a duration of swimming of 15–20 minutes, these three types of dolphins have relative elastic parameters that are about the same. Apparently, in these speed ranges of these dolphins, the skin has a certain hydrodynamic value. At speeds corresponding to the duration of swimming for 1 day, the relative parameters of elasticity of the dolphins studied significantly differ from each other. This result probably indicates that the characteristics of the skin are tuned to a specific swimming speed. In Ref. [224], the Froude numbers are denoted as the relative maximum speed, which characterizes the different types of dolphins.

Measurements were made of the elastic characteristics of the skin and a number of elastomers in order to compare them and determine the effectiveness of the interaction of these materials with the boundary layer, whose structure is a system of longitudinal vortices. Fig. 6.21 shows the results of measuring the elastic properties of the skin of three bottlenose dolphins and three elastomer samples. Measurements on dolphins were carried out in three places: the head, middle, and tail parts of the body. The abscissa axis shows

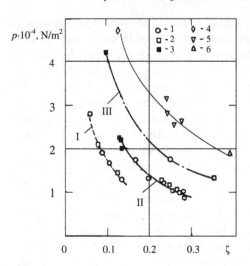

FIGURE 6.21 Dependence between stress and deformation of the skin of healthy (1), sick (2), and dead (3) dolphins in the head (*I*), middle (*II*) and tail (*III*) body parts, (4–6) measurements on elastomers (Table 6.2).

TABLE 6.2 Elasticity elastomer samples.

Elastomer sample no.	ρ (kg/m^3)	Number of pores per 1 cm	Thickness (cm)	E (MPa; megaPascal)	Measuring device
1	42	20–26	1	0.116–0.023	BH-5704
				0.09–0.05	Elastic meter
2	300	30–40	0.8	0.035–0.031	BH-5704
3	1045	6	0.8	0.45	Elastic meter
4	1050	6	0.8	0.45	Elastic meter
5	107	30–40	0.8	0.16–0.06	BH-5704
				0.13–0.09	Elastic meter
6	107	32	1	0.049	Elastic meter
Dolphin Type of rubber	1030	–	1.5–0.6	0.21–0.34	Elastic meter
2959	–	–	–	0.98	[145]
1847	–	–	–	2.4	[145]

the relative deformation $\zeta = (l-l_o)/d$, where l_o is the displacement of the measuring rod relative to the inductance coil of the elastic sensor (Section 6.4) on a rigid surface, l is on an elastic surface; d is the thickness of the sample or the upper deformable layers of the skin; on the ordinate axis the tension p (N/m^2) is plotted.

Characteristics of elastomer samples, whose elasticities are shown in Fig. 6.21, are indicated in Table 6.2 with the same numbers 4–6. In addition, in the table for comparison with the characteristics of the skin of live dolphins, the properties of other elastomer samples are given.

The static (conditionally balanced) modulus of elasticity of elastomers E is usually defined as the tangent of the angle of inclination to the abscissa axis of the $p(\zeta)$ curve in the region of linear deformations at very low loading rates, or with a significant holding under load [239]. Considering that this condition was fulfilled in the present measurements, the elastic moduli of the skin and the elastomer samples were determined using the p/ζ ratio.

Measurements on dolphins were carried out along the lateral line. The thickness of the skin was taken as the distance from its surface to the skin muscle, ranging from $15 \cdot 10^{-3}$ m in the head to $6 \cdot 10^{-3}$ m in the tail of the body (Section 4.3, Fig. 4.26). The modulus of elasticity in the head part of the body of a healthy dolphin varied within $E = (2.1 - 0.96) \cdot 10^5 \, \text{N/m}^2$, in a sick dolphin, $E = (4.7 - 1.2) \cdot 10^5 \, \text{N/m}^2$; in the middle part of the body, in a healthy dolphin, $E = (1.03 - 0.34) \cdot 10^5 \, \text{N/m}^2$, in a sick dolphin, $E = (0.54 - 0.37) \cdot 10^5 \, \text{N/m}^2$, in a dead one, $E = (1.7 - 1.4) \cdot 10^5 \, \text{N/m}^2$, and in the tail part of the body, respectively: $0.7 \cdot 10^5 \, \text{N/m}^2$, $0.37 \cdot 10^5 \, \text{N/m}^2$, $4.2 \cdot 10^5 \, \text{N/m}^2$. The external covers of the dead dolphin had a smaller range of changes in skin elasticity throughout the body and significantly larger E values than in a healthy dolphin. At the same time, it was found that in a living healthy dolphin, the ductility of the cover in the dorsal and ventral directions decreased, while in the dead one they increased. The dependence of the parameters of the outer covers on the state of the animal is largely due to the different properties of the substrate, determined by the tone of the skeletal and dermal muscles [29].

As can be seen from Table 6.2, an elastomer with a large number of pores on elastic properties is closer to the skin of dolphins than an elastomer with a smaller number of pores, and elastomers, having a density close to the density of the skin of dolphins, differ significantly from it on elastic characteristics. The elasticity of elastomer samples was measured simultaneously by standard instruments and developed elastic meters (Section 6.4, Figs. 6.13–6.15). The results of measurements by the developed elastic meters shown in Table 6.2 showed that with the help of elastic meters, adequate results were obtained for the elasticity of the dolphin's skin.

6.6 Investigations into the skin elasticity of live dolphins

When changing the modes of movement in animals, the shape of the body and its geometry, flexural rigidity of the whole body, rigidity and mechanical properties of the skin, elasticity and shape of the propeller, body and blood temperature, pressure in the circulatory system, thickness of the mucous membrane playing the role of a dampening surface automatically regulated the production rate of mucous substances, etc.

The effect of the automatic regulation of the elasticity E of the skin of live dolphins on the swimming speed U has been investigated. To understand the mechanism of interaction of the skin of live dolphins with the characteristics of the flow of the environment, it is necessary to develop a physical model of the skin and, on this basis, a mathematical analogue of the skin. In Ref. [298], the differential equation of an analogue of the skin of dolphins is presented in the form of a linear model of a Voigt–Kelvin viscoelastic medium, and in Ref. [17], in the form of a so-called standard or generalized model of a linear viscoelastic medium. In the same paper, a simplified mathematical analogue of skin is given, which takes into account, in contrast to the standard model, tension in the skin layers and bending deformations. The scheme of a more complete analogue of the skin of dolphins will differ significantly from the simplified scheme. Indeed, if we consider the structure of the skin of dolphins [14,16,17,278−282], then the skin scheme given in Refs. [16,50,53,62] (Section 4.3, Fig. 4.24; Section 6.3, Fig. 6.12) should be supplemented in the following way. The skin layer (4) is schematized not only by the elastic element, but also by the dampener. Skin layer (5) should be schematized with an elastic element, and layer (60, like layer (3), with the Voigt−Kelvin element. In addition to layers (5) and (6), it is necessary to add layer (7), which is schematized by the Voigt−Kelvin element. Between layers (3) and (6), and layers (3) and (7), elastic connections are inserted parallel to the indicated elements, schematizing connective membranes. It is known from the theory of polymers [239] that the mechanical properties of rubber can be most completely described using nonlinear integro-differential equations of the Volterra type. There are currently no mathematical methods for describing the mechanical characteristics of filled rubbers. Considering that the tension in the layers of the skin and its dampening properties substantially depend on the swimming speed of dolphins and are reflex-regulated; that all skin is multilayered, with the upper layers having a longitudinal lamination; that the deformation of only the individual collagen fibers that make up the skin layers is described by a complex nonlinear equation [108], that skin oscillations correlate with the work of the caudal fin and are nonstationary; that all skin can be modeled only by a nonlinear compositional viscoelastic medium, it becomes obvious that the mathematical skin model of live dolphins currently cannot be solved analytically. The solution of a simplified mathematical analogue can only be found by numerical methods. For example, the studies given in Ref. [212] can be applied in the first approximation to the analysis of nonstationary skin oscillations without taking into account the tension in the layers. In this work, a conclusion was obtained that testifies to the positive role of the dynamic interaction of skin layers with different elasticities and oscillating masses, which lead to a significant dampening of nonstationary oscillations.

Thus, it becomes clear that it is necessary to accumulate various experimental data on the mechanical characteristics of dolphin skin. In addition to the dynamic characteristics, the definition of the static elastic modulus of the skin of dolphins is also of interest.

With small deformations, the elasticity obeys Hooke's law, and the regularity remains the same under tension and compression [239]. Therefore the elasticity measured on dolphins under compression, taking into account the smallness of the deformation, characterizes the static elastic modulus, and the measurements made by the elastic meter shown in Fig. 6.13. Section 6.4 investigates the behavior of the skin described in static by the Voigt−Kelvin model. In this regard, the results given below characterize the general static state of the entire skin without taking into account its dynamic features noted above.

In Section 6.5 some results of measuring the value of E are presented, made with the help of a specially manufactured instrument such as a Shore hardness meter. According to the results of these measurements, the elastic parameters k_s were calculated for one cross section of the animal body in the area of the mid-section, which turned out to be approximately equal for the three types of dolphins at their cruising speeds. The elastic modulus was determined using the Hertz formula, which introduced the error, which was reduced in accordance with the dependence of the elastic modulus, expressed in Shore units, on the elastic modulus, expressed in dimensional units [260].

When developing methods for measuring the E value, three versions of the device were made, one of which with a weak load spring measured the elasticity of the outer layers of the short-beaked common dolphin skin (presumably the epidermis and papillary layer), which amounted to $1.4 \cdot 10^5 \, \text{N/m}^2$. The modulus of elasticity of a thicker layer of skin, measured when it was stretched in Cramer's work, was $1.03 \cdot 10^6 \, \text{N/m}^2$ (Section 6.5). The discrepancy between the measurement results is explained by the fact that when measuring with the above device it is difficult to accurately estimate the skin thickness t, which

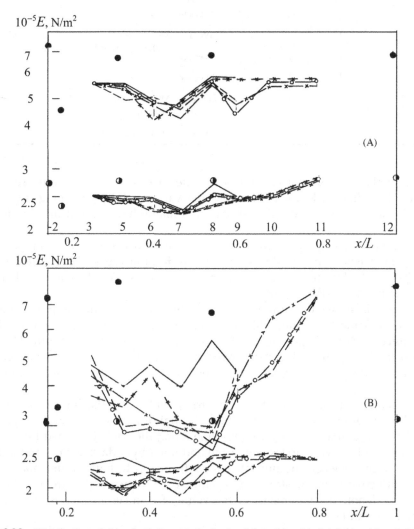

FIGURE 6.22 Distribution of skin elasticity over the body of the white-sided dolphins (a) and the bottle-nosed dolphin. (b) at various swimming speeds [28].

TABLE 6.3 Dolphin measurement conditions

Dolphin species	Number of dolphins	Length body (m)	Number of measurements	Effort springs	Note
Short-beaked common	4	1.4	17	Weak	The measurements were carried out at different times of the year with an interval of 6 months under different living conditions at different times of the day
				Average	
				Large	
Bottlenose	3	1.5	1	Weak	Measurements were taken in the morning and in the evening
		1.37	1	Average	
		1.3	2	Average	
		1.9	3	Average	Measurements were taken at weekly intervals under different habitat conditions
		1.79	1	Average	
		1.7	1	Average	
Harbor porpoise	2	0.7	2	Large	Measurements were taken in the morning and evening
		0.6	1	Large	Dolphin was sick

the pressure of the measuring rod applies to, due to possible reflex contractions of the skin muscle. Note that M. Kramer measured the value of E by stretching the preserved samples of the skin of another dolphin species, not taking into account the size and age of the animal.

In Ref. [28], with the help of a proven instrument and measurement technique (Sections 6.4 and 6.5), the elasticity of the skin of three live dolphin types was studied: the short-beaked (*Delphinus delphis*), the bottlenose dolphins (*Tursiops truncatus*) and the harbor porpoise (*Phocaena phocaena*) (Fig. 6.22).

The abscissa axis shows the numbers of cross sections along the dolphin body (at the top of the x/L axis) and the distance along the body in the dimensionless form x/L, where x is the distance from the beginning of the body and L is the body length. The location map of these and longitudinal sections over the dolphin body is given in Section 6.5, Fig. 6.19. Data on the size of animals and statistics of measurement results are shown in Table 6.3.

The measurement of the value of E was carried out along the lines of cross sections from top to bottom at points lying on the lines of longitudinal sections evenly spaced from each other. The measurements began with a line along the back, indicated by a solid line, along the side by a line with circles, along the abdomen by a dashed line, and along lines located respectively between the indicated lines from top to bottom and marked with a dashed line with crosses on the strokes and with crosses between the strokes. Half-shaded circles and dots denote measurements of the value of E in certain places along the body, indicated by circles in Fig. 6.20.

Approximate calculations showed that pressure pulsations in the turbulent boundary layer when swimming at a speed of 10 m/s have a value of about $3.3 \cdot 10^3$ dyn/cm^2, and in the laminar boundary layer an order of magnitude less. In a transitional flow regime in the boundary layer, pressure pulsations can be several times higher than the magnitude of turbulent pulsations [175]. Analyzing the graph of changes in dolphin skin thickness along the body (Section 4.3, Fig. 4.26), we assume that at low swimming speed, the intensity of disturbances in the boundary layer allows them to spread to a small amount of t in the region of the epidermis and papillary layer. As the velocity increases, the kinetic energy of the disturbances of the boundary layer increases, and the oscillations caused by them in the skin can spread deeper, penetrating also into the subpapillary layer. With an increase in the intensity of pressure pulsations in the boundary layer, for example, during a transitional flow regime or with an increase in swimming speed, oscillations can spread deeper in the skin of dolphins. Figs. 6.22A and B show the results of measuring the value of E, made by the device with normal (average) spring stiffness under the assumption that these values correspond to the average swimming speeds U_{av} and characterize the elastic properties of the upper layers of the skin, including the subpapillary layer.

We assume that with U_{cr}, increased pressure pulsations will penetrate into the underlying layers of the skin, and skin elasticity will increase to maintain its optimal stabilizing properties in accordance with the similarity criteria (Sections 1.4 and 6.3). In this connection, measurements of the value of E with a device with double (high) spring stiffness were carried out. These results are shown at the top of Fig. 6.22A. The measurement of the E value of the short-beaked common dolphin was carried out on a trained animal; therefore the measurement errors caused by the dolphin's reaction to skin irritation were insignificant. The bottom of Fig. 6.22B shows the results of measurements of the E value of a trained bottlenose dolphin, and the upper part of this figure shows when it had just been captured. The animal was agitated, and its muscles were tense. Therefore the results of measurements performed by the same instrument differ. It was assumed that the increased elasticity of the skin corresponded to U_{cr} and therefore measurements with a double spring instrument on the bottlenose dolphins were not carried out. Measurements have shown that, in the short-beaked common and bottlenose dolphins, the elasticity of the skin varies throughout the body from top to bottom and from nose to tail, so that the softest part of the skin lies on the sides in the middle part of the body. This pattern is observed in both cases—when measured with normal and double spring stiffness. However, as the pressure of the measuring rod increases, the body elasticity changes more dramatically (Fig. 6.22A). The same is noticeable in Fig. 6.22B, which shows that the contraction of skeletal muscles and skin muscle leads to an even greater increase in skin

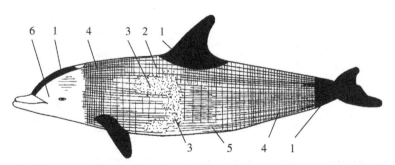

FIGURE 6.23 The topography of skin elasticity of the white-sided dolphin: (1) 1.9–2.0 g/mm²; (2) 1.8–1.85; (3) 1.75–1.77; (4) 1.85–1.9; (5) 1.77–1.8; (6) not measured [50,53].

elasticity with the same pressure of the measuring rod and that the simulation of an increase in E value with increasing speed by measuring with a device with double spring rigidity was performed correctly.

In dolphins, the harbor porpoise followed approximately the same patterns, but it was difficult to carry out the required number of measurements, as these dolphins did not respond well to training. In addition, it was found that the sick dolphin's skin elasticity was much less than that of healthy animals. This is due to the fact that the animal was relaxed, so the elastic characteristics of its body and skin were changed.

Fig. 6.23 shows a diagram of the distribution of the value of E over the body of a short-beaked common dolphin, obtained on the basis of the measurements made with a device with normal spring stiffness. The numerical values of elasticity are expressed in terms of the pressure P of the spring of the measuring rod, related to the area of its tip. Attention should be paid to areas with maximum skin elasticity. In accordance with the morpho-functional value, these areas experience the maximum hydrodynamic pressure [26]. The tail stem and the front part of the body in the region of the lateral fins experience less hydrodynamic loads, so the elasticity of these areas is decreased. The middle part of the body experiences the smallest hydrodynamic loads, so the value of E in this part of the body is the smallest.

The skin of dolphins, in its mechanical properties and structure, is similar to elastomers, the elastic properties of which depend not only on the structure, but also on the thickness and on the mechanical properties of the substrate [239,249]. Based on the measurements made, and also on the basis of the structure of the dolphin's body and its skin as a whole (Sections 4.1−4.3), by analogy with elastomers, it can be assumed that the patterns obtained for the distribution of skin elasticity along the body of three dolphin types are due to the difference in skin structure along the body, its thickness, and the structure of the skin substrate.

Since pressure pulsations in the boundary layer can be compared with a compressive load, and when swimming at depth, on a dolphin's body there is significant hydrostatic

TABLE 6.4 Dolphin skin thickness t for determining stiffness S

| Section no. | $t \cdot 10^3$ m | | | |
| | Short-beaked common, $L = 1.4$ m | | Bottlenose, $L = 1.9$ m | |
	$U_{av} = 6$ м/c	$U_{cr} = 7.8$ м/c	$U_{av} = 6.9$ м/c	$U_{cr} = 9$ м/c
2	1.9	13	2.2.	17
3	2.4	17	2.7	21
5	2.4	25	2.7	28
6	2.8	26	3.1	30
7	3.0	26	3.3	28
9	2.8	17	3.1	21
10	2.5	10	2.8	14
11	1.8	5	2.1	9

$10^{-7}S$, N/m³

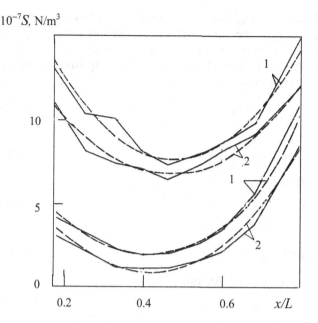

FIGURE 6.24 Distribution of skin stiffness along the body of the short-beaked common dolphin (1) and the bottlenose dolphin (2) at different swimming speeds [17,50,62].

pressure compressing the body, taking into account the effect of skin thickness on the results of measurements of its elasticity, the elastic properties of the skin can also be characterized by S determined from the ratio $S = E/t$ (t is the skin thickness, m).

The value of S was calculated by the averaged values of the values of E and t. The value of t for the short-beaked common dolphin was determined according to Fig. 4.26 (Section 4.3) at U_{av} taking into account the three upper layers of the skin, and at U_{cr} by taking into account the entire thickness of the skin, except for the layer of subcutaneous fat. The topography of the bottlenose dolphin skin thickness is not precisely measured. With known individual areas of the skin, the distribution of average values of the bottlenose is interpolated in accordance with Fig. 4.26 (Section 4.3). Table 6.4 shows the values of t for these layers.

Fig. 6.24 shows the results of calculating the value of S at U_{av} (top) and at U_{cr} (bottom) [17,50,62]. The solid lines indicate the results of the calculations, and the dashed lines indicate the approximations of these calculations by the formulas for the upper curves:

$$\text{(curve 1)} \qquad S \cdot 10^{-7} = 67x/L(x/L - 0.96) + 23, \tag{6.15}$$

$$\text{(curve 2)} \qquad S \cdot 10^{-7} = 48.7x/L(x/L - 0/94) + 17.5, \tag{6.16}$$

for lower curves:

$$\text{(curve 1)} \qquad S \cdot 10^{-7} = 49.4x/L(x/L - 0.84) + 10.5, \tag{6.17}$$

$$\text{(curve 2)} \qquad S \cdot 10^{-7} = 53x/L(x/L - 0.84) + 10, \tag{6.18}$$

The value of S along the dolphin's body changes with the same regularity as the elasticity, but more dramatically, and at different swimming speeds the rigidity changes according to the same type of parabolic dependencies. At U_{cr}, the S value of the short-beaked common and bottlenose dolphins changes with almost the same regularity. Formulas (6.15)–(6.18) limit, apparently, the optimal range of variation of S with characteristic swimming speeds.

For simplicity, the curves located at the bottom of Fig. 6.24 can be approximated by straight lines on the left for the short-beaked common dolphin:

$$S \cdot 10^{-7} = -11 \, x/L + 6, \tag{6.19}$$

for the bottlenose dolphin

$$S \cdot 10^{-7} = -10.5 \, x/L + 4\,8, \tag{6.20}$$

and on the right for short-beaked common

$$S \cdot 10^{-7} = 32 \, x/L - 16, \tag{6.21}$$

for the bottlenose dolphin

$$S \cdot 10^{-7} = 31 \, x/L - 17. \tag{6.22}$$

Averaging expressions (6.19) and (6.20) gives the formula

$$S \cdot 10^{-7} = -10.8 \, x/L + 5.4, \tag{6.23}$$

and expressions (6.21) and (6.22) give

$$S \cdot 10^{-7} = 31.5 \, x/L - 16.3. \tag{6.24}$$

An experimental study of the hydrodynamic stability of the flow around dampening surfaces simulating the skin of dolphins showed [170] that the stabilization of the boundary layer depends on the design of the surfaces and their mechanical properties, characterized by similarity criteria. One of the criteria is the elasticity parameter.

$$k_s = S \frac{\nu}{\rho} \frac{1}{U^3}, \tag{6.25}$$

where ν, ρ, respectively, are the coefficient of kinematic viscosity and density of the liquid. Using data from Fig. 6.23 and Table 6.3, the distribution of k_s along the body (Fig. 6.25) of

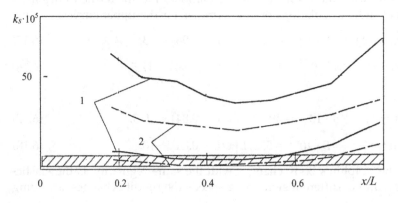

FIGURE 6.25 Distribution of the elastic parameter over the body of short-beaked common dolphins (1) and bottlenosed dolphin (2) [17,50,62].

the short-beaked common dolphins (solid lines) and the bottlenose dolphins (dashed lines) for U_{av} (top) and U_{cr} (below) is calculated using the formula (6.25). In Fig. 6.25, the region of optimal values of k_s obtained by calculation in Ref. [120] is hatched. When swimming dolphins with U_{av}, the value of k_s along the entire length of the body has nonoptimal values, and when swimming dolphins with U_{cr} in the middle part of the body the optimum values.

At the start of the body, the value k_s of the short-beaked common dolphin is outside the optimal zone, but in this place a negative pressure gradient on the dolphin's body plays a significant role in stabilizing the boundary layer (Section 2.3, Fig. 2.4). The dolphin stalk, being a part of the mover, makes an oscillatory motion and its flow around it is different from the flow around the middle part of the body, because when the tail fin oscillates, a sucking force appears in the tail part of the body. It is seen that with an increase in swimming speed up to the limiting speeds, the value of k_s has an optimal value along the entire length of the body. The relation (6.24) shows that when the swimming speed changes, in order to preserve the value of k_s in the region of optimal values, the value S should be changed. To check this position, we perform the following calculations for section (7) of the white-side dolphin. According to the corresponding lower curve in Fig. 6.24 for this section, the optimal value of k_s will be $3.18 \cdot 10^5$. According to the formula (6.24) for $U_{av} = 6$ m/s, the optimal value is $S = S = 6.8 \cdot 10^6$ N/m^3, while in Fig. 6.22 it is $7.3 \cdot 10^7$ N/m^3. The adjusted optimal value of S taking into account the difference in skin according to the Table. 6.3 will be $5.67 \cdot 10^7$ N/m^3. The difference between the measured and the optimal value of S [119] is equal to:

$$\Delta = \frac{7.3 \cdot 10^7 - 5.67 . 10^7}{7.3 \cdot 10^7} \cdot 100\% = 22.4\%.$$

Similar calculations for the bottlenose dolphin give a difference of 40%.

Kramer in Ref. [178] transformed Eq. (6.25) as a dependence on the Reynolds number, which made it possible to determine the pattern of changes in stiffness along the body. The obtained relations differ from formulas (6.15)–(6.24) in that the value of S monotonously decreases with increasing x in the Cramer relations. The resulting discrepancy with the measurements made is due to the fact that the tail part of the dolphin body is mobile and flows around a nonstationary flow. Calculations made using Kramer's formulas for sections (2) and (7) of the short-beaked common and bottlenose dolphins showed that with U_{av} the distribution of S across the body (Fig. 6.23) does not satisfy the ratios proposed by Kramer and which U_{cr} satisfies.

In analyzing the investigations performed, one should take into account the hydrodynamic patterns considered in Ref. [26], on the basis of which the quasistationary approach should be extended only to the body length in sections 2–8. Indeed, the nasal part of the body experiences large dynamic and tangential stresses, has a specific frontal protrusion that affects the thickness of the skin, and makes small oscillatory movements. The tail part of the body is part of the mover with increased dynamic loads and variable flow speeds.

The results of the investigations performed allowed drawing the following conclusions:

1. In the structure of the dolphin skin laid automation of regulating the stabilizing properties of the skin. As the speed increases, pressure pulsations increase, which

penetrate into the deeper layers of the skin. At the same time this increases the tension of skeletal, motor, and skin muscles.

Due to the specific structure of the skin (Section 4.3), the elasticity of the skin increases and its thickness decreases in the region of the skin muscle, and its diameter decreases in the middle part of the body. Thus, measurements carried out using a belt worn on the dolphin's body in front of the dorsal fin showed that when moving at high speeds or accelerations, the *body diameter* in this area *decreased* by 4—8 mm, depending on the size and type of dolphins. If we assume that the elasticity of the skin does not change with an increase in speed, then a decrease in the thickness of the skin by about 2 mm will lead to an increase in its rigidity by 10—15%. When measuring the elasticity of a device with single and double springs, the difference in the depth of the core of the measuring device into the skin of a dolphin averaged 3 mm. If we assume that the decrease in body diameter is associated with both a decrease in skin thickness and skeletal compliance, then the results of measuring skin elasticity are in good agreement with the data for reducing body diameter with increasing swimming speeds. Calculations show that to maintain the optimal value of k_s in a wide range of cruising swimming speeds, dolphins need only change the rigidity of their skin by 20—40%. The change in the elasticity of the body, which is the substrate of the skin, has an effect on the rigidity of the skin.

2. During slow swimming, when the dolphin's body is relaxed, assuming that the dampening properties of the skin are evenly distributed along the body, even with suboptimal values of the parameter k_s, the regularity of the distribution of the rigidity of the skin along the body is optimal for stabilizing the boundary layer.

Rheological analysis of soft collagen tissue was performed in Ref. [108].

The results of an experimental study of the characteristics of the boundary layer in the flow of elastic coatings that simulate the structure of the dolphins skin are given in Refs. [132,136,154]. In Ref. [137] a new method was proposed for determining the elastic-damping mechanical characteristics of materials simulating the structure of the skin. In Refs. [31,214] some results of a theoretical investigations into the problem of the flow of elastic coatings performed by Voropaev and Nikishova are presented.

In Parts II and III, the results of modeling the structure and functioning of hydrobionts and some results of practical application are given in detail.

6.7 Investigations into the oscillating mass parameter of dolphin skin

Under the oscillating mass M of the outer covers of aquatic animals is meant the mass of those skin layers, to a depth of which spread boundary layer disturbances at a given speed of movement, causing elastic waves of small amplitude in the skin. Obviously, the depth of penetration of oscillations into the skin will depend on the thickness of the boundary layer δ, the energy of the disturbing motion E, the magnitude of the pressure pulsations p of the boundary layer on the skin surface and the spectrum of velocity pulsations in the boundary layer (Section 2.3, Fig. 2.5). All these quantities are interrelated and are determined by the speed of motion U and the characteristics of the flow of the outer covers of aquatic animals. The magnitude of the δ laminar boundary layer or the thickness of the viscous sublayer of the turbulent boundary layer determines the amplitude of the oscillations of the skin and the pattern of distribution of the thickness of the skin along the body. The energy of the disturbing motion is determined by the values of the pulsating velocities in the boundary layer and their distribution in magnitude δ. As shown in Refs. [44,70,111,170], the analogues of the outer covers of aquatic animals reduce the energy of the disturbing motion and redistribute the maximums of the pulsating velocities by δ.

The value of p calculated for a rigid surface [29,256] can reach large values and indicate that vibrations can propagate to a considerable depth in the skin of dolphins. Depending on the flow regime in the boundary layer, the spectrum of the pulsation velocity of the boundary layer will be different and differ from the spectrum on a hard surface [26,44,70]. It is known from the mechanics of elastomers that the oscillations propagate deeper into about $1/2$–$1/4$ of the wavelength of the disturbing motion. Calculated on the basis of experimental studies of hydrodynamic stability on a viscoelastic surface [170], the wavelength of a disturbing motion in a laminar boundary layer on a body of revolution with an elongation of 4 at $\alpha\delta^* = 0.6$ will be for $x = 0.1$ m and $U = 8$ m/s, $\lambda \approx 1.04 \cdot 10^{-3}$ m. Such a wave can spread to the depth of the first two or three layers of skin. With an increase in energy, thickness of the boundary layer, and size λ, oscillations can penetrate deeper into the skin of dolphins. Based on the values of velocity and pressure pulsations known in hydromechanics in the laminar, transitional, and turbulent boundary layers, it is shown (Sections 6.5 and 6.6) which dolphin skin layers can participate in oscillatory motion at different swimming speeds.

Since, at present, the physics of the processes of interaction of the flow with elastic surfaces has not been completely investigated, it is not possible to accurately determine the thickness of the skin layers, the magnitude of which penetrates the vibrations of the boundary layer. Only the simplified problem of the absorption of the pulsation energy of the boundary layer by a monolithic viscoelastic layer was solved theoretically [111]. In this regard, the coefficient of oscillating mass of the skin can be approximately calculated by the following formula [44,70,170]:

$$k_{\check{i}} = \frac{UM}{\mu}, \tag{6.26}$$

where μ is the dynamic coefficient of viscosity of water. Determining the $k_{\scriptscriptstyle M}$ value using formula (6.26) is based on modeling the skin using the Voigt–Kelvin viscoelastic model (Section 6.3) in the form of a membrane surface, which introduces some error compared to the monolithic layer model [44,70,111,170]. However, the qualitative picture will be reflected with sufficient certainty.

The oscillating mass parameter can also be defined as follows (Section 6.3):

$$K_{\check{i}} = \frac{Re_1}{\rho} M_{\Sigma}, \tag{6.27}$$

where Re_1 is the unit Reynolds number, $Re_1 = U_{\infty}/\nu$ m^{-1}; ρ is the density of seawater (kg/m^3); and M_{Σ} is the total oscillating mass, $M_{\Sigma} = M_1 + M_2$ (kg $\cdot$ s^2/m^3). According to the mechanical model of the skin of dolphins (6.3), the oscillating mass M_1 is the mass per unit area of the epidermis (Section 6.3, Fig. 6.12, layer 1), and M_2 is the mass per unit area of the underpapillary layer (layer 3). The values of M_1 and M_2 were determined by measuring the thickness of the skin layers of dolphins (Section 4.3, Fig. 4.26). Pressure pulsations in a turbulent boundary layer can be determined according to the expression [120]:

$$(\overline{p'^2})^{1/2}/q = 0.008,$$

where $(\overline{p'^2})^{1/2}$ is the (rms) value of pressure pulsations, and q is the hydrodynamic pushing, $q = \rho U_\infty^2/2$. When a dolphin is swimming with a speed of $U_\infty = 10$ m/s, magnitude $\overline{p} = 40$ dyn/cm^2. For a laminar boundary layer, the magnitude $\overline{p}$ is an order of magnitude smaller. The parameter of the oscillating mass may depend on the number Re and on the values of disturbances wavelength [120]. It can be assumed that under the condition of laminar flow under the oscillating mass of the skin of live dolphins, it is necessary to understand the mass M_1 of the epidermis with a thickness of 0.2 mm. With an increase in the speed and values of the pressure pulsations, the oscillating mass should be understood as the mass $M_\Sigma = M_1 + M_2$, and the mass M_1 includes the first and second skin layers. The mass M_2, which takes into account the third layer of the skin, generally depends on the p value. The k_M parameter is calculated for both cases; in the second case, the M_2 value is fully calculated.

The values of the oscillating mass were determined on a fixed material. Rectangles of the skin were cut out so that all layers of the skin were in sections. The mass of the entire skin section, and then the mass of the individual skin layers, was calculated with an accuracy of 0.01 g. Accordingly, the volume of the entire section and individual skin layers was determined in a measuring cylinder with a division value of 1 mL. Then, as a control, the total measurements of skin layers were repeated. In determining the volume, due to the dissolution of fat in water, the measuring cylinder was washed with warm water before each measurement. The error in determining the density of layers of the skin, caused by an increase in the mass of the skin due to its stay in formalin, is approximately 10% of the true value. The calculated density values of the skin layers were reduced by

TABLE 6.5   Density and mass of individual layers of dolphin skin

Dolphin species	ρ_m кг/м^3 skin layers 1–3	ρ_m кг/м^3 skin layer 4	ρ_m кг/м^3 skin layer 6	ρ_m кг/м^3 skin layer 7	Thickness 1–3 layers $t_m \cdot 10^3$ m, сеч. No. 5	The mass of the unit surface of the skin, 1 layer kg/m^2 thickness $2 \cdot 10^{-4}$ m	M_Σ 1–3 layers (kg/m^2)	$k_M \cdot 10^{-4}$ 1 layer	$k_M \cdot 10^{-4}$ 1–3 layers
Short-beaked common	1020	920	1200	870	2.2	0.204	2.24	0.122	1.73
Bottlenose	1030	910	1220	860	1.7	0.206	1.75	0.141	1.56
Harbor porpoise	1050	970	1180	900	0.8	0.21	0.84	0.094	0.42

this value. The averaged densities of individual layers of the skin thus obtained the values of the oscillating masses, and the values of the parameters of the oscillating mass are summarized in Table 6.5.

The dolphin swimming speed values necessary for calculating the parameters of the oscillating mass are taken from Section 2.5. To calculate the parameters k_{M_1}, the averaged speed was taken with the duration of a voyage for 1 day, and for calculating the parameter k_{M_Σ} with a duration of swimming of 15–20 minutes. The density and kinematic viscosity of water are taken at a temperature of 20°C.

In Ref. [120], for the case of air flowing around the body at a speed of 10^2 m/s, the following optimum values of the elastic surface characteristics are given: membrane density $\rho_m = 1.8$ lb · s^2/ft^4 (930 kg/m^3), thickness $t_m = 0.001$ inch (25.4 · 10^{-6}m), oscillating mass $M = 1.5 \cdot 10^{-4}$ lb · s^2/ft^3 (0.0236 kg/m^2), and the parameter of oscillating mass $k_M = 13.6 \cdot 10^4$. It is shown that the smaller the value of k_M, the greater the transition of the laminar boundary layer to the turbulent one is delayed. For the case of water flowing around the body, the parameter $k_M = 2.5 \cdot 10^4$.

Comparison of the obtained results with the data from Ref. [120] with air flow shows that the values of the oscillating mass, the thickness of the membrane, and the velocity of the flow are different, and the value of the density of the membrane and the dimensionless parameter of the oscillating mass are close. This is natural, as compared with two cases of flow around the body with water and air.

Calculations show that for the values of the k_M of dolphin skin, the short-beaked common and the bottlenose dolphin values are close, while in the porpoise the k_M values are somewhat smaller. Since the size and speed of swimming of a porpoise is less than that of other dolphins, it can be assumed that it swims only in the laminar flow regime, and in this case the pressure pulsations are very small: it is sufficient to take into account only the oscillating mass of the epidermis. Therefore, the values of k_M are smaller by an order of value $k_{M\Sigma}$. The oscillating mass also characterizes the density coefficient $m = \rho_m t_m / \rho \delta$, where ρ_M is the density of the oscillating mass (kg/m^3); t_M is the thickness of the oscillating mass (m); and δ is the boundary layer thickness (m). With a decrease in the value of m, the flow stabilizes and the instability zone shifts to lower frequencies, the degree of increase in the velocities of the disturbing motion ($Re_\delta = 8775$) with $m = 10$; four; 2 decreases by 10; 15; 38%. The calculated coefficient of density m of the skin of dolphins in the region of one-third of body length according to Section 6.5 shows (Table 6.6) that at high swimming speeds, when three layers of skin fluctuate, the density coefficient m is close to the values

TABLE 6.6 Dolphin skin density coefficient

Dolphin species	$Re_\delta \cdot 10^{-3}$ at U		Thickness $\delta \cdot 10^3$, m at U		Coefficient m at U	
	Small	Large	Small	Large	Small	Large
Short-beaked common	8.7	9.9	1.47	1.29	0.135	1.69
Bottlenose	10.8	12.3	1.59	1.38	0.126	1.23
Harbor porpoise	5.7	6.3	1.29	1.14	0.158	0.72

of the coefficient m given in Ref. [120]. Obviously, to obtain small values of m, it is desirable to have a material with the lowest possible density value. This is especially important for the case of movement in air, as when water flows around the body, it is easier to obtain a material with a density commensurate with the density of water.

From Table 6.5 it can be seen that the measured density of layers 1, 2, and 3 is very close to the density of seawater. Measurements of oscillating mass values according to the mechanical model of dolphin skin (Section 6.3) show that the values of k_M and m are in the optimal range. The above results were obtained using formula (6.26) for a single cross section of a dolphin body. Below are the results of the distribution of oscillating mass along the body of various dolphins, obtained using the formula (6.26). Calculations of the k_M value were performed for short-beaked common and bottlenose dolphins for two swimming speeds. Animal body lengths (L) and swimming speeds are the same as in Section 6.6. Note that the value of U is chosen somewhat arbitrarily, since it is believed that the short-beaked common dolphin is a faster dolphin than the bottlenose dolphin, while the latter has a length and body weight that are greater and, therefore, theoretically can develop a higher speed (Section 2.5).

Table 6.7 shows the results of calculating the k_M value for a short-beaked common dolphin. The first column of Table 6.7 shows the numbers of cross sections along the dolphin body, in which skin thicknesses were determined. A diagram of these sections is given in Section 6.5 (Fig. 6.19). In the second column, Table 6.7 shows the averaged values of the thickness of the epidermis, papillary, and papillary layers of the skin, indicated by h_{1-3}. Table 6.7 shows the averaged values of the corresponding skin layers (Section 4.3, Fig. 4.16). Knowing the density and thickness of the layers, we can determine the values of M. The indexes (1–3) indicate the first three layers of skin, (4) the mesh layer of the dermis, (1–4) the top four layers of the skin, (6) the layers of skin muscles, and Σ, the sum of all layers of the skin without a layer of subcutaneous adipose tissue. Accordingly, these indices denote the values of M and k_M.

TABLE 6.7 Averaged Dolphin Skin Layers

Section no.	$h_{1-3}\,10^2$ m	M_{1-3} kg/m^2	k_M^{1-3} 10^{-4}	$h_4 10^2$ m	M_4 kg/m^2	M_{1-4} kg/m^2	k_M^{1-4} 10^{-4}	$h_6 10^2$ m	M_6 kg/m^2	$h_\Sigma 10^2$ m	M_Σ kg/m^2	k_M^Σ 10^{-4}
2	0.19	1.94	1.15	0.91	8.38	10.32	8.35	0.2	2.4	1.3	12.72	9.82
3	0.24	2.45	1.45	0.66	6.07	8.52	6.6	0.8	9.6	1.7	18.12	14.0
5	0.24	2.45	1.45	0.66	6.07	8.52	6.6	1.6	19.2	2.5	27.72	21.4
6	0.28	2.86	1.7	0.72	6.63	9.49	7.33	1.6	19.2	2.6	28.69	22.2
7	0.3	3.06	1.82	0.7	6.44	9.5	7.34	1.5	18.0	2.5	27.5	21.2
9	0.28	2.86	1.7	0.72	6.63	9.49	7.33	0.7	8.4	1.7	17.89	13.8
10	0.25	2.55	1.52	0.65	5.98	8.53	6.6	0.1	1.2	1.0	9.73	7.5
11	0.18	1.84	1.09	0.32	2.94	4.78	3.7	0	0	0.5	4.78	3.7

The distribution of the thickness of the h_6 layer along the body is characterized by the fact that, depending on the speed of navigation, the value of h_6 varies. When sailing at small and medium speeds, h_6 is of small value. When swimming at maximum speeds, the tear-off and tangential loads on the dolphin's body [26] increase, according to which the tension and thickness of the skin muscles increase, which is clearly seen in the G. Cousteau films, shown in Section 2.7 in Fig. 2.22. In these films and, according to our observations, during fast swimming, the skin muscle is strained so much that two large longitudinal folds form along the sides of the dolphin's body. A slow-moving dolphin has no such folds. Thus, depending on U, not only the thickness of the skin covers, but also the cross-sectional shape of the body changes. Therefore, the value of h_6 when calculating the coefficients k_M^{1-4} is determined from the results of measuring the distribution of the skin muscle in the area of the longitudinal folds. In connection with the above, we assume that fluctuations at medium swimming speeds (for the short-beaked common dolphin $U_{av} = 6.0$ m/s) extend only in the upper layers of the skin (h_{1-3}). With increasing U, the perturbing motion spreads to the deeper layers of the skin, therefore the calculation of the k_M^{1-4} and k_M^{Σ} values for the short-beaked common dolphin is made at $U = 7.8$ m/s. For the bottlenose dolphin, the k_M^i values were calculated, accordingly, for $U = 6.9$ and 9.0 m/s. In those places where the skin muscles are thin, when swimming at maximum speeds, the muscle tenses less than in the area of longitudinal folds. For these places, the values of k_M^{1-4} and k_M^{1-4} were calculated taking into account changes in the values of h_{1-4} and h_{Σ}.

The results of the calculations are shown in Fig. 6.26 [32], which also shaded the region of optimal values of k_M [120]. The values of k_M^{1-3} as well as the values of k_M^{1-4} are in the region of optimal values, and the values of k_M^{1-4} in the region of the longitudinal folds of the skin muscle are more optimal than in the locations of its thin layers. The values of k_M^{Σ} in the first and in the tail part of the body of dolphins have optimal values, while in the middle part of the body they are in the region of nonoptimal values. This indicates, at

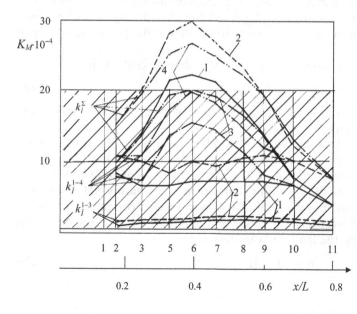

FIGURE 6.26 Distribution along the body of the coefficient of oscillating mass of the skin of dolphins of the white-sided (1, 3) and bottlenosed dolphins (2, 4) in the region of large (1, 2) and small (3, 4) skin muscle layers [32].

least, the boundaries of the distribution of fluctuations deep into the thickness of the skin layer, depending on the speed of swimming. Taking the obtained values of the k_M values of dolphin skin (Fig. 6.26) as optimal, one can determine how much the change in the U value follows from expression (6.26) to change the skin thickness, to the depth of which the disturbing oscillation extends. The calculation made for cross section (7) of the short-beaked common dolphin showed that when U is changed from 6 to 7 m/s, h_{1-3} should increase by 23% with the same skin elasticity. However, it is known (Section 6.6) that with increasing U skin elasticity increases. The thickness of the skin in the area of the longitudinal folds of the skin muscle increases, while in other parts of the body it decreases so that the overall diameter of the body D decreases and the value of h/D increases. With an increase in skin elasticity, by analogy with elastomers, the depth of vibrations entering the skin and their phase velocity increase. Given the redistribution of h_Σ in D, the value of k_M^j changes accordingly.

However, the interconnected change of the stiffness parameter k_s (Section 6.6, Fig. 6.23) and k_M with increasing speed can also occur with changes in the density or viscosity of the skin. As shown in Section 5.2, with an increase of U, the blood flow in the skin and its blood supply increase. The temperature on the surface of the body, apparently, does not significantly change (Section 5.4), which leads to a decrease in the density and viscosity of the skin, and, therefore, to an increase in the depth of propagation of oscillations and phase velocity in the skin. The simulation of such a process [170] made it possible to experimentally investigate the flow around an analogue of the skin of dolphins. Measurements showed that the velocity profile during flow around a heated elastic plate differs significantly from a rigid standard and an elastic plate without heating, both in the laminar flow in the boundary layer [170] and in the turbulent [44,70], indicating an increase in the laminar sublayer.

The same mechanism is also observed in fast-swimming fish, in which body temperature can reach 37°C [305,307]. An increase in skin temperature during fast swimming can also occur due to an increase in friction in the layers of the skin during its deformation caused by the work of the caudal fin, as well as skin dissipation of the pulsating energy of the boundary layer.

The oscillating mass of the skin is also characterized by a density coefficient

$$m = \rho_m \cdot h_i / \rho \delta, \qquad (6.28)$$

where ρ_m and h_i are, respectively, the density and thickness of the oscillating mass of the skin; and ρ is the density of seawater.

In Refs. [120,170], it was shown that the laminar boundary layer is stabilized by an elastic surface the better, the smaller the value of m, the optimal value of which is no more than $1-4$, and ρ_m must be of the same order as ρ. The above calculations showed that with a low swimming speed of dolphins, the resulting skin size m is practically not implemented, and at high swimming speeds, the value of m for short-beaked common and bottlenose dolphins was 1.69 and 1.23, respectively. Since dolphins have $\rho_m \approx \rho$, the following condition is optimal for stabilizing the laminar boundary layer:

$$h_i \approx (1 - 4)\delta \qquad (6.29)$$

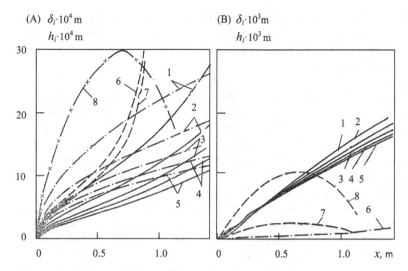

FIGURE 6.27 The distribution of the thickness of the laminar (A) and turbulent (B) boundary layers along the body at different swimming speeds. (A): (1) $U = 5$ m/s; (2) 10; (3) 15; (4) 20; (5) 25; (6) 6; (7) 7.8 m/s, (8) distribution along the body of the skin thickness h_{1-3}; (B): (1–5) the same as for (A), (6) distribution of the laminar boundary layer for the body of revolution, with $U = 5$ m/s; (7, 8) skin thickness of dolphins h_{1-3} and h_{1-4}, respectively.

Fig. 6.27 shows the dependences of the thickness of the laminar and turbulent boundary layers, as well as the dependences of the quantities h_{1-3} and h_{1-4} on the length of the body. Fig. 6.25A shows solid lines for the calculation of the δ_{lam} for the body of revolution with an extension of 4 [256], the dashed lines for the body of rotation with an extension of 5 (model No. 32) [9], and the dash-dotted lines for the plate [256]. Comparison of the regularities of the distribution of the quantities h_{1-3} and δ_{lam} for curves (1), (6), and (7) shows that condition (6.29) is fulfilled with an admissible error. Comparison of δ_{lam} for a plate and a body of revolution shows that δ_{lam} at the body of rotation is less due to the presence of a negative pressure gradient in the nose of the body. In Fig. 6.25B, the solid lines show the values $\delta_{turb.}$ calculated for the plate according to Ref. [256], taking into account the initial laminar region. Just as for the laminar boundary layer, curves (1–5) for the body of revolution, by analogy with Fig. 6.26A, will be located lower, and then, compared to curve (8), it becomes obvious that in this case condition Eq. (6.29) is also satisfied.

Thus, the results shown in Fig. 6.26 fully confirm the data in Fig. 6.27. Note that experimental investigations into skin analogues showed the best results in cases where the analogue in structure and physics of interaction with the flow was closest to nature. For example, the best results were obtained on elastic plates, in which the coefficient m was 0.5 and 0.8.

The relation (6.29) determines the region of the integuments interacting with the pulsation field of the boundary layer. It is known that the pulsating energy of the flow is concentrated in the near-wall region of the boundary layer. The most energy-carrying eddies are located on the outer region of the boundary layer. In this connection, it is necessary to investigate the characteristic value of the thickness δ of the boundary layer, with which h_i can be dimensioned, as well as the dependence of this parameter on x/L in various species of fast-swimming aquatic organisms.

We also measured the thickness h_i of the skin of fast-swimming fish, which either do not have any, or have a small area of mucous or scaly layer of the skin. The material for the study was frozen fish caught during expeditionary trips to the Atlantic Ocean. The following species of fish were studied: pelamid ($L = 0.6$ m), big-eyed ($L = 0.86$ m), striped ($L = 0.63$ m), long-eyed ($L = 0.83$ m), and spotted ($L = 0.76$ m and $L = 0.49$ m) tuna, and swordfish ($L = 1.41$ m without the sword).

To study the thickness of the skin of fish, the specimens examined were cut (Pirogov cuts) in various fixed areas of the body. Skin thickness measurements were carried out using a measuring ruler in MBS binocular around the entire perimeter of the incision at 0.01 m, which made it possible to obtain material on the nature of changes in the thickness of the skin itself (corium) depending on body length. Unfortunately, with this method of research it is impossible to measure the thickness of the epidermis, which was peeled off and lost in the process of freezing and transportation. In addition, the thickness of the mucous membrane in the relevant areas of the body was not taken into account.

Fig. 4.49 (Section 4.6) shows the distribution of skin thickness along the body of fast-swimming fish. It is noted that swordfish have a completely different change in skin thickness along the body. For fast-swimming dolphins and fish, the dependence of skin thickness and character of its distribution throughout the body on the swimming speed U and, as a result, on the hydrodynamic load of the flow medium at the interface, that is, on the skin of fast-swimming aquatic animals, is notable. Since the body sizes of the considered animal species are approximately the same, then at different speeds, a different distribution of the boundary layer thickness is formed so that the ratio:

$$h_i / \delta_i = \text{const.} \tag{6.30}$$

It can be assumed that in the process of evolution under the influence of the medium and depending on its load (the pulsating field of pressure and friction of the boundary layer), such thickness of the skin was developed that prevented the painful influence of the medium on the body (the principle of receptor sensitivity: Section 1.2), and depending

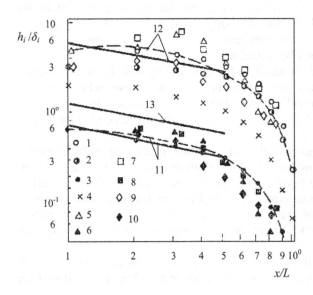

FIGURE 6.28 The distribution of the parameter of the oscillating mass along the body of fast-swimming aquatic organisms: short-beaked common dolphin $L = 1.4$ m; (1) $h_{1-3}/\delta_{\text{lam}}$, $U = 10$ m/s; (2) $h_{1-3}/\delta_{\text{lam}}$, $U = 6$ m/s; (3) $h_{1-3}/\delta_{\text{turb}}$, $U = 10$ m/s; (4) $h_{1-4}/\delta_{\text{turb}}$, $U = 10$ m/s; spotted tuna $L = 0.76$ m, $U = 15$ m/s; (5) $h_{\Sigma}/\delta_{\text{lam}}$; (6) $h_{\Sigma}/\delta_{\text{turb}}$; big-eyed tuna $L = 0.86$ m, $U = 15$ m/s; (7) $h_{\Sigma}/\delta_{\text{lam}}$; (8) $h_{\Sigma}/\delta_{\text{turb}}$; pelamida $L = 0.6$ m, $U = 10$ m/s; (9) $h_{\Sigma}/\delta_{\text{lam}}$; (10) $h_{\Sigma}/\delta_{\text{turb}}$; (11, 12) averaged curves at turbulent and laminar boundary layers, respectively; (13) theoretical curve.

on the mechanical parameters of the skin stabilizes the boundary layer. In accordance with this hypothesis, it can be expected that in high-speed species of aquatic animals the parameter Eq. (6.30) will have a universal pattern.

To test the hypothesis put forward, on the basis of Section 2.5 data, the averaged velocities of the considered animal species were determined, and the corresponding δ values were calculated from Refs. [9,256].

The values of h_i correspond to the curves (1–3) shown in Fig. 4.49 (Section 4.6). The results of calculations of relations (6.30) are shown in Fig. 6.28. The character of the dimensionless parameter h_i/δ_i along the body of the long-finned tuna, starting with $x/L = 0.5$, is the same as that of the spotted one, and the latter ($L = 0.49$ m long) is the same as that of pelamida and dolphin at $U = 0.6$ m/s. If we calculate the flow rate inside the gills of the spotted tuna, then the values of the parameter h_i/δ_i will completely coincide with those of the dolphin. The pattern of distribution of the h_i/δ_i parameter along the body is the same for all considered species of hydrobionts, however, the numerical values differ in the case of a laminar or turbulent flow around the body (Fig. 6.28).

The scatter of numerical values with a laminar flow is greater than with turbulent flow. The flow regime in the boundary layer in the fish species examined, unlike dolphins, was not measured. A transient flow regime between the laminar and turbulent boundary layers was detected. Curve (11) in Fig. 6.28 denotes the averaged dependence $h_i/\delta_i = f(x/L)$ for a turbulent flow around a body, and curve (12) for a laminar flow. In Sections 4.3–4.6 it is shown that dolphins and fish have a specific structure of the skin that promotes drag reduction. At fast swimming speeds in fish, the boundary layer becomes thinner and more turbulized than dolphins, therefore intermediate curve (13) is plotted closer to curve (11). Based on these methods [119] for curve (13), an empirical formula is proposed for $x/L = 0–0.5$:

$$lg(h_i/\delta_i) = lg\ 1.28 - 2.4^{-1}lg(x/L)\ \text{or}\ lg(h_i/\delta_i) = lg\ 1.28 - 2.4^{-1}lg(x/L). \quad (6.31)$$

At $x/L > 0.5$, the regularity Eq. (6.31) changes, as the stem (tail fin drive) moves in the oscillatory mode and the boundary layer on this part of the body becomes nonstationary. In this regard, the skin is experiencing not only the effects of a nonstationary boundary layer with a large dynamic load, but also alternating tensile forces caused by the work of the lateral muscles. Since in the caudal part of the body a suction force acts, caused by the work of the caudal fin, the mechanism for drag reduction in this area is different.

Thus, a unified pattern of the distribution of the parameter h_i/δ_i along the body was obtained for all considered fast-swimming species of animals. Numerical values of the parameter h_i/δ_i in the range $x/L = 0–0.5$ do not exceed the previously obtained values Eq. (6.29):

$$h_i/\delta_i = 0.6 - 1.28. \quad (6.32)$$

6.8 Some mechanical characteristics of dolphin skin

In Section 6.3–6.7, some criteria were considered that are necessary for determining the basic characteristics of the skin and creating optimal artificial surfaces. Parameters of elasticity of K_E and oscillating mass of κ_M dolphin skin are given in Sections 6.5–6.7. We define the parameters of limiting frequency, tension, and dampening (Section 6.3). The limit frequency parameter is characterized by the expression:

$$k_\omega = \left(\frac{E'_\Sigma}{M_\Sigma}\right)^{1/2} \frac{1}{U_\infty Re_1} = \left(\frac{k_E}{k_M}\right)^{1/2}. \tag{6.33}$$

where E'_Σ is the stiffness of the skin element (N/m^3); M is the oscillating mass (kg $\cdot$ s^2/m^3); U_∞ is the swimming speed of the dolphin (m/s); Re_1 is the unit Reynolds number, $Re_1 = U_\infty/\nu$ (m^{-1}), and ν is the kinematic viscosity of water (m^2/s).

The frequency limit parameter can be written as:

$$\kappa_\omega = \omega/Re_\delta, \tag{6.34}$$

where ω_0 is the dimensionless limiting frequency of oscillation of the dolphin's skin with zero dampening; Re_δ is the Reynolds number calculated from the thickness of the boundary layer.

Using formulas (6.33) and (6.34), the values of k_ω and ω_0 were calculated for three species of dolphins. Re_δ numbers are calculated for one-third of the body length using the formula for laminar flow around a smooth plate according to Section 6.5. The calculation results (Table 6.8) show that, at slow swimming speeds, the values of κ_ω and ω_0 are different for different species of dolphins. At fast swimming speeds for all species of dolphins, these values are close.

In Refs. [120,170], the dependence of the increase in speeds on the disturbing motion is given for different values of ω_0. When the value of ω_0 decreases, the increase in speed occurs at a slower rate. However, at $\omega_0 \leq 0.06$, class C instability arises, and at $\omega_0 \geq 1.4$, the effect of ω_0 is negligible. Comparison of these data with the calculated values of ω_0 shows that at high swimming speeds, the values of κ_ω and ω_0 of the dolphins' skin are in the range of optimal values; at low swimming speeds, these values are significantly overestimated.

Tension parameter The dimensionless velocity of propagation of an elastic wave in a flexible coating, caused by a disturbing motion in a laminar boundary layer, can be defined by the expression [120]:

$$c_{0m} = C_m/U_\infty, \tag{6.35}$$

TABLE 6.8   Parameters of the frequency of fluctuation of the skin of dolphins

Dolphin species	$Re_\delta \cdot 10^{-3}$ at U		Parameter limiting frequency $10^4 \cdot \kappa_\omega$ at U		Dimensionless limiting frequency ω_0 at U	
	Small	Large	Small	Large	Small	Large
Short-beaked common	8.7	9.9	1.14	0.222	0.99	0.22
Bottlenose	10.8	12.3	0.76	0.169	0.82	0.208
Harbor porpoise	5.7	6.3	1.72	0.275	0.98	0.173

TABLE 6.9 Dolphin skin tension parameters

Dolphin species	Tension parameter $10^{-3} \cdot \hat{e}_T$ at U		Skin tension cover T, N/m at U	
	Small	Large	Small	Large
Short-beaked common	0.69	9.7	4.2	76.4
Bottlenose	0.79	8.8	5.5	80.1
Harbor porpoise	0.53	2.4	2.4	13.6

where C_m is the wave propagation velocity in the coating, $C_m = (T/M)^{1/2}$ (m/s); and T is the value of the tension (N/m). Substituting into formula (6.34) the expressions for T and M (Sections 6.3 and 6.7), we get:

$$\hat{e}_{\dot{O}} = c_{0m}^2 \cdot \hat{e}_{\hat{I}} \tag{6.36}$$

Substituting in the expression (6.36) the values c_{0m}^2 and $\hat{e}_{\hat{I}}$, we will have

$$\hat{e}_{\dot{O}} = \dot{O}/\mu \, U_\infty \tag{6.37}$$

The increasing speeds of the disturbing motion decrease with decreasing c_{0m} [120]. However, the c_{0m} value cannot be lower than the phase velocity of Tollmien-Schlichting waves, and a significant decrease in the increasing speeds is possible only at $c_{0m} < 1.0$. Therefore, the optimal range of values of c_{0m} is in the range $0.48 \leq c_{0m} \leq 1$.

The value of tension T on the dolphin was not measured. Calculations showed that the values of the parameters $\hat{e}_E$, $\hat{e}_{\hat{I}}$, and κ_ω are optimal. Given that the parameter $\hat{e}_T$ is directly dependent on the parameter $\hat{e}_E$, we assume that the c_{0m} value for dolphin skin is optimal. Values $\hat{e}_T$ and T, calculated at $c_{0m} = 0.75$, are given in Table 6.9. Small T values are due to the fact that the direction of hydrostatic forces contributes to a decrease in T values. With diving and a significant increase in hydrostatic pressure in dolphins, the value of T practically does not change due to some physical laws (Section 5.1).

The dampening parameter can be determined from dynamic tests according to the expression:

$$\kappa_\eta = 2\zeta = \Delta/\pi, \tag{6.38}$$

where ζ is the dampening factor; and Δ is the logarithmic attenuation coefficient.

Since dynamic studies of the mechanical characteristics of the skin of live dolphins were not carried out due to the lack of necessary equipment, and the prepared skin significantly changes its properties, the dampening parameter can be determined by the formula [120] (Table 6.9):

$$\kappa_\eta = d \cdot Re_\delta / \kappa_M \alpha_0 c_0 \tag{6.39}$$

Here d is the dampening coefficient:

$$d = \eta/\rho U_\infty, \tag{6.40}$$

where η is the viscosity or dampening of the skin element ($N \cdot s/m^3$), ρ is the density of seawater (kg/m^3), and α_0 is a dimensionless wave number:

$$\alpha_0 = 2\pi\delta/\lambda, \tag{6.41}$$

where δ is the boundary layer thickness; and λ is the wavelength of the disturbing motion.

The dimensionless propagation velocity of the disturbance wave over the elastic surface c_0 was calculated according to the expression:

$$c_0 = \left(c_{0i}^2 + \omega_0^2/\alpha_0^2\right)^{1/2} = \left(c_{0i}^2 + \kappa_\omega^2 \cdot Re_\delta^2/\alpha_0^2\right)^{1/2}. \tag{6.42}$$

The dampening parameter can be defined in two ways.

1. The aperiodic membrane motion law

For the aperiodic membrane motion law we can write the equation:

$$\varepsilon_1/\varepsilon_0 = \exp\left(-\tau \cdot \eta/2M\right) \cdot \left[(1/2 + \eta/4M \cdot k_1)\exp(k_1 \cdot \tau) + (1/2 - \eta/4M \cdot k_1) \times \exp(-k_1 \cdot \tau)\right], \tag{6.43}$$

where ε_1 and ε_0 are the final and initial values of the deformation of the flexible surface (m); and τ is the relaxation time (s);

$$k_1 = \left[\left(\eta/2M\right)^2 - E'_\Sigma/M\right]^{1/2}.$$

It can be shown that the following relationship is performed:

$$\eta/2M \approx (1 - 3)\sqrt{\left(E'_\Sigma/M\right)} \tag{6.44}$$

Substituting the value d according to Eq. (6.40) and the value $\hat{e}_i$ from Section 6.7 into expression (6.39) and taking into account formula (6.44), we have:

$$k_\eta = \eta/2\dot{I} \cdot (2\delta/U_\infty \cdot \alpha_0 \cdot c_0) \approx \sqrt{\left(\mathring{A}'_\Sigma/\dot{I}\right)} \cdot \delta/(U_\infty \cdot \alpha_0 \cdot c_0) = (k_\omega \cdot Re_\delta/\alpha_0 \cdot c_0) \tag{6.45}$$

To calculate the values of the dampening parameter, the values $\mathring{A}'_\Sigma$, M, and U determined from Sections 6.6, 2.5, and 6.7, the value δ from Table 6.8, c_0 and α_0 according to the formulas (6.41) and (6.42), and we find λ according to the theory of stability and the data of experimental studies of stability on a plate [170] with the maximum increase value.

2. The damping parameter can also be determined as follows.

Experimentally, the value of the relaxation time for different species of dolphins was sought. Then, according to the expression $\eta = \tau \cdot E$ (§6.3), the value of d was determined by the formula (6.40), and the dampening parameter determined by formula (6.39). The values of Re_δ, k_M, c_0, and α_0 are found by the method described above.

The method for determining the relaxation time is as follows. When measuring the elasticity of the skin of live dolphins (Sections 6.6 and 6.7), after removing the measuring

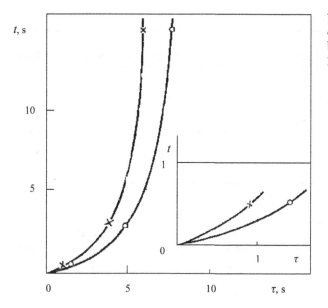

FIGURE 6.29 Interpolation graph for determining relaxation times: (1) short-beaked common dolphin, (2) harbor porpoise.

TABLE 6.10   Dolphin skin strain relaxation times

Dolphin species	Relaxation time $10^2 \cdot \tau$, s at U		Viscosity (dampening) $10^{-3} \cdot \eta$, N · s/m³ at U		Coefficient dampening d at U	
	Small	Large	Small	Large	Small	Large
Short-beaked common	1.3	1.0	1.09	1.01	0.176	0.126
Bottlenose Harbor porpoise	1.14	0.86	0.9	0.83	0.127	0.0897
	3.0	2.46	1.36	1.75	0.293	0.298

device from the body, a dent in the measuring rod remains on the skin. Let us assume that the relaxation time is a period of time during which the dent on the skin of the dolphin decreases by approximately "e" times. In this case, the load created by the measuring rod was maintained for 15, 3, and 0.5 seconds. Fig. 6.29 shows the results of measuring the relaxation time as a function of the load holding time for two types of dolphins.

Disregarding the pressure gradient and assuming that the disturbance band has a width of 0.01 m, we calculate the duration of the disturbances when dolphins are swimming with speeds equal to those accepted in Section 6.5. According to the schedule (Fig. 6.29) the corresponding relaxation times are determined, wherein the initial part of the relaxation time curves is approximated by straight lines, respectively, for the short-beaked common dolphin and harbor porpoise, at $t = 0.385\tau$ and $t = 0.228\tau$. For bottlenose dolphins,

TABLE 6.11 Dolphin skin damping coefficient

Dolphin species	α_0 at U Small	Large	c_0 at U Small	Large	κ_η at U Small	Large
Short-beaked common	0.54	0.534	1.38	1.1	1.68	0.122
Bottlenose	0.5	0.495	1.34	1.1	1.37	0.13
Harbor porpoise	0.585	0.57	1.36	1.05	2.24	0.746

TABLE 6.12 Dolphin skin damping parameters

Dolphin species	Viscosity (dampening) $10^{-3} \cdot \eta$, N·s/m^3 at U Small	Large	d at U Small	Large	κ_η at U Small	Large
Short-beaked common	0.83	3.0	0.135	0.374	1.32	0.376
Bottlenose	0.8	2.36	0.112	0.254	1.16	0.378
Harbor porpoise	0.73	1.81	0.158	0.314	1.24	0.644

the relaxation time is assumed to be the same as for the short-beaked common dolphin. The results of the calculation of relaxation times are summarized in Table 6.10 (the values of η and d are calculated by the second method). The calculated values of α_0, c_0, and k_η using formulas (6.41), (6.42), and (6.39) are presented in Table 6.11. For comparison, Table 6.12 shows the values of η, d, and k_η obtained by the first method. The discrepancy between the values of these parameters is explained by the fact that the first method was calculated for an aperiodic disturbance, and α_0 is taken from the condition of harmonic oscillation. The second method of calculation is also approximate due to the error in determining the relaxation times.

In Ref. [120], it was shown that with a decrease in the dampening coefficient, the rate of increase of the disturbing motion decreases.

The dampening coefficient should not be very small, since this may cause class B hydrodynamic instability. However, at $d \geq 0.3$, the increase in speeds becomes large compared to the flow around a rigid body. It is shown that the relative effect of dampening on the increase in speeds is significant if the values of M and c_M are small, otherwise the effect of dampening is insignificant. The results of calculations of the value of d by the second method showed that the dampening coefficient for the short-beaked common and bottlenose dolphins is in the optimal range, especially for high swimming speeds. The d values obtained by the first method are overestimated. A comparison of the performed calculations with the data from Ref. [120] shows that, when calculating by the second method for high swimming speeds, the dampening parameters have optimal values. When calculated by the first method, they are less favorable.

Measurement of the dampening on the prepared skin of a dolphin using the ball rebound method [257] showed that for skin, the relative dampening is 20%, and for rubber with the same thickness and rigidity it is 40%. Experimental data with flexible coatings

TABLE 6.13 Coefficients of stiffness, fluctuating mass and frequency of oscillation of the outer layers of the dolphin's skin

Parameter	Species	Average values Sections							
		2	3	4	5	7	9	10	11
$h_{1-3} \times 10^2$ (m)	Short-beaked common	0.19	0.24	0.24	0.28	0.3	0.28	0.25	0.18
	Bottlenose	0.22	0.27	0.27	0.31	0.33	0.31	0.28	0.21
$k_s^{1-3} \times 10^5$ (suboptimal)	Short-beaked common	61.8	48.8	47.4	38.4	34.2	39.8	45.8	70.3
	Bottlenose	342	25.1	22.8	21.8	19.7	25.0	28.0	36.7
$k_M^{1-3} \times 10^{-4}$ (optimal)	Short-beaked common	1.15	1.45	1.45	1.7	1.82	1.7	1.52	1.09
	Bottlenose	1.55	1.9	1.9	2.18	2.32	2.18	1.97	1.47
$k_\omega^{1-3} \times 10^5$ (suboptimal)	Short-beaked common	23.2	18.4	18.1	15.0	13.71	15.3	17.4	25.4
	Bottlenose	14.9	11.5	10.96	10.0	9.22	10.7	11.9	15.8

TABLE 6.14 Coefficients of stiffness, fluctuating mass and frequency of oscillation of the entire thickness of the dolphin's skin

Parameter	Dolphin species	Sections							
		2	3	4	6	7	9	10	11
$h_{1-4} \times 10^2$ (m)	Short-beaked common	1.1	0.9	0.9	1.0	1.0	1.0	1.0	0.5
	Bottlenose	1.3	1.2	1.0	1.2	1.1	1.3	1.2	0.9
$h_\Sigma \times 10^2$ (m)	Short-beaked common	1.3	1.7	2.5	2.6	2.5	1.7	1.0	0.5
	Bottlenose	1.7	2.1	2.8	3.0	2.8	2.1	1.4	0.9
$h_s^\Sigma \times 10^5$ (optimal)	Short-beaked common	9.0	6.9	4.74	3.86	3.92	6.9	11.7	23.5
	Bottlenose	4.37	2.92	1.55	1.39	1.39	2.64	5.06	11.9
$k_M^{1-4} \times 10^{-4}$ (optimal)	Short-beaked common	8.35	6.6	6.6	7.33	7.34	7.33	6.6	3.7
	Bottlenose	10.7	10.0	8.4	10.0	9.2	10.8	10.0	7.54
$k_M^\Sigma \times 10^{-4}$ (suboptimal)	Short-beaked common	9.82	14.0	21.4	22.2	21.25	13.8	7.5	3.7
	Bottlenose	15.1	19.8	27.9	29.6	26.8	19.5	12.2	7.5
$k_\omega^{1-4} \times 10^5$ (optimal)	Short-beaked common	3.28	3.24	2.68	2.3	2.31	3.06	4.2	7.97
	Bottlenose	2.02	1.71	1.36	1.18	1.23	1.56	2.25	3.97
$k_\omega^\Sigma \times 10^5$ (optimal)	Short-beaked common	3.03	2.22	1.49	1.32	1.36	2.24	3.95	7.97
	Bottlenose	1.7	1.21	0.75	0.69	0.72	1.16	2.04	3.98

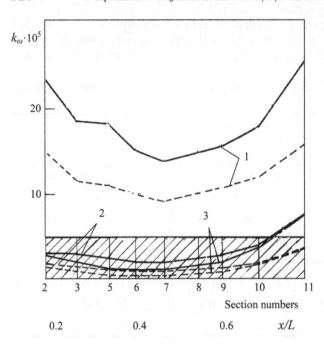

FIGURE 6.30 The distribution along the body of the parameter of the limiting frequency of oscillation of the skin of short-beaked common and bottlenose dolphins: (1) k_ω^{1-3}; (2) k_ω^{1-4}; (3) k_ω^{Σ}.

carried out with a relative dampening of these coatings of 44–47% are difficult to compare with the measurements obtained, but they also show that the d or κ_η values in dolphins are small [180].

The results of the investigation into the parameters characterizing the properties of the skin, indicate that the bottlenose and short-beaked common dolphin skin in all the studied characteristics are within optimal limits for stabilizing the laminar boundary layer.

The propagation of oscillations in the skin of hydrobionts is also characterized by the parameter of the limiting frequency (Section 6.3):

$$k_\omega = \left(k_s/k_{\hat{i}}\right)^{1/2} \tag{6.46}$$

The results of k_ω calculations for various swimming speeds of short-beaked common and bottlenose dolphins are summarized in Tables 6.13 and 6.14, and are illustrated in Fig. 6.30.

The region of optimal k_ω values is shaded for the data from Ref. [120]. The calculation showed that the values of k_ω^{1-3} during slow swimming are not optimal, while the values of k_ω^{1-4} and k_ω^{Σ} calculated for high swimming speeds are located in the region of optimal values, the values of which are in accordance with Fig. 6.25 (Section 6.7) and Fig. 6.30 are:

$$k_M = (1 - 20) \cdot 10^{-4}; k_\omega = (0.2 - 5)10^{-5}, k_s = (0.2 - 6.5)10^{-5} \tag{6.47}$$

The tension parameter $\hat{e}_T$ and the tension value T, calculated according to formula (6.36), the tension values along the dolphin body can be determined from other considerations. In Section 3.4, Figs. 3.4 and 3.5, the dependences of the drag coefficients on the

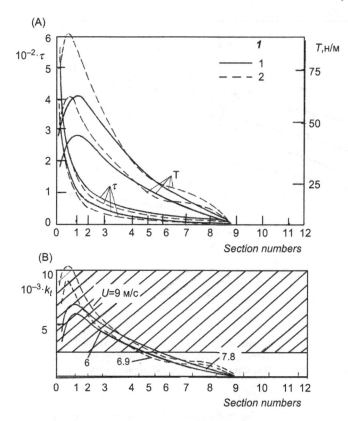

FIGURE 6.31 The distribution of the tangential stress τ and the tension T in the skin of dolphins (A), as well as the tension parameter k_T along the body of short-beaked common dolphins (1-solid lines) and bottle-nosed dolphins (2-dashed lines) at the swimming speed of the short-beaked common dolphin, $U = 7.8$ and 6 m/s, and bottlenose dolphin, $U = 9$ and 6.9 m/s. (B) - shaded area - optimal calculations in Ref. [120].

Reynolds number are given, experimentally obtained for different swimming regimes. The value of hydrobiont resistance R consists of three main components: the friction drag of the smooth surface of the body R_{friction}, the resistance of the shape R_{form}, and the inductive resistance of the fins of $R_{\text{inductive}}$ (Section 2.2). In the previous sections, it was shown that dolphins have a number of adaptations, which can significantly reduce the components of R_{form} and $R_{\text{inductive}}$.

It can be assumed that the resistance of dolphins by 90% is determined by the value $R_{\text{friction}} = C_x \cdot \Omega \cdot \rho\, U_0^2/2$, where C_x is the dimensionless coefficient of hydrodynamic resistance force related to the total area Ω of the wetted surface of the hydrobiont, including the surface area of the fins. At the chosen swimming speed, the Reynolds number is determined and the drag coefficient is determined from the experimental dependencies shown in Figs. 3.4 and 3.5. The area of the wetted body surface can be determined as shown in Section 5.1 (Fig. 5.2), and the area of the wetted surface of the fins according to the data given in Ref. [231]. Fig. 6.31 shows the calculated values of the tension T in the skin of a dolphin and the tension parameter k_T for short-beaked common dolphins and bottlenose dolphins. Fig. 6.31A also shows the distribution of shear stress τ along a flat rigid plate in a laminar boundary layer.

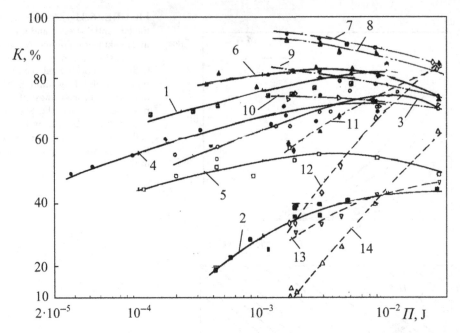

FIGURE 6.32 The dependence of the dampening coefficient on the disturbances energy (potential energy of the falling balls): (1–6) elastic materials (for notation, see Table 6.2, Section 6.5); (7, 8) healthy; (9) sick; (10) dead dolphins; (11) loose steel strip lying on a wooden plate; (12) silicate glass; (13) wooden plate, varnished; (14) plexiglass [18,50,62,70].

In Fig. 6.31B, the hatched area corresponds to the optimum values of the tension parameter k_T for elastic coatings in the air flow calculated in [120]. The shear stress of friction on the wall τ can be calculated by the formula $\tau = c_f \cdot \rho \cdot U^2/2$ [175,256].

Fig. 6.32 shows the results of measuring the absorbing properties of dolphin's skin using the ball rebound method, according to which steel balls of various mass m were dropped from a fixed height in a glass cylinder onto the surface of the test material when the holding electromagnet was disconnected [18,47,50,62,70]. The height of the rebound of the balls h from the material under study was determined with respect to the initial height of the fall of the balls h_0. The ordinate axis shows the value of the damping coefficient: $K = (1 - h/h_0) \cdot 100$, and the abscissa axis shows the potential energy of disturbance (falling balls) acting on the skin of a dolphin or on elastomer samples: $\Pi = mgh_0$.

It can be seen that the dependence of the dampening coefficient for the skin of live dolphins is significantly different from the characteristics of the skin of the sick or dead dolphins, as well as from elastomers and construction materials. The main difference is that the absorption coefficient of the skin of live dolphins approaches 100% at a disturbance energy value corresponding to the pressure pulsation energy of the turbulent or transition boundary layers. In the accepted range of energy change with increasing Π, the K value of dolphins monotonously decreases, the construction materials increase, and elastomer samples have a maximum in the region of $\Pi \approx 2 \cdot 10^{-3} - 2 \cdot 10^{-2}$ J. Because absolute energy dissipation should asymptotically approach the limiting value as the total energy increases, the relative scattering in the form of the dampening coefficient K should be at the maximum. Unlike elastomers, for the outer covers of dolphins, the maximum will be in the region $\Pi \approx 1.5 \cdot 10^{-3}$ J, and for structural materials in the region $\Pi \approx 3 \cdot 10^{-2}$ J.

As the integuments interact with velocity and pressure pulsations in the boundary layer, it is advisable to estimate the disturbance energy of the turbulent boundary layer by determining its position on the abscissa axis (Fig. 6.30). It is known that the energy of the pressure pulsations p' on the wall is related to the flow velocity U_∞ by the dependence

$$\sqrt{(\overline{p'})^2}/(\rho U_\infty^2/2) = \alpha\tau \cdot \lambda, \tag{6.48}$$

where α_τ is the proportionality coefficient (for a smooth wall, $\alpha_\tau \approx 2.1$), and the coefficient λ is determined from the relation

$$\lambda = \frac{2\tau_0/\rho}{U_\infty^2} = \frac{2\overline{u'v'}}{U_\infty^2}, \tag{6.49}$$

where τ_0 is the shear stress on the wall, $2\overline{u'v'}/U_\infty^2$ is the Reynolds stress, expressed in terms of the longitudinal u' and transverse v' components of the pulsation velocities of the boundary layer.

Measurements carried out in air [44,70] showed that $\lambda = 0.0046$ for a rigid plate and $\lambda = 0.0034$ for an elastic plate. Substituting these values in Eq. (6.48), we obtain for the flow velocity $U_\infty = 10$ m/s in the case of a rigid plate $\sqrt{(\overline{p'})^2} = 4.83 \cdot 10^2$ Pascal and elastic $\sqrt{(\overline{p'})^2} = 3.57 \cdot 10^2$ Pa. The same calculations in Section 6.7 determined for a rigid plate $\sqrt{(\overline{p'})^2} = 3.3 \cdot 10^2$ Pa. If we convert for this speed the results of measurements of the transformation ratio $\alpha_\tau \cdot \lambda$ of measurements on dolphins [251] with passive swimming $\sqrt{(\overline{p'})^2} = 4.45 \cdot 10^2$ Pa, and with active swimming $\sqrt{(\overline{p'})^2} = 1.95 \cdot 10^2$ Pa. In Refs. [251,269], the scale of the dissipative vortex during the flow over dolphins (1 sm) and the height from which the lumps of fluid "fall" on the wall ($y/\delta \approx 0.15$–0.25) are estimated, respectively. If we take the distribution of the boundary layer δ over the dolphin's body as calculated in Section 6.7, then the pressure pulsation energy on the wall for the middle part of the body at $\sqrt{(\overline{p'})^2} = 4.83 \cdot 10^2$ Pa will be $\Pi = 0.996 \cdot 10^{-4}$ J, and at $\sqrt{(\overline{p'})^2} = 1.95 \cdot 10^2$ Pa $\Pi = 0.39 \cdot 10^{-4}$ J.

Given the results of these calculations, as well as the value of the pulsations, the speed, which is 1.5–2.0 times higher in the transitional flow regime than in the turbulent boundary layer [44,70], we assume (Fig. 6.30) that the maximum absorption capacity of the skin has to the energy region corresponding to the transition boundary layer. In contrast to the skin, all elastomer samples are characterized by an absorption maximum for energy values that are several orders of value larger than those existing in the boundary layer.

The structure of the skin of dolphins illustrates the interconnectedness of body systems that affect the stabilizing properties of the skin. Increasing swimming speed increases muscle tension, including skin, blood flow and blood filling of the skin, changes in body geometry, which are interconnected, reflexively and automatically leading to optimal performance of the propulsion complex, energy metabolism in the body, and setting the stabilizing properties of the skin, characterized, in particular, by the parameters k_s, k_M, and k_ω. The thickness of the outer layers of the skin is determined by the thickness of the boundary layer, and that of the inner layers by the frequency spectrum and the energy of the pulsation speeds of the boundary layer.

The ordered longitudinal structure of the outer skin layers (Sections 4.1−4.3) with an increase in blood flow in the blood vessels of the dermal papillae stimulates the maintenance of stable longitudinal vortices in the boundary layer (Section 5.3). Such longitudinal vortices can also be stable due to the specific interaction of the flow with the skin. The longitudinal and transverse stratification and anisotropy of the properties of the skin allows us to consider the structure of the outer layers of the skin as waveguides [44,70]. Oscillations penetrating the skin, partially dissipate in the skin, and are partly reflected from the papillary layer of the dermis and spread along the body between the rows of dermal papillae, as in waveguides. Some of the energy of these oscillations dissipates, and the rest is dampened by the work of the caudal fin. Thus, most of the energy of disturbing oscillations is dissipated by layers in the skin, but some of the energy is reflected by the underpapillary layer and is formed in waveguides—rows of dermal papillae for stimulating stable longitudinal vortices like Görtler vortices in the viscous sublayer of the boundary layer.

6.9 Propagation speed of oscillations on dolphin skin

To determine the basic mechanical characteristics of the skin of hydrobionts, the important parameters and the regularities are shown in Fig. 6.33 [231].

The number of propulsive waves $m(Re_L)$, stacked on the body length of hydrobionts, depending on their Reynolds number, can be calculated using the formula:

$$m = k_4 \cdot Re^P, \quad \Phi = \text{idem},$$

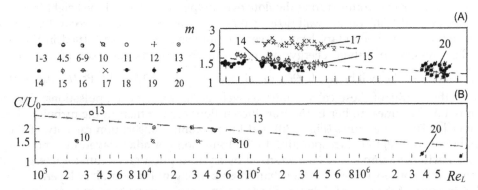

FIGURE 6.33 Dimensionless characteristics of various forms of swimming of hydrobionts with bending-oscillatory movements of the body and fin mover: (A and B) wave characteristics by the number of propulsive waves on the body m (Re_L) and relative phase velocity of propagation of these waves $c/U_0 = f(Re_L)$. Fish: (1−3) rainbow trout; (4, 5) church; (6−9) silver carp; (10) eel and black fish; (11) tropical fish; (12) different types of fish; (13) different types of combos fish; (14) lake trout; (15) mullet; (16) louvar; (17) garfish. Dolphins: (18) harbor porpoise; (19) short-beaked common; (20) bottlenose [231].

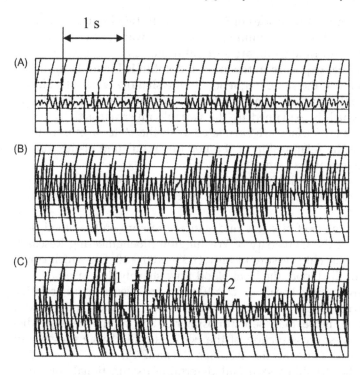

where $k_4(\Phi)$, $p \approx -0.074$. The characteristic is nonlinear, such as exponential function. In the cases considered, $m = 1.1-2.5$; the greatest number of waves in an eel-like fish (garfish), the smallest in a dolphin with a fin mover, and intermediate values of fish of the combroid type (trout and mullet).

The relative speed of the propulsion wave, $c/U_0 = f(Re)$, differs significantly depending on the Reynolds number and the type of hydrobiont, whether it has eel-like or combroid type of swimming. For eel-like fish

$$c/U_0 \approx \text{const} = 1.55, \quad 10^3 \leq Re \leq 2 \cdot 10^6,$$

that is, Reynolds self-similarity is characteristic. U_0 is the average swimming speed. For combo fish and dolphins there is approximately nonlinear type dependence of exponential function

$$c/U_0 = k_5 \cdot Re^q, k_5 \approx \text{const}, q \approx 0.075.$$

The results of some experiments performed in the United States at the Department of Psychology and Physiology at the University of California, where measurements were made to compare microvibration of the skin surface in humans and dolphins under identical conditions, are given in Ref. [231] (Fig. 6.34).

In humans, in air, microvibrations of the skin surface are regular, with an amplitude of $1-5\,\mu m$ and a frequency of 11 Hz. In a dolphin lying on foam in the air, microvibrations are also regular, but with much larger amplitudes of 5−20 μm, with two frequencies being distinguished: 2.5 and 13 Hz. In a dolphin swimming in shallow water, microvibrations are not as regular. They change with time (compare sections 1 and 2 of Fig. 6.34C). Thus, the sensitivity of the skin to external exposure in the air of a dolphin is higher than that of a human, by half an order of value at the same frequency. In the shallow water of a slowly floating dolphin, the experimental recording of microvibrations of the skin was variable in amplitude and less rhythmic, which is explained by the variable muscular activity of the dolphin. Microvibrations of the skin surface are correlated with the state of muscle tone, thereby experimentally showing the relationship of the nervous regulation of the state of the epidermis and the deeper layers of the skin.

The results of studies conducted on dolphins located in the air on foam rubber are not correct. Such studies should be carried out in a box filled with water so that the dolphin's body is in a suspended state.

Special experiments to study the effect of stress under local pressure on the blood flow in human skin were carried out at the University of Washington Bioengineering Research Center. With an increase in the external pressure of only $(1.33−2.00) \cdot 10^3$ Pa, which corresponds to the venous pressure of the blood, the local blood flow sharply decreased.

With an average dolphin swimming speed of $U = 10$ m/s, an overpressure (relative to the local hydrostatic pressure) is created in the nose of the body about $\rho \cdot U_0^2/2 = 5.1 \cdot 10^4$ N/m^2, that is, more than semiatmospheric, and at the peak of discharge, approximately in the largest cross section of the body, is a discharge pressure of approximately 20% of the indicated value (Section 2.3, Figs. 2.5 and 2.15).

To study the properties of the dolphin's skin and determine its functional role, it is advisable to study the parameters of the propagation of surface oscillations over the skin and to study the behavior of the cover when the dolphin moves freely in water. The main parameter of the surface fluctuations of the skin is the speed of propagation and attenuation of oscillations with distance. Kidun carried out experimental investigations into the velocity of propagation and attenuation of forced oscillations on the skin surface of a live dolphin, while at rest and while swimming in water [146].

The propagation velocity was measured by a variable base phase method [93]. Fluctuations on the skin of a dolphin were excited by a rod with a disc hole-like tip of an electrodynamic pathogen. Phase interval multiples of π were measured with a probe vibration sensor with a preamplifier and a distance meter and recorded with a phase matching indicator. The measurements were carried out in the frequency range of 50−400 Hz on the skin surface of a bottlenose dolphin.

It has been established that in the studied frequency range, surface oscillations are excited on the skin of the dolphin, spreading in all directions from the point source pathogen in the form of traveling waves. The traveling waves appear on the cover of both living and dead dolphins. The speed of propagation of C_m fluctuations is different in different areas of the skin. Fig. 6.35 presents a graph of the velocity of propagation of C_m oscillations on the skin of one side of a dolphin, where the axes of the projection of the skin of one side of a dolphin are plotted along the x and y axes and the value of the velocity C_m along the z axis [146]. The scheme of the measurement points is the same as in Section 4.3 (Fig. 4.26) and in Section 5.4 (Fig. 5.4). It can be seen (Fig. 6.35) that the speed of propagation of C_m oscillations on the skin of the head (Section 6.4) is relatively high, in the neck

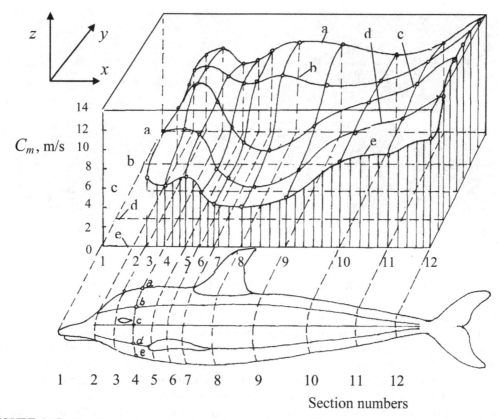

FIGURE 6.35 The distribution of the phase velocity of forced oscillations on the skin of a bottlenose dolphin. (1–12) - the measurement scheme is the same as in Section 4.3 Fig.4.17 and Section 5.4 Fig.5.18 [70,146].

(Section 6.6) it falls, from the middle of the body (Sections 6.8 and 6.9) it rises again, and then further falls slightly. At the beginning of the caudal stem and further along the stem, C_m rises. The change in speed along lines a, b (Fig. 6.35) is smaller compared with the change in speed along lines c, d, and e located closer to the dolphin's belly.

For some dolphins, this picture was somewhat disturbed due to the presence over the skin of defects in the form of scars, wounds, or cuts. A reflection of the oscillations was observed in these places, which led to the formation of standing waves. The velocity of propagation of surface oscillations, as shown in Fig. 6.35, varies from 4 to 12 m/s, that is, it lies in the range of swimming speeds of dolphins. For comparison, we point out that the speed of oscillation propagation on a tense muscle of a human forearm is about 25–30 m/s. The speed of propagation of oscillations does not depend on the direction of propagation. The speed measurement was carried out at a frequency of 200 Hz. The relative error in measuring the velocity, which consisted of the error in counting intervals of multiples of n and the error in indicating the coincidence of the phases, did not exceed 7%–10%. The data on the speed of propagation of oscillations over the skin shown in Fig. 6.35 represent the

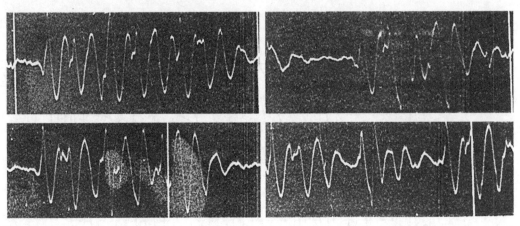

FIGURE 6.36   Photos of oscillograms of oscillations of the surface of a dolphin's skin during slow swimming and at rest (time stamps correspond to 0.1 s) [70,146].

averaged picture for six adult bottlenose dolphins. The lengths of their bodies were within 1.8—2.5 m.

The same method was used to measure the attenuation of oscillations on the skin of live dolphins in the same frequency range. Oscillations were excited on the skin with an electrodynamic vibrator. The wavelength was measured by the phase method, and the dampening of the oscillations by the amplitude comparison method. Measurements showed that in the frequency range of 50—400 Hz, the value of the logarithmic dampening decrement depends little on the frequency and for the skin of a stationary live dolphin ranges from 0.71 to 0.81, that is, at a distance equal to one wavelength, the amplitude of oscillations decays by 2—2.5 times.

In addition to measuring the parameters of forced vibrations on the cover of a motion-less dolphin, oscillations were recorded on the surface of the skin of a dolphin moving freely in water. The transmission of oscillations in the form of electrical signals from a dolphin was carried out over a flexible thin shielded cable. The dolphin was placed in a coastal aviary with a size of 6×30 m. A vibration sensor was attached to it, from which signals were transmitted via a flexible cable to receiving and recording equipment. A belt was fixed on the dolphin in front of the dorsal fin. On one side in the square between Sections 6 and 7 and the longitudinal lines b and c (Fig. 6.35) a probe vibration sensor was installed so that the sensor vibration probe fit to the dolphin cover and was in front of the belt. On the second side, on the opposite side, a communication cable was attached to the belt. Dolphin speed was estimated visually. There were areas of free swimming, acceleration, and deceleration of the dolphin in the water. When the dolphin was moving in water at low speed and at rest, oscillations of the skin were recorded (Fig. 6.36). The oscillations were pulsed, single, or in groups, following each other with a period of 0.1—0.12 seconds, a duration of 0.035—0.09 seconds, and a filling frequency of 130—140 Hz. During the recording, there were no jerks in the movement of the animal or twitching of the communication cable.

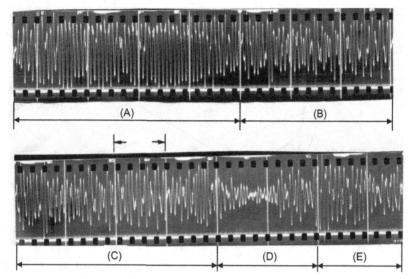

FIGURE 6.37 Photos of oscillograms of oscillations of the dolphin's skin surface during a jump through the net: (A) slow swimming, (B) acceleration, (C) inertial motion, (D) movement in air, (E) entrance to water. Jump cycle: (B − E) (marks, 1 s) [70,146].

To ensure the rapid movement of the dolphin in the water, it was decided to use its acceleration in the aviary during a jump through the movable network partition. To this end, at the beginning of the experiment, the partition in the aviary moved in the direction that the dolphin was. Seeing the diminishing space in the aviary, the dolphin, with acceleration, made a spike through the partition. Then the partition was moved in the opposite direction and the dolphin made a second jump back. Later, the dolphin learned the task and, at the command of the experimenter, made jumps. The height surmounted by a dolphin during a jump above the water reached 1.5−2 m. During the jump, the dolphin emerged from the water at about a 45−50 degrees angle at a speed of about 8−9 m/s. From the oscillogram in Fig. 6.37 it can be seen that the duration of the entire jump was ∼ 1.0 seconds. The frequency of oscillations on the skin was variable and lies in the range from 115 to 230 Hz. The frequency of oscillations during acceleration increased, was maintained relatively constant, and then decreased. On the oscillogram there are special places where the oscillation phase changes by 180 degrees. These places correspond approximately to the time the animal entered the water in the process of jumping. During the experiment, about 10 jumps were recorded. To verify the reliability of the data obtained, after the end of the experiment, the effects of jerks, bends of the supply cable, and the flow stream on the vibration probe were investigated. Studies have shown that cable bends and the effect of water flow by the sensor were not recorded.

Occurrence of the detected oscillations can occur with the help of the central nervous system or locally due to self-oscillations of the cover during movement. Since the speed of propagation of electrical impulses in the muscle tissues of animals lies within 15−18 m/s, then with the step of alternating the structures of the skin cover 5−6 cm, the time of

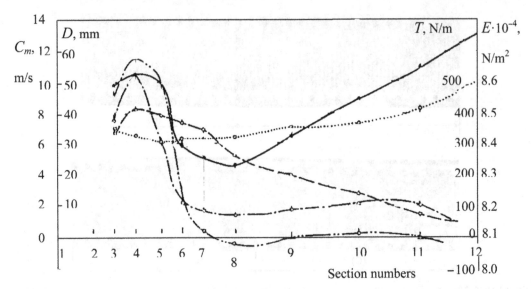

FIGURE 6.38 Propagation of the phase velocity of forced oscillations of the skin and other mechanical characteristics along the dolphin body: (1) C_m value; (2) modulus of elasticity E; (3) skin thickness D; (4) static tension T; (5) the total tension of the skin during movement [70,146].

passage of two structures, that is, one wavelength, will be equal to 0.008–0.0055 seconds, which corresponds to a frequency of about 125–180 Hz. This suggests that the detected oscillations can be excited by the nervous system. On the other hand, it should be noted that the frequency of energy-carrying velocity pulsations in a turbulent flow lies within 100–200 Hz. It is possible that such pulsations can create surface oscillations on the skin. Measurements were also performed using wireless equipment that measures skin vibrations at two points and the speed of movement in the open sea ($U = 3$–4 m/s) and in a long hydrodynamic channel in which the speed of the dolphin was $U = 7$ m/s. In the experiments an accelerometer and vibration sensor were applied. At $U = 7$ m/s, oscillations of the skin surface were recorded in the form of pulses with a duration of 0.1–0.2 seconds at a filling frequency of 240–270 Hz.

Fig. 6.38 shows a comparison of the experimental results obtained for the distribution of the velocity of surface fluctuations on the skin along the dolphin body (Fig. 6.35), the elastic modulus of the skin (Section 6.6), and the thickness distribution of the skin along the dolphin body (Section 4.3, Fig. 4.27). It can be seen that the velocity of propagation of oscillations on the surface of the dolphin's body along line c (Fig. 6.34) correlates with the elastic modulus and the total thickness of the skin. With an increase in elasticity and a decrease in the thickness of the cover, the velocity of propagation of oscillations increases. Using data from the total thickness of the skin cover and determining its average density from the density of individual layers (Section 6.7, Table 6.5), we find for line c (Fig. 6.35) the distributed mass of the cover.

In accordance with expression (6.8) for the velocity of propagation of oscillations, it is possible to calculate the tension of the cover from the relationship:

$$T = C_m^2 M, \tag{6.50}$$

where T is the tension of the skin (N/m); C_m is the speed of propagation of oscillations on the skin (m/s); and M is the distributed mass of the skin (kg/m^2).

The tension of the skin was calculated using experimental data on the velocity of propagation of C_m fluctuations and the distributed mass of the skin M (curve 4, Fig. 6.38).

The tension of the skin was calculated using experimental data on the velocity of propagation of C_m fluctuations and the distributed mass of the skin M (curve 4, Fig. 6.38). The calculated tension is static, as it characterizes the dolphin cover at rest. It is defined for line c on one side of the skin (Fig. 6.35). The tension value is in the range from 54 to 502 N/m. The magnitude of the dynamic tension that occurs when moving in water at high speed, according to Section 6.8 (Table 6.9), for a bottlenose dolphin is 80.1 N/m.

For the analysis of the deformation of the skin of greatest interest is the total tension, consisting of static and dynamic tension. Under the action of dynamic tension, the skin, being fixed around the head, front fins, and chest [287], is subject to tension, which shifts the skin relative to the skeleton and skeletal muscles due to the presence of subcutaneous fat. However, in the region of the dorsal fin and on the caudal stem, the cover is attached to the skeletal muscles and cannot shift [287]. This leads to its compression in the area between the dorsal fin and the caudal stem. Muscle bundles of skin muscles located in this part of the body (Section 4.1, Fig. 4.8; Section 4.3, Fig. 4.24) continue into bundles of connective tissue fibers, which are attached in the skin, mainly in its dermal layer or form connective tissue fasciae and aponeuroses located on border dermis and subcutaneous adipose tissue. Therefore the skin muscles prevent the shear stresses of the ambient flow, and its stress corresponds to the tension in the skin T. From Fig. 6.38, it can be seen that in the cross section (7–10) there is a decrease in the total tension.

Compare the magnitude of the measured logarithmic dampening decrement with the data given in Section 6.8. In accordance with the expression (6.38) (Section 6.8), the logarithmic dampening decrement Δ can be determined from the relation:

$$\Delta = \pi \kappa_\eta, \tag{6.51}$$

where κ_η is the dampening parameter. For the bottlenose dolphin according to Section 6.8, at a low navigation speed, $\kappa_\eta = 1.37$ and $\Delta = 4.3$, and at a high swimming speed, $\kappa_\eta = 0.13$ and $\Delta = 0.4$, that is, at a low swimming speed, at a distance equal to one wavelength, the oscillation dampening is 73 times, and with a high swimming speed it is 1.5 times.

Investigations conducted by us showed that in a dolphin in a state of rest, the oscillations on the cover spread with attenuation no more than 2–2.5 times. From this it follows that the properties of dampening do not change wide limits when the speed of movement changes, and already at rest they are close to the values corresponding to a high swimming speed.

Elastic surfaces reduce the spectrum of unstable oscillations, are shifted downstream, and stretch the transition region from laminar to turbulent flow in the boundary layer. This reduces adverse gradient loads on the body of hydrobionts. It can be assumed that the shear rate pulsations are dampened due to the elastic-dampening properties of the transverse microfolds. According to measurements on hydrobionts, in Ref. [173] it is stated that oscillations in the boundary layer correlate with the work of the caudal fin.

Oscillations of the second type were also recorded in Ref. [173]; they are equal to twice the frequency of the tail fin and noticeably affect the character of the flow in the boundary layer beyond the mid-section. Above are the measurements obtained by Kidun for pressure pulsations on the hydrobiont surface. During slow swimming, the frequency of pressure pulsations on the skin was 130–140 Hz, and during jumps it was 115–230 Hz. According to our data, when applying a single pulse on the surface of plexiglass oscillations spread with a frequency of 600 Hz, on the surface of a hydrobiont skin analogue, it was 130 Hz, and when temperature controlled it was 157 Hz. By changing the mechanical parameters of the skin, hydrobionts are able to influence the field of pressure pulsations.

During the movement of hydrobionts in their boundary layer, three types of disturbing motion are generated: from the action of transverse microfolds on the skin surface (Section 4.3, Fig. 4.12), from skin vibration as an elastic-viscous material, and due to the work of the tail mover.

The spectrum of generated disturbances is significantly different (for example, at $U = 10$ m/s in the first case, tens of kilohertz; in the second, hundreds of hertz; and in the third, tens of Hertz), therefore, according to the mechanism of complex interactions of disturbing movements in the first case, the skin-generated disturbances interact with the disturbances of the boundary layer, and in the second and third cases the impact will be on the energy spectrum of the boundary layer, especially on its low energy-carrying frequencies.

The effect of the transverse microfolds on the boundary layer, which is maximum at the beginning of the body, decreases as the microfolds move to the midsection, since the movement of the body of a hydrobiont under a sinusoidal law can be considered as occurring in the same sinusoidal trail of fluid caused by oscillation of the head. Therefore the body is moved around a plane-parallel unsteady flow. The boundary layer thickens, and the role of microfolds weakens, while the second and third disturbance factors of the boundary layer act constantly along the whole body. At $0.4\,l$, the subsequent stages of the transition in the boundary layer develop, where l is the body length. At the beginning of the body, the formation of longitudinal vortex disturbances occurs rapidly in the boundary layer under the influence of the indicated receptivity phenomena. These vortex systems subsequently develop and stabilize under the influence of centrifugal forces caused by the curvature of the body during its oscillatory motion.

Thus, the results of experimental investigations indicate a correlation between the structure of the skin and parameters that determine the state of the cover in the process of movement of the dolphin. The resulting material does not exclude the possibility of the formation of an appropriate mechanism on the dolphin cover, which can help reduce its resistance to movement in water. In the future, it will be important to compare the results obtained with the characteristics of the boundary layer, in particular, experimental investigations of the boundary layer on the plate surface with a deformable surface in the form of a traveling wave with certain parameters performed [255]. In addition, it is important to take into account modern ideas about the structure of the boundary layer, cited in Refs. [70,200,201,256,269].

References

[1] G.B. Agarkov, V.Y. Lukhanin, On the issue of the muscular muscles of the tail section of the white-sided dolphin, Bionica 4 (1970) 61–64.

[2] G.B. Agarkov, V.G. Khadzhinsky, To the question of the structure and innervation of the skin of the Black Sea dolphins in connection with their protective function, Bionica 4 (1970) 64–70.

[3] G.B. Agarkov, V.V. Babenko, Z.I. Ferents, On the innervation of the skin and skin muscles of the dolphin in connection with the hypothesis of stabilization of the flow in the boundary layer, Problems of Bionics, Science, Moscow, 1973, pp. 478–483.

[4] G.B. Agarkov, B.G. Khomenko, V.G. Khadzhinsky, Dolphin Morphology, Naukova Dumka, Kiev, 1974, p. 167.

[5] G.B. Agarkov, V.F. Sych, Morpho-functional analysis of the epaxial muscles of the propulsive complex of the porpoise, Bionica 8 (1974) 122–127.

[6] G.B. Agarkov, A.A. Vishnyakov, On the implantation of sensors of average and pulsating pressure into the skin of marine animals, Bionica 9 (1975) 131–134.

[7] D. Adelung, "Gläserne" Pinguine – Anwendung telemetrischer Methoden bei der Erforschung dieser Meeresvogel. Meer und Museum, 1996, Band 12: 26–32.

[8] U.R. Adie, Frequency and energy windows when exposed to weak electromagnetic fields on living tissue, TIIER 1 (1980) 140–148.

[9] T.E. Alekseeva, V.P. Gromov, A.F. Dmitrieva, et al., Calculation of the Characteristics of a Laminar Boundary Layer on Rotation Bodies, Science Siberian Branch, Novosibirsk, 1968, p. 220.

[10] J.M. Anderson, The vorticity control unmanned undersea vehicle – a biologically inspired autonomous vehicle, Proceedings of the International Symposium on Seawater Drag Reduction, Newport, RI, 1998, pp. 479–483.

[11] M.A. Arakelyan, A device for determining the radiant and convective components of heat transfer in a production environment, Thermometry, Energy and Resource Saving, Institute for Problems of Energy Saving, Kiev, 1989, pp. 17–27.

[12] V.V. Babenko, D.A. Morozov, Some physical laws during dolphin diving, Mechanisms of Movement and Orientation of Animals, Sciences. Dumka, Kiev, 1968, pp. 49–57.

[13] V.V. Babenko, N.A. Gnitetsky, L.F. Kozlov, Preliminary results of investigation of the elastic properties of the skin of living dolphins, Bionica 3 (1969) 12–19.

[14] V.V. Babenko, R.M. Surkina, Some hydrodynamic features of dolphin swimming, Bionica 3 (1969) 19–26.

[15] V.V. Babenko, N.A. Gnitetsky, L.F. Kozlov, Preliminary results of the investigation of the temperature distribution on the surface of a dolphin body, Bionica 4 (1970) 83–88.

[16] V.V. Babenko, R.M. Surkina, Determination of the parameter of the oscillating skin mass covers of some marine animals, Bionica 5 (1970) 94–98.

[17] V.V. Babenko, Main characteristics of flexible coatings and similarity criteria, Bionica 5 (1971) 73–75.

[18] V.V. Babenko, Some mechanical characteristics of the skin of dolphins, Bionica 5 (1971) 76–81.

[19] V.V. Babenko, L.F. Kozlov, S.V. Pershin, On alternating skin damping of dolphins at various swimming speeds, Bionica 6 (1972) 42–52.

[20] V.V. Babenko, D.A. Morosov, Einige physikalische Gesetzmäßigkeiten beim Tauschen der Delphine, in: G. Tembrock Dr (Ed.), Mechanismen der Bewegung und Orientierung der Tiere, Akademie-Verlag, Berlin, 1973, pp. 49–56.

[21] V.V. Babenko, L.F. Kozlov, Experimental investigation of hydrodynamic stability on rigid and elastic-damping surfaces, in: Proceedings of the Academy of Sciences of the USSR. Fluid Gas Mech. (1) (1973): 122–127.

[22] V.V. Babenko, L.F. Kozlov, S.V. Pershin, Damping coating. USSR Author's Certificate 413286, 1974.

[23] V.V. Babenko, L.F. Kozlov, V.I. Korobov, Damping coating. USSR Author's Certificate 483538, 1975.

[24] V.V. Babenko, L.F. Kozlov, V.P. Kayan, Waxing fin mover. USSR Author's Certificate 484129, 1976.

[25] V.V. Babenko, The fin propulsor. USSR Author's Certificate 529104, 1976.

[26] V.V. Babenko, O.D. Nikishova, Some hydrodynamic regularities of the structure of the skin of marine animals, Bionica 10 (1976) 27−33.

[27] V.V. Babenko, L.F. Kozlov, V.I. Korobov, Adjustable damping coating. USSR Author's Certificate 597866, 1978.

[28] V.V. Babenko, Investigation of skin elasticity of live dolphins, Bionica 13 (1979) 43−52.

[29] V.V. Babenko, To the interaction of flow with an elastic surface, Mechanics of Turbulent Flows, Science, Moscow, 1980, pp. 292−301.

[30] V.V. Babenko, N.F. Yurchenko, Experimental investigation of Görtler stability on rigid and elastic flat plates, Hydromechanics 41 (1980) 103−108.

[31] V.V. Babenko, G.A. Voropaev, N.F. Yurchenko, On the problem of modeling the interaction of the outer covers of aquatic animals with the boundary layer, Hydromechanics 41 (1980) 73−81.

[32] V.V. Babenko, On the oscillating mass of the skin of dolphins, Bionica 14 (1980) 57−64.

[33] V.V. Babenko, On the methodology of experimental studies in hydrobionics, Bionica 15 (1981) 88−98.

[34] V.V. Babenko, V.A. Tarasenko, V.A. Gaponenko, V.V. Davydov, Fin mover for boats. USSR Author's Certificate 796074, 1981.

[35] V.V. Babenko, Damping coating. USSR author's certificate 802672, 1981.

[36] V.V. Babenko, L.F. Kozlov, S.V. Pershin, Self-tuning of cetacean skin damping with active swimming, Bionica 16 (1982) 3−10.

[37] V.V. Babenko, Institute of Hydromechanics of NASU. The cetacean skin's property is to actively regulate the hydrodynamic resistance to swimming by controlling the local interaction of the skin with the flowing stream. Diploma number 265, 4.XI.1982. Published 05/15/1983. Bulletin N17. Science and Innovation. Ukrainian Magazine Review of the Future 5 (2009): 39−41.

[38] V.V. Babenko, Some features of heat regulation of external covers of aquatic animals, Bionica 17 (1983) 35−39.

[39] V.V. Babenko, N.F. Yurchenko, On the modeling of the hydrodynamic functions of the outer covers of aquatic animals, Hydrodynamic Issues of Bionics, Naukova Dumka, Kiev, 1983, pp. 37−46.

[40] V.V. Babenko, L.F. Kozlov, S.A. Dovgj, et al., The influence of the outflow generated vortex structures on the boundary layer characteristics, Proceedings Second IUTAM Symposium on Laminar-turbulent Transition, Springer-Verlag, Novosibirsk: Berlin, 1985, pp. 509−513.

[41] V.V. Babenko, The problem of boundary layer receptivity to various disturbances, Bionica 22 (1988) 15−23.

[42] V.V. Babenko, A.P. Koval, On the hydrodynamic properties of the skin of aquatic animals, Bionica 23 (1989) 38−42.

[43] V.V. Babenko, On the interaction of hydrobionts with flow, Bionics 25 (1992) 3−11.

[44] V.V. Babenko, M.V. Kanarsky, V.I. Korobov, The Boundary Layer on Elastic Plates, Naukova Dumka, Kiev, 1993, p. 263.

[45] V.V. Babenko, V.P. Musienko, V.I. Korobov, Y.A. Ptukha, Experimental investigation of spherical groove influence on the intensification of heat and mass transfer in the boundary layer. Euromech Colloquium 327, Effects of Organised Vortex Motion on Heat and Mass Transfer. August 25−27, 1994, Institute of Hydromechanics, National Academy of Sciences of Ukraine, Kiev. Book of Abstracts, 1994, pp. 23−24.

[46] V.V. Babenko, Interaction of biological systems during the motion of marine animals, Review Meeting on "Bio-Locomotion and Rotational Flow Over Compliant Surfaces", Johns Hopkins University, Baltimore, MD, 1995.

[47] V.V. Babenko, General principles of hydrobionics investigations and perspectives of research in hydrobionics, Review Meeting on "Bio-Locomotion and Rotational Flow Over Compliant Surfaces", Johns Hopkins University, Baltimore, MD, 1995.

[48] V.V. Babenko, General principles of hydrobionic investigations and perspectives of research in hydrobionics, in. Workshop of the Society for Technology of Biology and Bionics, Jena, Germany, 1997, pp. 167.

[49] V.V. Babenko, Polymers submission optimization with the help of sword-shaped tips, in: Proceedings of the 10th European Drag Reduction Working Meeting, Berlin, Germany, 1997, pp. 19−21.

[50] V.V. Babenko, Hydrobionic principles of drag reduction, in: AGARD FDP Workshop on "High Speed Body Motion in Water," Kiev, Ukraine, 1998, pp. 3 (1–14).

[51] V.V. Babenko, Method of influence on coherent vortices structures of a boundary layer, in: Proceedings of the International Symposium on Seawater Drag Reduction, Newport, RI. Workshop on "High Speed Body Motion in Water," 1998, pp. 113–120.

[52] V.V. Babenko, A.A. Yaremchuk, On biological foundations of dolphin's control of hydrodynamic resistance reduction, in: Proceedings of the International Symposium on Seawater Drag Reduction, Newport, RI. Workshop on "High Speed Body Motion in Water," 1998, pp. 451–452.

[53] V.V. Babenko, Hydrobionic principles of drag reduction, in: Proceedings of the International Symposium on Seawater Drag Reduction, Newport, RI. Workshop on "High Speed Body Motion in Water," 1998, pp. 453–455.

[54] V.V. Babenko, Combined method of drag reduction. AGENDA, in: ONR Workshop on Gas Based Surface Ship Drag Reduction, Newport, RI, 1999.

[55] V.V. Babenko, V.I. Korobov, V.P. Musienko, Formation of vortex structure on curvilinear surfaces and semi-spherical cavities, in: 11th International Couette-Taylor Workshop. Bremen, Germany, 1999, p. 103.

[56] V.V. Babenko, Hydrobionic principles of resistance reduction, Appl. Fluid Mech. 2 (74) (2000) No. 2: 3–17.

[57] V.V. Babenko, Method for reducing dissipation rate of fluid ejected into boundary layer. Patent US 6,138,704, 2000.

[58] V.V. Babenko, V.V. Moros, I.I. Martynenko, Experience of creation of underwater remotely operated vehicles in Ukraine and perspective of their application in geological oceanography, Miner. Resour. Ukr. 1 (2000) 21–25.

[59] V.V. Babenko, Control of the coherent vortical structures of a boundary layer, Aerodynamic Drag Reduction Technologies, Proceedings of the CEAS/DragNet European Drag Reduction Conference, 2000, Potsdam, Germany, Springer–Verlag Berlin Heidelberg, 2001, pp. 341–350.

[60] V.V. Babenko, E.A. Shkvar, Combined method of drag reduction, in: Proceedings of the International Summer Scientific School "High Speed Hydrodynamics," 2002, pp. 321–326.

[61] V.V. Babenko, P.W. Carpenter, Dolphin hydrodynamics. IUTAM symposium on flow past highly compliant boundaries and in collapsible tubes. 26–30 Mach, 2001, University of Warwick Coventry England, in: Peter W. Carpenter, Timothy J. Pedley (Eds.), Proceedings of the IUTAM Symposium, Kluwer Academic Publishers, Dordrecht/Boston/London, 2003, pp. 293–323.

[62] V.V. Babenko, Hydrobionic principles of drag reduction, Int. J. Fluid Mech. Res. 30 (2) (2003) 125–146.

[63] V.V. Babenko, A.I. Kuznetsov, A.I. Kuznetsov, V.V. Moroz, Technique of towing tests in the test basin with the help of two models of a gliding vessel, Appl. Fluid Mech. 5 (77) (2003) No. 4: 5–11.

[64] V.V. Babenko, A.P. Koval, Structure and hydrodynamic properties of sword-fish skin, Appl. Fluid Mech. 6 (78) (2004) No. 3: 3–19.

[65] V.V. Babenko, V.A. Voskoboynik, A.V. Voskoboinik, V.T. Turik, Speed profiles in the boundary layer of a plate with grooves, Acoust. Bull. 7 (3) (2004) 14–27.

[66] V.V. Babenko, Interaction of quickly floating hydrobionts with flow, in: First International Industrial Conference Bionic 2004, Hannover, Germany, Reihe 15, Umwelttechnik, Nr. 249. 2004, pp. 153–159.

[67] V.V. Babenko, Interaction of quickly swimming hydrobionts with flow, in: Proceedings of the 2nd International Symposium on Seawater Drag Reduction, ISSDR 2005, Busan, Korea, 2005, pp. 579–592.

[68] V.V. Babenko, H.H. Chun, I. Lee. Coherent vortical structures and methods of their control for drag reduction of bodies, in: Proceedings of the 9th International Conference on Hydrodynamics, ICHD-2010, Shanghai, China, 2010, pp. 45–50.

[69] V.V. Babenko, F.M. Abbas, N.A. Gnitetsky, The interaction of the boundary layer with three-dimensional disturbances, Appl. Fluid Mech. 13 (85) (2011) No. 3: 3–22.

[70] V.V. Babenko, H.H. Chun, I. Lee, Boundary layer flow over elastic surfaces, Compliant Surfaces and Combined Methods for Marine Vessel Drag Reduction, Elsevier Publishers. Butterworth-Heinemann, Amsterdam, Boston, Heidelberg, London and others, 2012, p. 613.

[71] R. Bainbridge, The speed of swimming of fish as related to size and to the frequency and amplitude of tail beat, J. Exp. Biol. 5 (1) (1958) 17–33.

[72] P.R. Bandyopadhyay, Viscous drag reduction of a nose body, AIAA J. 27 (3) (1989) 274–282.

[73] P.R. Bandyopadhyay, M. Gad-el-Hak, Reynolds number effects in wall-bounded turbulent flows, Appl. Mech. Rev. 47 (28) (1994) 139.

[74] P.R. Bandyopadhyay, W.H. Nedderman, M.J. Donelly, J.M. Castano, A dual flapping foil maneuvering device for low-speed rigid bodies, Proceedings of Third International Symposium on Performance Enhancement for Marine Applications, Newport, RI, 1997, pp. 257−261.

[75] P.R. Bandyopadhyay, J.M. Castano, D. Thivierge, W. Nedderman, Drag reduction experiments on a small axisymmetric body in salt water using electromagnetic micro tiles, Proceedings of the International Symposium on Seawater Drag Reduction, Newport, RI, 1998, pp. 373−378.

[76] P.R. Bandyopadhyay, J.M. Castano, W. Nedderman, D. Thivierge, Phased vortex seeding for thrust modulation in a rigid cylinder with flapping foil thrusters, Proceedings of the International Symposium on Seawater Drag Reduction, Newport, RI, 1998, pp. 457−461.

[77] R. Bannasch, <http://www.festo.com/INetDomino/coorp_sites/de/544ca592e1ff5d58c12572b9006e05bd.htm>.

[78] R. Bannasch, J. Fiebig, Herstellung von Pinguin modellen für hydrodynamische Untersuchungen, Der Präparator. Boch. 38 (1) (1982) 1−5.

[79] R. Bannasch, Schwimm- und Tauchleistungen der Pinguine, Milu 6 (1985) 295−308.

[80] R. Bannasch, Functional anatomy of the "flight" apparatus in penguins. Ch 12 in: L. Maddock, Q. Bone, J.M. Rayner (Eds.), Mechanics and Physiology of Animal Swimming, Cambridge University Press, 1994, pp. 163−197.

[81] R. Bannasch, Hydrodynamic aspects of design and attachment of a back-mounted device in penguins, J. Exp. Biol. 194 (1994) 83−96.

[82] R. Bannasch, Hydrodynamics of penguins − an experimental approach, in: P. Dann, I. Norman, P. Reilly (Eds.), Advances in Penguin Biology, Surrey Beatty, Sydney, 1995.

[83] R. Bannasch, Widerstandsarme Strömung Körper Optimalformen nach Patenten der Natur. Bionareport 10, in: Werner Nachtigal, Alfred Wisser (Eds.), Technische Biologie and Bionik Kongress, Mannheim 1996, Akademie der Wissenschaften und der Literatur, Mainz, Gustav Fischer Verlag, Stuttgart, Jena, Lubeck, Ulm, New York, 1996, pp. 151−176.

[84] B.A. Barbanel, V.G. Bogdevich, L.I. Maltsev, A.G. Malyuga, Some Practical Applications of Boundary Layer Control Theory, Institute of Thermal Physics, Academy of Sciences of Russia, Novosibirsk, 1994, p. 48.

[85] A.M. Bassin, A.I. Korotkin, L.F. Kozlov, Management of the Boundary Layer of the Vessel, Shipbuilding, Leningrad, 1968, p. 491.

[86] V.G. Belinsky, P.I. Zinchuk, Resistance of a disk and a sphere during accelerated motion from a state of rest, Bionics 27−28 (1998) 88−99.

[87] V.M. Belkovich, About physical regulation of the white whale, in: Q: Proceedings of the Meeting: Ichthyological Commission of the USSR Academy of Sciences, 1961.

[88] A.G. Belousov, An experimental study of the accelerated motion of bodies in water, Bionics 27−28 (1998) 100−103.

[89] V.A. Berezovsky, N.N. Kolotilov, Biophysical characteristics of human tissues, Directory, Naukova Dumka, Kiev, 1990, p. 224.

[90] H. Bippes, H. Görtler, Dreidimensionale Störungen in der Grenzschicht an einer konkaven Wänden, Actamechanics 13 (14) (1972) 251−267.

[91] R. Blackwelder, Analogies between transitional and turbulent boundary layers, Phys. Fluids 26 (10) (1983) 2807−2816.

[92] Q. Bone, On the function of the types of myotomal muscle fiber in elastomobranch fish, J. Mar. Biology. Assoc. UK 46 (2) (1966) 321−349.

[93] N.N. Brazhnikov, Ultrasonic Phasometry, Energy, Moscow, 1968, p. 248.

[94] T.R. Brett, The swimming energetic of Salmon, Sci. Amer. 213 (2) (1965) 80−85.

[95] P.W. Carpenter, Biology-based drag reduction, in: Proceedings of the 2nd International Symposium on Seawater Drag Reduction, ISSDR 2005, Busan, Korea, 2005, pp. 45−52.

[96] P.K. Chang, Separated Flows, Pergamon Pess, Oxford, 1970, p. 229.

[97] O.B. Chernyshov, A.P. Koval, A.A. Drobakha, Some Features of the Morphology of the Gill Apparatus of Fish, Associated with the Speed of their Swimming, 12, National Academy of Sciences of Ukraine, Kyiv, Hydromechanics, 1978, pp. 103−108.

[98] A.V. Chepurnov, The body shape of some cetaceans in connection with their swimming speed, Mechanisms of Movement and Orientation of Animals, Naukova Dumka, Kiev, 1968.

[99] O.B. Chernyshov, V.A. Zayets, Some features of the structure of shark fish skin, Bionica 4 (1970) 77–83.

[100] O.B. Chernyshov, V.A. Zayets, Change in the keels of the placoid scale of sharks depending on the speed of swimming, Bionica 8 (1974) 82–86.

[101] B.D. Clark, W. Bernis, Kinematics of swimming of penguins at Detroit Zoo, J. Zool. Lond. 188 (1979) 411–428.

[102] S. Deutsch, M. Money, et al. Microbubble drag reduction in rough wall turbulent boundary layer, in: Proceedings ASME Fluids Engineering, 2003, pp. 1–9.

[103] S.A. Dovgy, V.V. Moros, V.V. Babenko, S.V. Polishchuk, Underwater apparatus. Patent of Ukraine 65250A, 2004.

[104] B.M. Egidis, V.M. Shakalo, B.M. Egidis, V.M. Shakalo, The use of electrolytic micro rotors for measuring the speed of movement in sea water, Bionica 5 (1971) 128–131.

[105] Y.V. Ergin, Magnetic Properties and Structure of Electrolyte Solutions, Science, Moscow, 1983, p. 183.

[106] N.Y. Fabricant, Aerodynamics, General Course, Science, Moscow, 1964, p. 814.

[107] F. Saeed, R. Taghavi Ray, M. Brett Ronald, M. Saeid, S. Tom, Active vortex management concepts for maneuvering submarines, Proceedings of Third International Symposium on Performance Enhancement for Marine Applications, Newport, RI, 1997, pp. 55–60.

[108] M. Fliesen, M. Magi, L. Sonnerup, A. Viidik, Rheological analysis of soft collagenous tissue, J. Biomech. 2 (1) (1969) 13–20.

[109] H. Fok, Über die Ursachen der hohen Schwimmgeschwindigkeiten der Delphine, Z. für Flügwissenschaft 2 (1965).

[110] A.A. Fontaine, S. Deutsch, T.A. Brungart, H.L. Petrie, M. Fenstermacker, Drag reduction by coupled systems: microbubble injection with homogeneous polymer and surfactant solutions, Exp. Fluids 26 (1999) 397–403.

[111] L.F. Kozlov, A.I. Tsyganyuk, V.V. Babenko, et al., Formation of Turbulence in Shear Flows, Naukova Dumka, Kiev, 1985, p. 281.

[112] T.G. Genedict, Vital energetics, A Study in Comparative Basal Metabolism, Carnegie Institute of Washington Publicaions, 1938.

[113] O.A. Gerashchenko, Fundamentals of Thermometry, Naukova Dumka, Kiev, 1971, p. 191.

[114] L.S. Glikman, Sharks of the Paleogene and their Stratigraphic Significance, Science, Moscow, 1964.

[115] V.N. Glushko, V.P. Kayan, V.A. Kochin, On the optimization of the propulsive characteristics of a flapping propulsion device, Bionica 25 (1992) 75–80.

[116] H. Görtler, Über eine dreidimensionale Instabilität laminarer Grenzschichten an konkaven Wänden, Nachr. Wiss. Göttingen, II Math-Phys. Kl. 2 (15) (1940) 1–26.

[117] J. Gray, Animal Locomotion, Norton, New York, 1968, p. 479.

[118] L. Greiner, Hydrodynamics and power engineering of underwater vehicles, Translation from English, Shipbuilding, Leningrad, 1978, p. 384.

[119] R.S. Guter, B.V. Ovchinsky, Elements of Numerical Analysis and Mathematical Processing of Experimental Results, Science, Moscow, 1970, p. 432.

[120] D. Gyorgyfalvy, The possibilities of drag reduction by the use of flexible skin, in: AIAA 4-th Aerospace Science Meeting, Los Angeles, CA, AIAA Paper N 66–30, 1966, p. 37.

[121] V.G. Hadzhinsky, Some morpho-functional features of dolphin skin, Bionica 6 (1972) 58–66.

[122] A. Mojetta, M. Falcone, E. Vandone, Haie: Biografie eines Raubers, Jahr Verlag Hamburg, 1997, p. 168.

[123] Hassan Y.A. and Ortiz-Villafuerte. J. Experimental Study of Micro-bubble Drag Reduction Using Particle Image Velocimetry. Department of Nuclear Engineering Texas & M. University.

[124] Y.A. Hassan, C.C. Gutierrez-Torres, J.A. Jimenez-Bernal, Temporal correlation modification by microbubbles injection in a boundary layer channel flow, Int. Commun. Heat. Mass. Transf. 32 (8) (2005) 1009–1015.

[125] V.N. Hatuntsev, V.V. Babenko, V.P. Musienko, O.A. Mertechechenko, Investigation of non-stationary flow around bodies in the case of an arbitrary plane motion, Bionica 27–28 (1998) 104–108.

[126] H. Hertel, Structure, Form and Movement (Biology and Technology), Reinhold Publishing Corporation, 1966, p. 251.

[127] H. Hertel, Hydrodynamics of swimming and wave riding dolphins, in: H. Andersen (Ed.), The Biology of Marine Mammals, Acad. Press, New York, 1969, pp. 31–93.

[128] S.F. Hoerner Fluid-Dynamic Drag. 2nd ed. Hoerner S.F., Bricktown, NJ, 1965.

[129] B.G. Homenko, V.G. Hadzhinsky, Morphological and functional bases of dermal reception in dolphins, Bionica 8 (1974) 106–113.

[130] J.W. Hoyt, Hydrodynamic drag reduction due to fish slimes, in: T.Y. Wu, C.J. Brokaw (Eds.), Swimming and Flying in Nature, Vol. 2, Plenum Press, New York, 1975.

[131] C.A. Hui, Swimming in penguins. Dissertation, University of California, Los Arlgeles, 1983.

[132] Investigations on Bionics. Miscellany of Articles. Kiev: Naukova Dumka, 1965.

[133] V.P. Ivanov, V.V. Babenko, V.A. Blokhin, L.F. Kozlov, V.I. Korobov, Investigation of the velocity field in a hydrodynamic bench of small turbulence using a laser Doppler velocity meter, Eng. Phys. J. 37 (5) (1979) 818–824.

[134] S.F. Ivanova, On the role of the circulatory system of the gills in ensuring high efficiency of their work, Bionica 17 (1983) 76–80.

[135] B. Jacob, A. Olivieri, M. Miozzi, E.F. Campana, R. Piva, Drag reduction by microbubbles in a turbulent boundary layer, Phys. Fluids 22 (2010) 1151–1154.

[136] M.V. Kanarsky, V.V. Babenko, L.F. Kozlov, Experimental investigation of a turbulent boundary layer on an elastic surface, Stratified and Turbulent Flows, Naukova Dumka, Kiev, 1979, pp. 59–67.

[137] V.V. Kanarsky, V.V. Babenko, Method of determining the elastic-damping mechanical characteristics of materials. Patent of the USSR 1183864, 1985.

[138] O.G. Karandeeva, I.A. Protasov, N.P. Semenov, On the question of the physiological rationale for the Gray paradox, Bionica 4 (1970) 36–43.

[139] V.P. Kayan, V.E. Pyatetsky, A closed-type bio-hydrodynamic installation for studying the hydrodynamics of swimming of marine animals, Bionica 5 (1971) 121–124.

[140] V.P. Kayan, On the drag coefficient of a dolphin, Bionica 8 (1974) 31–35.

[141] V.P. Kayan, V.E. Pyatetsky, Hydrodynamic characteristics of a bottlenose dolphin at different acceleration modes, Bionica 8 (1974) 48–55.

[142] V.P. Kayan, V.E. Pyatetsky, Kinematics of the bottlenose dolphin swimming depending on the acceleration mode, Bionica 11 (1977) 36–41.

[143] V.P. Kayan, L.F. Kozlov, V.E. Pyatetsky, Kinematic characteristics of swimming some aquatic animals. News of the USSR Academy of Sciences, Fluid Gas. Mech. 5 (1978) 3–9.

[144] V.P. Kayan, On the hydrodynamic characteristics of a fin mover dolphin, Bionica 13 (1979) 9–15.

[145] V.P. Kayan, Experimental investigation of the hydrodynamic thrust created by an oscillating wing, Bionica 17 (1983) 45–49.

[146] S.M. Kidun, Investigation of the velocity of oscillation propagation on the dolphin cover, Bionica 13 (1979) 52–58.

[147] V.I. Klassen, The Magnetization of Water Systems, Chemistry, Moscow, 1982, p. 296.

[148] P.S. Klebanoff, K.D. Tidstrom, L.M. Sargent, J. Fluid Mech. 12 (1962) 1–34. pt. 1.

[149] M. Kleiber, The Fire of Life. An Introduction to Animal Energetic, New York, London, 1961.

[150] S.E. Kleinberg, Success. Modern Biol. 41 (1956) 3.

[151] S.E. Kleinberg, et al., Beluha, Science, Moscow, 1964.

[152] S.I. Kline, Similitude and Approximation Theory, McGraw-Hill, New York, 1965, p. 167.

[153] S.J. Kline, The structure of turbulent boundary layers, J. Fluid Mech. 18 (7) (1967) 223–231.

[154] V.I. Korobov, V.V. Babenko, L.F. Kozlov, Interaction of a turbulent boundary layer with an elastic plate, Phys. Eng. J. 56 (2) (1989) 220–225.

[155] V.I. Korobov, V.V. Babenko, V.G. Belinsky, Fin propeller. Patent of Ukraine 1671515, 1991.

[156] V.I. Korobov, Complex effect of compliance of the surface and high-molecular polymer additives on turbulent friction, Appl. Fluid Mech. 2 (74) (2000) No. 2: 59–63.

[157] A.I. Korotkin, Attached Masses of the Vessel. Reference Book, Shipbuilding, Leningrad, 1986, p. 312.

[158] A.P. Koval, Roughness and some structural features of sword-fish skin, Bionica 6 (1972) 73–77.

[159] A.P. Koval, On the question of the functional significance of certain skin derivatives in a sailboat, Bionica 8 (1974) 88–93.

[160] A.P. Koval, Ampoules of sword-fish skin and their possible functional significance, Bionica 11 (1977) 86–91.

[161] A.P. Koval, Crypto-like mucus-forming structures of the skin and gill covers of swordfish, Bionica 12 (1978) 108–111.

[162] A.P. Koval, Morphology of the mucous-forming apparatus of sword-fish skin, Bionica 14 (1980) 90−96.

[163] A.P. Koval, V.A. Zayats, T.A. Kalyuzhnaya, Morphological peculiarities of the structure of the skin of fish of various speed groups, Bionica 21 (1987) 77−84.

[164] L.F. Kozlov, On the biological efficiency of some marine fish, Bionica 4 (1970) 44−46.

[165] L.F. Kozlov, V.M. Shakalo, Telemetry equipment for recording flow regimes in the boundary layer when moving in the aquatic environment, Bionica 4 (1970) 55−60.

[166] L.F. Kozlov, V.M. Shakalo, Some results of measurements of the velocity pulsations in the boundary layer of dolphins, Bionica 7 (1973) 50−52.

[167] L.F. Kozlov, O.D. Nikishova, On the issue of hydrodynamics of dolphin swimming, Bionica 8 (1974) 3−9.

[168] L.F. Kozlov, V.M. Shakalo, L.D. Buryanova, N.N. Vorobyev, On the effect of nonstationarity on the flow regime in the boundary layer of the Black Sea bottlenose dolphin, Bionica 8 (1974) 13−16.

[169] L.F. Kozlov, New Science − Hydrobionics, Society "Knowledge" of the Ukrainian SSR, Kiev, 1977, p. 45.

[170] L.F. Kozlov, V.V. Babenko, Experimental Investigations of the Boundary Layer, Naukova Dumka, Kiev, 1978, p. 184.

[171] L.F. Kozlov, Hydrodynamic problems of biomechanics. Biomech. Sofia 7: 88−91.

[172] L.F. Kozlov, Hydrodynamics of aquatic animals with a semi-lunar tail fin, Bionica 13 (1979) 3−9.

[173] L.F. Kozlov, V.M. Shakalo, On the flow regime in the quasistationary boundary layer of some cetaceans, Bionica 14 (1980) 74−81.

[174] L.F. Kozlov, S.A. Dovgy, Sliding of cetaceans on swell waves, Bionica 15 (1981) 49−55.

[175] L.F. Kozlov, Theoretical Biohydrodynamics, Higher School, Kiev, 1983, p. 240.

[176] L.F. Kozlov, S.V. Pershin, Complex studies of the active regulation of dolphin's skin by reducing hydrodynamic resistance, Bionica 17 (1983) 3−12.

[177] M.O. Kramer, Boundary layer stabilization by distributed damping, J. Am. Soc. Nav. Eng. 72 (1) (1960) 25−33.

[178] M.O. Kramer, Boundary layer stabilization by distributed damping, J. Am. Soc. Nav. Eng. 74 (2) (1962) 341−348.

[179] M.O. Kramer, Means and method stabilizing laminar boundary layer flow. Patent USA 3.161.385, 1964.

[180] M.O. Kramer, Hydrodynamics of the dolphin, Advances in Hydro Science, vol. 2, Academic Press, New York and London, 1965.

[181] N.F. Krasnov, V.F. Zakharchenko, V.N. Koshevoi, Basics of Aerodynamic Calculation, High School, Moscow, 1984, p. 264.

[182] K.J. Muir, M.S. Triantafullou, A fast-starting and maneuvering vehicle, the ROBOPIKE. Drag reduction and turbulence control in swimming fish-like bodies, Proceedings of the International Symposium on Seawater Drag Reduction, Newport, RI, 1998, pp. 485−490.

[183] T.G. Lang, P. Karen, Hydrodynamic performance of porpoises (Stenella attenuata), Science 152 (3721) (1966).

[184] T.G. Lang, Hydrodynamic analysis of dolphin fin profiles, Nature 104 (9) (1971) 1110−1111.

[185] R. Latorre, V.V. Babenko, Role of bubble injection technique drags reduction, Proceedings of the International Symposium on Seawater Drag Reduction, Newport, RI, 1998, pp. 319−325.

[186] R. Latorre, A. Miller, R. Philips, Ship hull drag reduction using bottom air injection, Ocean. Eng. 30 (2003) 161−176.

[187] L.M. Lebedev, Machines and Instruments for Testing Polymers, Mechanical Engineering, Moscow, 1967, p. 212.

[188] M.J. Lighthill, Hydrodynamics of aquatic animal propulsion, Ann. Rev. Fluid Mech 1 (1969) 413−446.

[189] M.J. Lighthill, Aerodynamic aspects of animal flight, in: T.Y.T. Wu, C.J. Brokaw, C. Brennen (Eds.), Swimming and Flying in Nature, vol.1, Plenum, New York, 1975, pp. 423−491.

[190] J. Lighthill, Mathematical Biofluid Dynamics, Societe for Industrial and Applied Matheatiks, Philadelphia, 1983, p. 281.

[191] J. Lilly, Man and Dolphin, Doubleday: Doubleday, Garden City, NY, 1961.

[192] G.V. Logvinovich, Hydrodynamics of Flows with Free Boundaries, Naukova Dumka, Kiev, 1969, p. 208.

[193] G.V. Logvinovich, Hydrodynamics of a thin flexible body, Bionica 4 (1970) 3−5.

[194] G.V. Logvinovich, Hydrodynamics of swimming fish, Bionica 7 (1973) 3−8.

[195] L.D. Lukyanova, B.S. Balmukhanov, A.T. Ugolev, Oxygen-Dependent Processes in the Cell and its Functional State, Science, Moscow, 1982, p. 301.

[196] N.K. Madavan, S. Deutsch, C.L. Merkle, Reduction of turbulent skin friction by microbubbles, Phys. Fluid 27 (1984) 356–363.

[197] N.K. Madavan, S. Deutsch, C.L. Merkle, Measurements of local skin friction in a microbubble-modified turbulent boundary layer, J. Fluid Mech. 156 (1985) 237–256.

[198] A.K. Martynov, Applied Aerodynamics, Mechanical Engineering, Moscow, 1972, p. 447.

[199] V.A Matyuhin, Bioenergy and Physiology of Swimming Fish, Science, Siberian Branch, Novosibirsk, 1973, p. 154.

[200] J.C.S. Meng, Wall layer microturbulence phenomenological model and a semi-marcov probability model for active control of turbulent boundary layers, in: R.L. Panton (Ed.), Self-Sustaining Mechanism of Wall Turbulence, Computational Mechanics Publications, Southampton, UK and Boston, MA, 1997, pp. 201–252.

[201] J.C.S. Meng, Engineering insight of near-wall micro turbulence for drag reduction and derivation of a design map for seawater electromagnetic turbulence control, Proceedings of the International Symposium on Seawater Drag Reduction, Newport, RI, 1998, pp. 359–367.

[202] M. Pang, J. Wei, Experimental investigation on the turbulence channel flow laden with small bubbles by PIV, Chem. Eng. Sci. 94 (2013) 302–315.

[203] K.J. Moore, T. Rajan, V.A. Gorban, V.V. Babenko, Method and apparatus for increasing the effectiveness and efficiency of multiple boundary layer control techniques. United States Patent 6,357,374, 2002.

[204] K.J. Moore, Engineering an efficient shipboard friction drag reduction system, in: Proceedings of the 2nd International Symposium on Seawater Drag Reduction, ISSDR 2005, Busan, Korea, 2005, pp. 345–358.

[205] K.J. Moore, C.M. Moore, M.A. Stern, S. Deutch, Design and test of a polymer drag reduction system on sea flyer, in: 26th Symposium on Naval Hydrodynamics, Rome, Italy, 2006.

[206] V.V. Moroz, Experimental study of the dynamics of an underwater towed system, Bionica 27–28 (1998) 115–119.

[207] D.A. Morozov, A.G. Tomilin, Elements of hydrostatics of dolphins, Bionica 4 (1970) 50–54.

[208] W. Nachtigall, Bionik. Grundlagen und Beispiele für Ingenieure und Naturwissenschaftler, Springer-Verlag, Berlin, 1998, p. 331.

[209] W. Nachtigall, D. Bilo, Strömungsanpassung des Pinguins beim Schwimmen unter Wasser, J. Corp. Physiol. 137 (1980) 17–26.

[210] Narhov A. S. – Moscow, Russia: Zool. J. 1937. 16. 4.

[211] O.V. Nechaeva, V.N. Plekhanov, V.G. Kadzhinsky, Features of the distribution of potentials on the skin of dolphins, Bionica 7 (1973) 79–83.

[212] N. Van Dao, Unsteady oscillations of a dynamic system with a damper. News of the Academy of Sciences of the USSR, Fluid Gas. Mech. 4 (1965) 92–96.

[213] E.E. Niemi, Renner Jr., J. Michael, Development of a two-person, human-powered submarine, Proceedings of Third International Symposium on Performance Enhancement for Marine Applications, Newport, RI, 1997, pp. 97–103.

[214] O.D. Nikishova, V.V. Babenko, Flow of an elastic body by a fluid flow, Bionica 9 (1975) 55–60.

[215] V.V. Ovchinnikov, Swordfish and Sailfish, AtlantNIRO, Kaliningrad, 1970, p. 269.

[216] V.V. Pavlov, Wing design and morphology of the harbor porpoise dorsal fin, J. Morphol. 258 (2003) 284–295.

[217] V.V. Pavlov, Dolphin skin as a natural anisotropic compliant wall, Bioinspir. Biomim. 1 (2006): 31–40.

[218] V.V. Pavlov, A.M. Rashad, A non-invasive dolphin telemetry tag: computer design and numerical flow simulation, Mar. Mammal. Sci. (2011) 1–12.

[219] V.V. Pavlov, A.M. Rashad, A non-invasive dolphin telemetry tag: computer design and numerical flow simulation, Mar. Mammal. Sci. 28 (1) (2012) E 16–E 27.

[220] V.V. Pavlov, D. Riedeberger, U. Rist, U. Siebert, Analysis of the relation between skin morphology and local flow conditions for a fast-swimming dolphin, Notes Numer. Fluid Mech. Multidiscip. Des. 119 (2012) 239–253.

[221] D.A. Parry, The Structure of Whale Blubber and a Discussion of its Thermal Properties, Quart. J. Microsc. Sc. 90 (1) (1949) 273–279.

[222] S.V. Pershin, Biological and hydrodynamic studies of high-speed aquatic animals in nature and oceanariums, Mechanisms of Movement and Orientation of Animals, Naukova Dumka, Kiev, 1967, pp. 29–38.

[223] S.V. Pershin, Biohydrodynamic patterns of swimming of aquatic animals as principles for optimizing the motion of immersed bodies, Questions of Bionics, Science, Moscow, 1967, pp. 555–560.

[224] S.V. Pershin, Hydrodynamic characteristics of cetaceans and normalized swimming speed of dolphins in natural conditions and in captivity, Bionica 3 (1969) 5–12.

[225] S.V. Pershin, Optimization of the aft fin mover in nature on the example of cetaceans, Bionica 3 (1969) 26–34.

[226] S.V. Pershin, A.S. Sokolov, A.G. Tomilin, The phenomenon of self-regulation of the hydroelasticity of cetacean fins. Discovery 95 (USSR), 1971.

[227] S.V. Pershin, Rationing of the layer-by-layer model of damping skin of dolphin, Bionica 7 (1973) 66–71.

[228] S.V. Pershin, O.B. Chernyshov, L.F. Kozlov, et al., Pattern in the covers of high-speed fish, Bionica 10 (1976) 3–21.

[229] S.V. Pershin, Self-tuning of skin damping and reduction of hydrodynamic resistance during active swimming of cetaceans, Bionica 10 (1976) 33–40.

[230] S.V. Pershin, Biogidrodynamic phenomenon of swordfish as a limiting case of high-speed hydrobionts, Bionica 12 (1978) 40–48.

[231] S.V. Pershin, Basics of Hydrobionics, Shipbuilding, Leningrad, 1988, p. 263.

[232] A.A. Petrov, Interaction of resonant fin movers with the medium, Bionica 26 (1993) 108–117.

[233] I.M. Petrova, Hydrobionics in Shipbuilding, TSNIITEI, Leningrad, 1970.

[234] V.I. Pinchuk, The Determinant of World Ocean Sharks, Science, Moscow, 1976, p. 122.

[235] R. Plonsi, R. Barr, Bioelectricity: A Quantitative Approach, World, Moscow, 1991, p. 366.

[236] S.V. Polishchuk, V.V. Babenko, Fin mover. Patent of Russia 2033938, 1995.

[237] S.V. Polishchuk, V.V. Babenko, Apparatus with a fin mover. Patent of Ukraine 95073406, 1995.

[238] S.V. Polishchuk, V.V. Babenko, Apparatus with fin mover. Patent of Ukraine 96020765, 1996.

[239] V.N. Poturaev, V.I. Dyrda, I.I. Krush, Applied Rubber Mechanics, Naukova Dumka, Kiev, 1975, p. 215.

[240] I.L. Povh, N.V. Finoshin, Calculation of impedances in pipes of variable cross section, Theoretical and Applied Mechanics, 21, Higher School, Kharkov, 1990, pp. 120–124.

[241] L.P. Prosser, F. Brown, Comparative physiology of animals. M. Sci. (1967).

[242] P.E. Purves, Locomotion of the whales, Nature 197 (1963) 48–65.

[243] V.E. Pyatetsky, V.P. Kayan, Some kinematic characteristics of swimming of a dolphin-Azovka, Bionics 6 (1972) 18–21.

[244] V.E. Pyatetsky, V.P. Kayan, A.M. Kravchenko, Experimental device, apparatus and methods for studying the hydrodynamics of swimming of aquatic animals, Bionica 7 (1973) 91–101.

[245] V.E. Pyatetsky, V.P. Kayan, On the kinematics of swimming a bottlenose dolphin, Bionica 9 (1975) 41–46.

[246] V.E. Pyatetsky, V.M. Shakalo, Flow regime in the boundary layer of a dolphin model, Bionica 9 (1975) 46–50.

[247] V.E. Pyatetsky, V.M. Shakalo, A.I. Tsyganyuk, I.I. Sizov, Investigation of the flow pattern of aquatic animals, Bionica 16 (1982) 31–36.

[248] R. Reiner, Selbstorganisation: Anwendungen eines biologischen Prinzips, Biona-Report 8 (1998) 13–25.

[249] M.M. Reznikovskiy, A.I. Lukomskaya, Mechanical Testing of Rubber and Rubber, Chemistry, Moscow-Leningrad, 1964, p. 184.

[250] E.V. Romanenko, V.G. Yanov, Results of experiments on the study of hydrodynamics of dolphins, Bionica 7 (1973) 52–56.

[251] E.V. Romanenko, Fundamentals of Statistical Biohydrodynamics, Science, Moscow, 1976, p. 167.

[252] E.V. Romanenko, Theory of Swimming of Fish and Dolphins, Science, Moscow, 1986, p. 152.

[253] E.V. Romanenko, Hydrodynamics of Fish and Dolphins, KMK Publishing, Moscow, 2001, p. 411.

[254] N.N. Savitsky, Biophysical Basis of Blood Circulation and Clinical Methods for Studying Hemodynamics, Main Publishing House of Medical Literature, Leningrad, 1963.

[255] W. Schilz, Untersuchungen über den Einfluss biegeförmiger Wandschwindungen auf die entwicklung der Strömungs grenschicht, Acustica 15 (21) (1965) 27–36.

[256] H. Schlichting, K. Gersten, Boundary-Layer Theory, eighth revised enlarger ed, Sringer-Verlag, Berlin Heidelberg NewYork, 2000, p. 801.

[257] F.S. Schmidt, Zu beschleunigter Bewegung des Sphäre körpers in aufsetzenden sich Mitten, Annu. Phys. LPZ (1920) 633–664. Bd. 61.

[258] G.B. Schubauer, H.K.U.S. Skramstad, Department of Commerce National Bureau of Standards 1947; Research Paper RP 1722, p. 38.
[259] G.B. Schubauer, W.G. Spangerberg, Forced mixing in boundary layers, J. Fluid Mech. 8 (1960) 10–32. Part 1.
[260] J.R. Scott, Physical Testing of Rubber and Rubber, Chemistry, Moscow, 1968, p. 315.
[261] L.I. Sedov, Methods of Similarity and Dimension in Mechanics, Science, Moscow, 1977, p. 438.
[262] R.L. Panton (Ed.), Self–Sustaining Mechanism of Wall Turbulence, Computational Mechanics Publications, Southampton, UK and Boston, MA, 1997, p. 422.
[263] B.N. Semenov, On the existence of the hydrodynamic phenomenon of the bottlenose dolphin, Bionica 3 (1969) 54–61.
[264] N.P. Semenov, V.V. Babenko, V.P. Kayan, An experimental study of some features of the hydrodynamics of dolphin swimming, Bionica 8 (1974) 23–31.
[265] V.M. Shakalo, One method of measuring the rms value of velocity pulsations in a nonstationary turbulent flow, Bionica 6 (1972) 105–110.
[266] V.M. Shakalo, Cable telemetry equipment for measuring the velocity pulsations in the boundary layer of marine animals using a wire anemometer, Bionica 14 (1980) 67–74.
[267] A.N. Shebalov, Some questions of the influence of non-stationarity on the mechanism of formation of resistance, Bionica 3 (1969) 61–66.
[268] K. Shimoni, Theoretical Electrical Engineering, World, Moscow, 1964, p. 773.
[269] A.A. Shlanchyauskas, N.I. Vegite, A model of a turbulent boundary layer composed of large-scale mixing, Wall Turbulent Flows Part 2, Novosibirsk, 1975, pp. 203–208.
[270] N.G. Shpet, Features of the shape of the torso and tail fin of whales, Bionica 9 (1975) 36–41.
[271] V.V. Shuleikin, Physics of the Sea, Publishing House of the Academy of Sciences of the USSR, Moscow, 1953.
[272] K.P. Shvan, K.R. Foster, Impact of high-frequency fields on biological system, TIER 1 (1980) 121–132.
[273] J. Siekmann, Theoretical studies of sea animal locomotion, Part I. Ing. Arch. 31 (1962) 214–228. 1963. Theoretical studies of sea animal locomotion. Part 2. Ing. Arch. 32: 40–50.
[274] M.M. Sleptsov, Cetaceans of the Far Eastern Seas, Primorskoe Book Publishing, Vladivostok, 1955.
[275] E.J. Slijper, Organ weights and symmetry problems in porpoises and seals, Arch. Neerl. Zool. 13 (1958) 97–113.
[276] A.M.O. Smith, On the growth of Taylor – Görtler vortices along highly concave walls, Q. Appl. Math. 13 (3) (1955) 233–262.
[277] V.E. Sokolov, The structure of the skin of some cetaceans, Bull. Mosc. Soc. Nat 60 (6) (1955) 45–60. Department of Biology.
[278] V.E. Sokolov, The structure of the skin of some cetaceans, Rep. 2. Sci. Rep. High. School. Biol. Sci. 3 (1962) 47–53.
[279] V.E. Sokolov, Adaptive features of the skin of aquatic mammals, Marine Mammals, Science, Moscow, 1965, p. 272.
[280] V.E. Sokolov, V.D. Burlakov, Z.Y. Grushanskaya, Theoretical drawing of the Black Sea dolphin bottlenose dolphin, Bull. Mosc. Soc. Nat 77 (6) (1972) 45–53.
[281] V.E. Sokolov, Skin of Mammals, Science, Moscow, 1973, p. 487.
[282] V.E. Sokolov, Morphology and Ecology of Marine Mammals, Science, Moscow, 1987, p. 453.
[283] W.R. Stahl, The analysis of biological similarity, Adv. Biol. Med. Phys. 9 (1963) 355–464.
[284] L. Steven, Friction drag reduction of external flows with bubble and gas injection, Annu. Rev. Fluid Mech. 42 (2010) 183–203.
[285] J.T. Stuart, Hydrodynamic stability, Appl. Mech. Rev. 18 (7) (1965) 223–231.
[286] R.M. Surkina, The structure of the connective tissue skeleton of a dolphin skin, Mechanisms of Movement and Orientation of Animals, Naukova Dumka, Kiev, 1968.
[287] R.M. Surkina, On the structure and function of the skin muscles of dolphins, Bionica 5 (1971) 81–87.
[288] R.M. Surkina, Location of dermal rolls on the body of the white-sided dolphin, Bionica 5 (1971) 88–94.
[289] Swimming and Flying in Nature, Plenum Press, New York, London, 1975, vol. 1, p. 421.
[290] I.Y. Tamm, Fundamentals Theory Electricity, Science, Moscow, 1989, p. 504.
[291] A.G. Tomilin, Thermoregulation and geographical races of cetaceans, Rep. Acad. Sci. USSR 54 (5) (1946) 469–472.

[292] A.G. Tomilin, Cetaceans. The Beasts of the USSR and Adjacent Countries, Publishing House of the Academy of Sciences of the USSR, Moscow, 1957, p. 756.

[293] A.G. Tomilin, Dolphins Serve Man, Science, Moscow, 1969, p. 246.

[294] Undersea Warrios: Submarines of the World. Captain Ernest Louis Schwab, USN (Ret.). Foreword by Vice Admiral Robert Y. Kaufman, USN (Ret.). Publications International, Ltd., 1991, p. 256.

[295] G.I. Vasilevskaya, To the peculiarities of the structure of the pectoral fin of dolphins, Bionica 8 (1974) 127–132.

[296] G.G. Vinberg, Success. Mod. Biol. 61 (1966) 2.

[297] Y.I. Voitkunsky, Water Resistance to Vessel Traffic, Shipbuilding, Leningrad, 1964, p. 412.

[298] G.A. Voropaev, V.V. Babenko, Absorption of pulsation energy by a damping coating, Bionica 9 (1975) 60–68.

[299] M.J. Wolfgang, S.W. Tolkoff, A.N. Techet, D.S. Barret, M.S. Triantafullou, D.K.P. Yue, et al., Drag reduction and turbulence control in swimming fish-like bodies, Proceedings of the International Symposium on Seawater Drag Reduction, Newport, RI, 1998, pp. 463–469.

[300] T.Y. Wu, Hydromechanics of swimming of fishes and cetaceans, Adv. Appl. Mech. 11 (1971) 1–63.

[301] V.A. Zayets, On the question of the variable roughness of shark covers, Bionica 6 (1972) 67–73.

[302] V.A. Zayets, The distribution of placoid scales over the body of sharks, Bionica 7 (1973) 83–87.

[303] V.A. Zayets, The relationship of the superficial layers of skeletal muscles with the skin shark fish, Bionica 8 (1974) 93–100.

[304] V. Zayets, A. Change in the structure of red muscles of shark fish depending on swimming speed, Bionica 11 (1977) 91–94.

[305] V.L. Zharov, The body temperature of tuna (Thunnidae) and some other fish of the order of the perci-formed tropical Atlantic, Ichthyol. Quest. 5 (1) (1965) 18–27.

[306] E.V. Zerbst, Bionik, biologische Funktionsprinzipien und ihre technischen Anwendungen, Teubner-Studienbucher, Berlin, 1987, p. 231.

[307] T.H. Zinewlaver, The hotbloods, Sea Frout 17 (2) (1971) 67–71.

[308] S.S. Zolotov, Y.S. Khodorkovsky, Features of frictional resistance to sword-shaped body, Bionics 7 (1973) 14–18.

[309] A.V. Yablokov, Functional morphology of the respiratory organs of toothed cetaceans, Proceedings of the Meeting on the Ecology and Fishing of Marine Mammals, Publishing House of the Academy of Sciences of the USSR, Moscow, 1961.

[310] A.V. Yablokov, The Key to the biological riddle. Whale at a depth of 2000 m, Nature 4 (1962).

[311] A.V. Yablokov, V.M. Belkovich, V.I. Borisov, Whales and Dolphins, Science, Moscow, 1972, p. 472.

[312] A.A. Yaremchuk, V.V. Babenko, On the question of comparing human and dolphin bioenergy, Bionica 27–28 (1998) 152–154.

[313] N.F. Yurchenko, V.V. Babenko, L.F. Kozlov, Experimental study of the Görtler instability in the boundary layer, Stratified and Turbulent Flows, Sciences. Dumka, Kiev, 1979, pp. 50–59.

[314] N.F. Yurchenko, V.V. Babenko, On the stabilization of longitudinal vortices by the skin of dolphins, Biophysics XXV (2) (1980) 299–304.

[315] N.F. Yurchenko, V.V. Babenko, On the modeling of hydrodynamic functions of the outer covers of aquatic animals, Hydrodynamic Issues of Bionics, Naukova Dumka, Kiev, 1983, pp. 37–46.

Index

Printed in the United States
By Bookmasters